W9-CKA-252

THE GREAT ENGINEERS

Published in association with
THE ROYAL COLLEGE OF ART

THE GREAT ENGINEERS

THE ART OF BRITISH ENGINEERS 1837-1987

DEREK WALKER

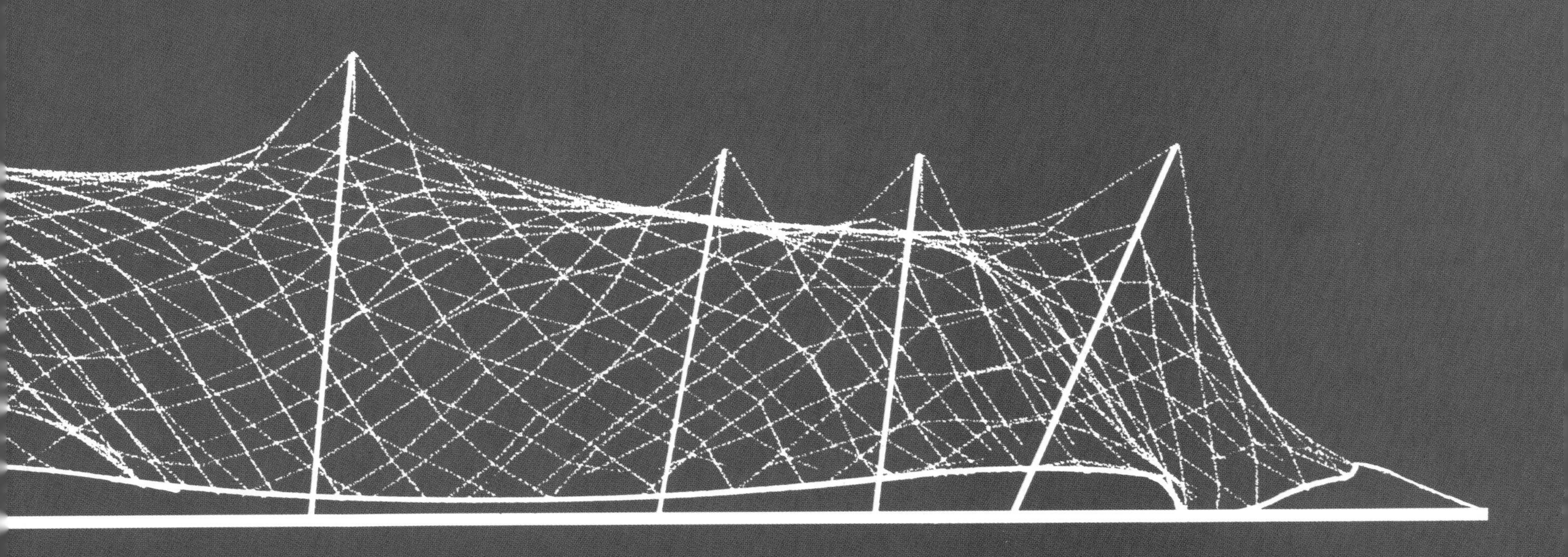

ACADEMY EDITIONS
LONDON
ST MARTINS PRESS
NEW YORK

This book is dedicated to
Jocelyn Stevens, Rector of the Royal College of Art,
on the occasion of the College's 150th Anniversary

Jacket: Ove Arup and Partners, Engineers; Foster Associates, Architects; Hong Kong and Shanghai Bank, 1986 (Martin Charles) with a contemporary photograph of I.K. Brunel inset (National Portrait Gallery). Design by Brian Tattersfield
Endpapers: an enlarged section of a 5mm Plessey microchip
Page 1: Sir Joseph Paxton's original sketch for a glass exhibition building, c. 1850
Pages 2-3: I.K. Brunel, Engineer; detail of the horizontal section of the SS Great Eastern, c. 1857-60 (Scott Russell drawing/Welsh Arts Council) and inset, Büro Happold, Engineers; Aviary, Munich, 1980, computer-generated elevation drawing
Page 5: Ove Arup and Partners, Engineers; Foster Associates, Architects; Hong Kong and Shanghai Bank at night, c. 1986
Page 6: Ove Arup and Partners, Engineers; Piano and Rogers, Architects; Centre Georges Pompidou, aerial view, c. 1976
Page 8: 'Making a cutting on the Glasgow, Paisley and Greenoch Railway', hand-coloured lithograph, c. 1841

Sponsored by
Balfour Beatty, British Steel and Plessey

Published in Great Britain in 1987 by
ACADEMY EDITIONS
an imprint of the Academy Group Ltd, 7 Holland Street, London W8 4NA

Copyright © 1987 Academy Editions
All rights reserved
No part of this publication may be
reproduced in any manner whatsoever without
permission in writing from the publishers

ISBN 0-85670-917-4 (UK)

Published in the United States of America in 1987 by
ST MARTIN'S PRESS
175 Fifth Avenue, New York, NY 10010

ISBN 0-312-01136-9 (USA)

Printed and bound in Hong Kong

Contents

H.R.H. THE DUKE OF EDINBURGH
Foreword 9

PROFESSOR DEREK WALKER
Introduction 11

PROFESSOR EDMUND HAPPOLD
A Personal Perception of Engineering 24

PROFESSOR WILLIAM GOSLING
A Short History of Electrical Communications 40

ROBERT THORNE
Paxton and Prefabrication 52

FRANCIS PUGH
Design, Engineering and the Art of Drawing 71

DR. R. ANGUS BUCHANAN
The Life-Style of the Victorian Engineers 83

PROFESSOR DEREK WALKER
The Railway Engineers 91

JAMES SUTHERLAND
Developments of the Use of Materials in Structures 108

DR. DENIS SMITH
Sir Joseph Bazalgette and
Public Health Engineering 119

PROFESSOR CHRISTOPHER FRAYLING
The Strange Case of the
Duke of Wellington's Funeral Car 128

DON HOLLAND
Construction as a Prime Export 136

PROFESSOR DEREK WALKER
The Sydney Opera House 145

JAMES GOWAN
The Engineering of Architecture 153

FRANK NEWBY AND DAVID COTTAM
The Engineer as Architect – Sir Owen Williams 159

DR. EDMUND C. HAMBLY
The North Sea Challenge 166

PROFESSOR DEREK WALKER
Services and Structure –
The Hong Kong and Shanghai Bank 178

DONALD HUNT
The Grim Tale of the Channel Tunnel 193

MAX FORDHAM
Intelligent Buildings for the Future 201

SIR ALASTAIR PILKINGTON
Research and Product Development – Glass 206

PROFESSOR DEREK WALKER
Today's Engineers – The Legacy Lives On 216

DR. JOHN C. BASS
Solid State Technology –
The Stimulus of Electronics 247

Biographies 260

Great British Inventions in their Time 272

Bibliography 274

Acknowledgements and Photographic Credits 281

Index 284

BUCKINGHAM PALACE.

The Industrial Revolution may have had a number of difficult consequences, but it was responsible for the creation of more wealth, better health and a higher standard of living for the people of this country than they had ever enjoyed before.

The Revolution was the creation of a number of remarkable engineers; many self-taught, whose energy and original ideas transformed life in this country and whose influence was to be felt throughout the world.

Just because engineering is a matter of manipulating materials does not mean that the great engineers of the eighteenth and nineteenth centuries were all unrepentant materialists. They were men of exceptional vision and imagination. Many of them made fortunes and most of them used their fortunes for the welfare of the community. Their names may be linked with their engineering masterpieces, but as individuals they were generous to charity, active in philanthropy, concerned with civic affairs and enthusiastic supporters of the arts.

Fortunately for this country, the talent for engineering has not dried up and every generation produces men whose ideas and enterprise provide opportunities to generate wealth for the benefit of the whole community.

I hope that this book, together with the "Great Engineers" Exhibition, will help people to understand the qualities and characters of these men as individuals and in the context of the times in which they lived.

1987

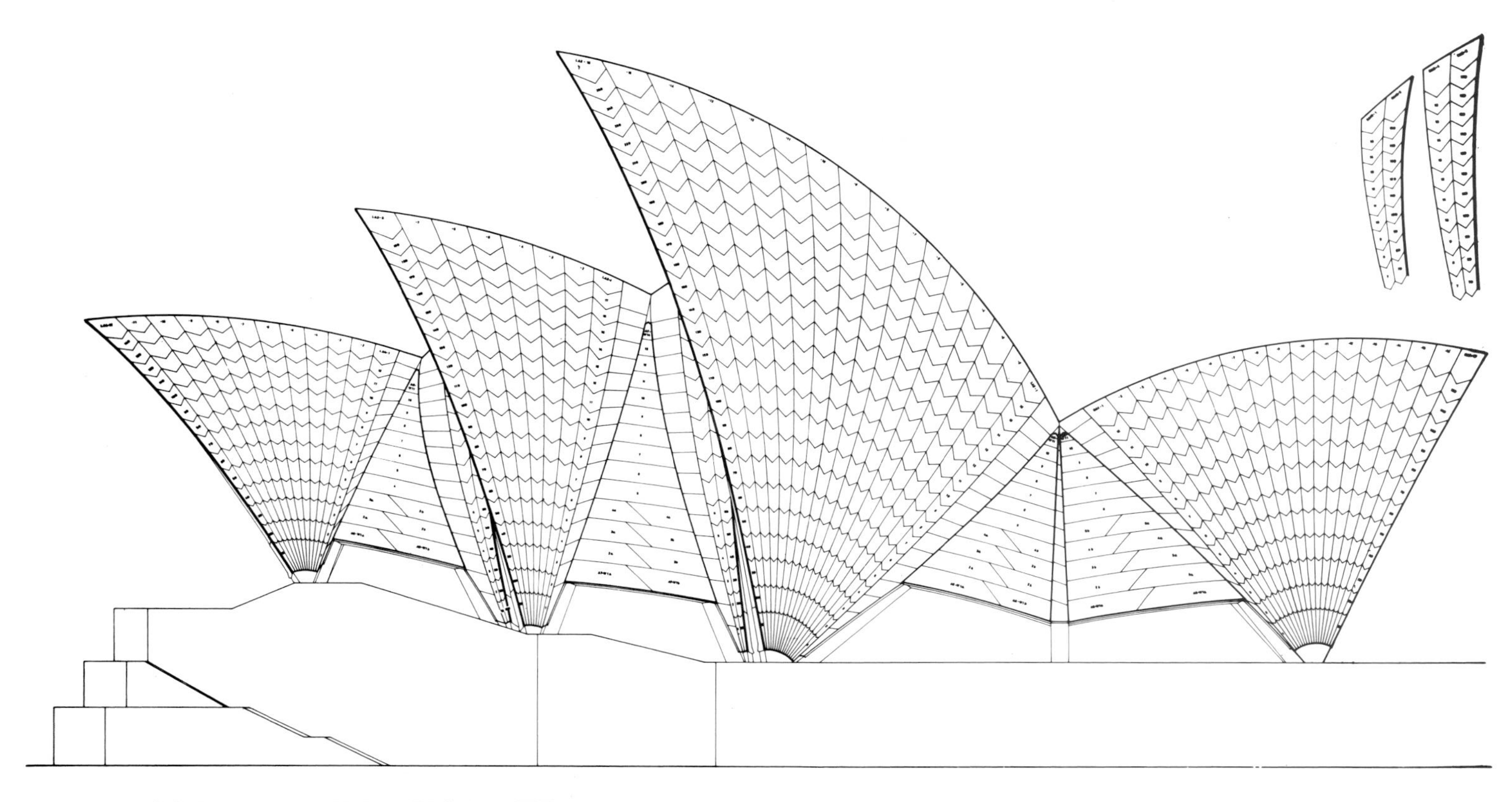

PROFESSOR DEREK WALKER

Introduction

I must take issue with Malcolm Quantrill, who, in naming his recent book 'The Environmental Memory', coined a phrase that I have long harboured as my own; using it to describe the portmanteau of ideas that replenish and drip feed the conceptual bloodstream like a visual memory bank. My environmental memory, like Quantrill's, is derived from the man-made environment which remembers and gives order to human action.

My earliest recollections in the northern countryside were the great structures – bridges and viaducts – that announced the 'railway age'. The magnificent sweep of York Station roof was much more exciting to me at five than was the Mansion House or Castle Howard. It wasn't just the romantic visions of a child impressed with the polished brass lettering and billowing smoke of a steam locomotive at rest in its natural home, or the smells and sounds as it moved slowly into life. It was the whole environment – the overall grandeur of the station. The bustle and movement of people and machines were so much more exciting to me than more civilised buildings.

A few years later I started to frequent a local book shop and, for very little money, amassed two libraries: one on the classical virtues of Yorkshire

Above:
Ove Arup and Partners, Engineers; Jorn Utzon, Architect; Sydney Opera House, elevation
Opposite:
Ove Arup and Partners, Engineers; Derek Walker Associates, Architects; 'WonderWorld' Theme Park, Corby, c. 1987-91: (*top*) theme building model, (*bottom*) axonometric view of site (David Reddick)

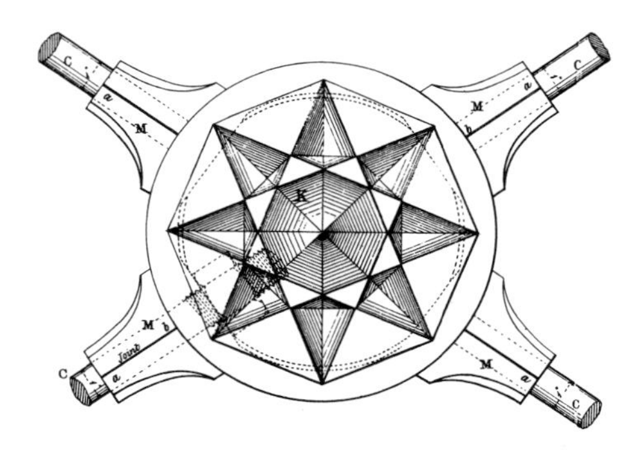

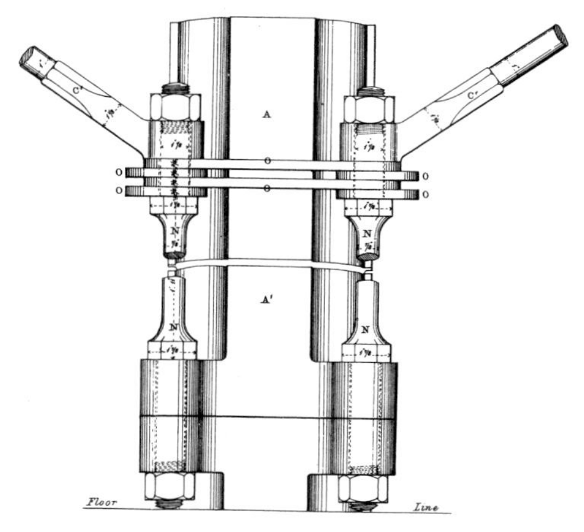

Above:
Sir Joseph Paxton/Sir Charles Fox, Engineers; The Crystal Palace, c. 1851: details of vertical diagonal bracing between columns
Opposite:
Crystal Palace, c. 1851: (*top*) 'View of the Interior of the Building in Hyde Park' (*Illustrated London News*), (*bottom*) 'Closing of the Great Exhibition' (drawing by Joseph Nash)

cricket, the other on northern industrialisation. The Arkwrights, Hargreaves, Cartwrights, Cayleys and Parsons of this world tripped easily from the lips of a Yorkshire child. I was impressed with the tales of engineering invention and ingenuity, and overwhelmed with pity at the privations and dangers experienced by the railway navvies as they built to impossible schedules in impossible conditions.

I started pinning up drawings in my room of strange machines, viaducts, bridges and station canopies, interspersed with my white flannelled heroes who held the county championship with a regularity bordering on divine right. The cricketers did not impress my uncle much, but the industrial images did, and on my twelfth birthday he presented me with two of his particular favourites – McDermott's *History of the Great Western Railway* and copies of the *Illustrated London News* relating to the Great Exhibition of 1851. I was captivated. I didn't know whether to become Brunel, Paxton or Leonard Hutton; a tremendous dilemma for a twelve year old with romantic illusions. The fascination with these boyhood preoccupations remains.

Although I have come to love Castle Howard I still find immense pleasure in examining buildings or structures demonstrating technical invention and structural integrity. It is equally predictable that my friendships in design have been drawn as much from the fields of engineering and product design as from architecture. I have always used architecture as a vehicle for expressing delight and feel that a true designer should be receptive to design ideas from a variety of sources, from product design and furniture to environmental planning. The more one is involved in design the greater one sees the opportunities of collaboration at a productive level.

When I work with Ted Happold, Frank Newby or Jack Zunz, the appraisal and solution to problems seem much more enjoyable, for one is drawing on new resources in working together towards the 'elegant solution'.

As buildings and environments become more complex they need to develop a wider appeal. A new intelligent solution to buildings is required which must, by its nature, be multi-disciplinary. This is not to say that architects and engineers cannot design alone, but simply that, in many cases, they design more efficiently together.

Certainly the shortcomings of separating the work of architects from that of engineers suggests that we should change the way in which these subjects are taught. Architects, structural engineers and environmental engineers must train together if skills are to be integrated so that we can come to a better understanding of the process and science of building.

My interest in technical education was not the only reason I wanted to celebrate the 150th anniversary of the Royal College of Art with an exhibition of 'Great Engineers'; it was a rare combination of circumstances. I re-read Martin Wiener's book *English Culture and the Decline of the Industrial Spirit 1850-1980* one weekend after a week's soul-searching with Ted Happold on the vagaries of interface between educational establishments and industry. He had just returned from the Aspen Design Conference

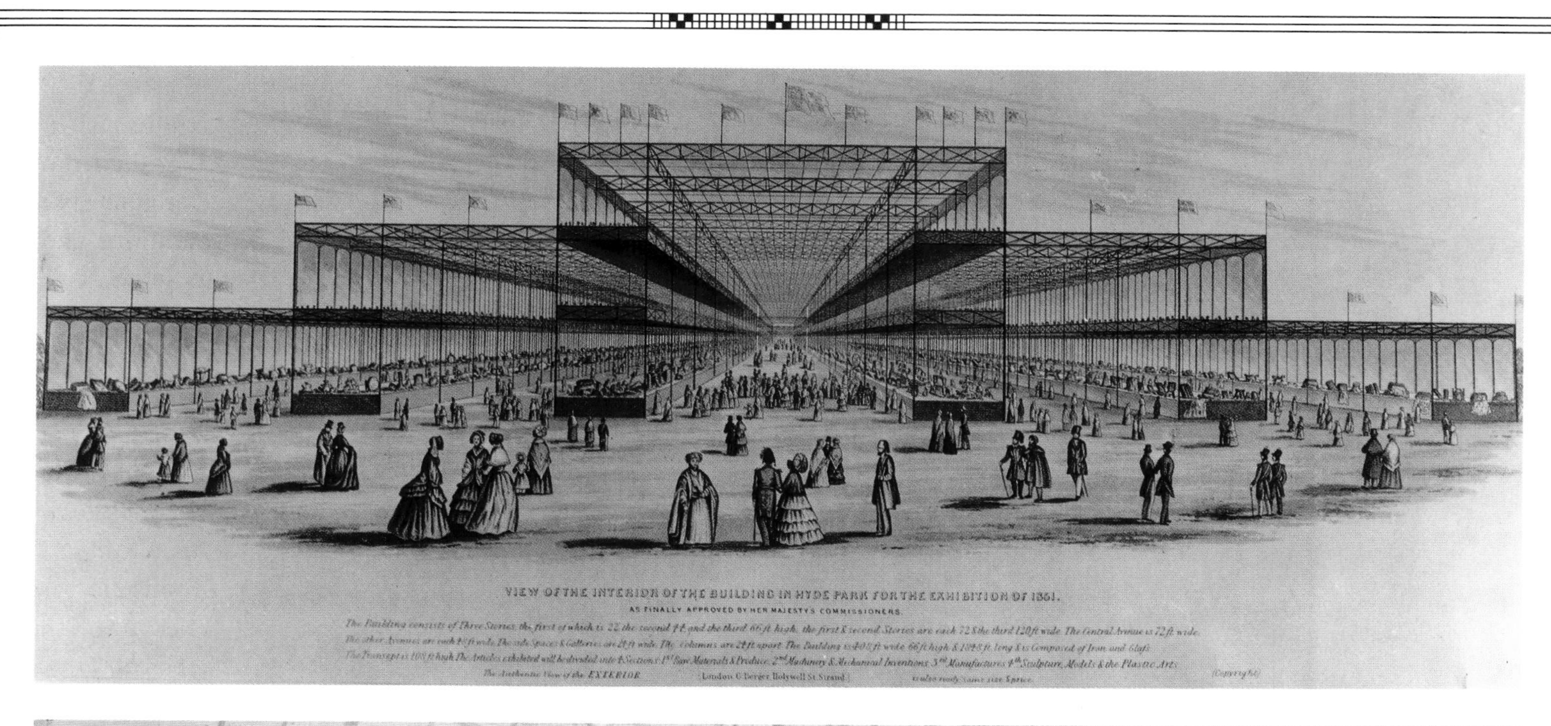
VIEW OF THE INTERIOR OF THE BUILDING IN HYDE PARK FOR THE EXHIBITION OF 1851.
AS FINALLY APPROVED BY HER MAJESTY'S COMMISSIONERS.
London G Berger Holywell St Strand

where he had witnessed a polished ritual 'dance of death' as the British contingent focused on ephemera and frou-frou. Although slick in presentation, the content was impoverished and the form of design necessary to provide enough satisfactory work for people in this country was conspicuous by its absence; the design of useful objects was neglected and instead a view of design concerned purely with fashion and marketing was celebrated. As the old joke goes: 'It's a great video but a lousy song'. The Conference was about packaging not product, style not function, art schools not engineering.

The Wiener Thesis is even more disconcerting because in a sense it chronicles the British ambivalence towards industry over the last 130 years and demonstrates how crippling this distaste for the capitalist ethos has been to our economic development.[1]

For a long time the English have not felt comfortable with 'progress'. As one social analyst has noted:

> 'Progress' is a word that in England has come to possess a curiously ambiguous emotive power. It connotes tendencies that we accept, even formally approve, yet of which we are privately suspicious.

It is an historic irony that a nation which sparked off the industrial revolution throughout the world, should have become embarrassed at the measure of its success. The English nation became so ill at ease with its prodigal progeny as to deny its legitimacy and fashion a conception of Englishness that virtually excluded industrialism.

This suspicion of material and technological development and the symbolic exclusion of industrialism were intimately related. They appeared in the course of the industrial revolution but, instead of fading away as the new society established itself, they persisted and indeed were extended and strengthened. In the later years of Victoria's reign they came to pervade 'educated opinion', forming a complex, entrenched cultural syndrome. The idealisation of material growth and technical innovation received a check, being more and more pushed back by the contrary ideals of stability, tranquility, closeness to the past and 'non-materialism'. An 'English way of life' was defined and widely accepted which stressed non-industrial, non-innovative and non-material qualities best encapsulated in rustic imagery – 'England is the country' as Stanley Baldwin put it (a phrase which was by this time already a *cliché*). This countryside of the mind was everything industrial society was not – ancient, slow moving, stable, cosy and 'spiritual'. The English genius, it proclaimed, was (despite appearances) not economic or technical but social and spiritual – it did not lie in inventing, producing or selling: but in preserving, harmonising and moralising. The English character was intrinsically not progressive but conservative; its greatest task – and achievement – lay in taming and 'civilising' the dangerous engine of progress it had unwittingly unleashed.

Throughout its history, the Royal College of Art has not always acknowledged the reason for its formation as a design establishment, which was to forge and maintain links with industry and to encourage an element of design

Above:
Top: Joseph Clement, drawing of a Beam Engine, 1830s. *Bottom:* James Nasmyth, Engineer; Steam Hammer, Patricroft, Manchester, c. 1851
Opposite:
Top: I.K. Brunel, Engineer; 'Raising the Tubes at Saltash', c. 1858. *Bottom:* Robert Stephenson, Engineer; 'Britannia Bridge under Construction', c. 1847

1. Martin J. Wiener, *English Culture and the decline of the Industrial Space*, Cambridge University Press, 1981.

Right:
'Banquet in the Reservoir of the Croydon Waterworks'
Opposite:
Top: Richard Turner, Engineer; Decimus Burton, Architect; 'The Palm House at Kew', c. 1844-48 (painting by Ben Johnson). *Bottom:* 'Metropolitan Line Station at King's Cross', c. 1863

within engineering by absorbing it into the process as a whole, rather than, like surface stylists, just applying a little polish to an object of overwhelming mediocrity. The climate of today's market does not warrant the mutual exclusivity of design and manufacturing at any level or scale and, because of industry's great need for invention, design, packaging and marketing, it is important we celebrate designers who really understand the process – intelligent building, corporate identity, appropriate use of materials and high levels of environmental comfort. The environment should not be a public lottery overseen by the blind. It demands concentration on quality and consistency.

This search for quality and innovation has flourished despite the tepid public response to inventive design and, though at times the line grows tenuous, it is possible to trace a family of solutions from the Crystal Palace right through to the Hong Kong and Shanghai Bank; giving us reason to suppose that the romantic breed of engineers who dominated the pre-eminence of British industrial growth in the 19th century have got their counterparts today and that the challenges remain equally crucial, though in many cases international corporate structures have replaced the isolated flashes of entrepreneurial skill prevalent in the 19th century. The quality product is still feasible and the key to success lies in the acceptance by designers and industry that good design markets the product and that the British disease of sublime prototypes and abysmal production can be solved.

The chapters that follow hopefully illustrate how a high quality of engineering and invention still permeates the British psyche. Edmund Happold in his survey of built engineering projects displays a dynamism

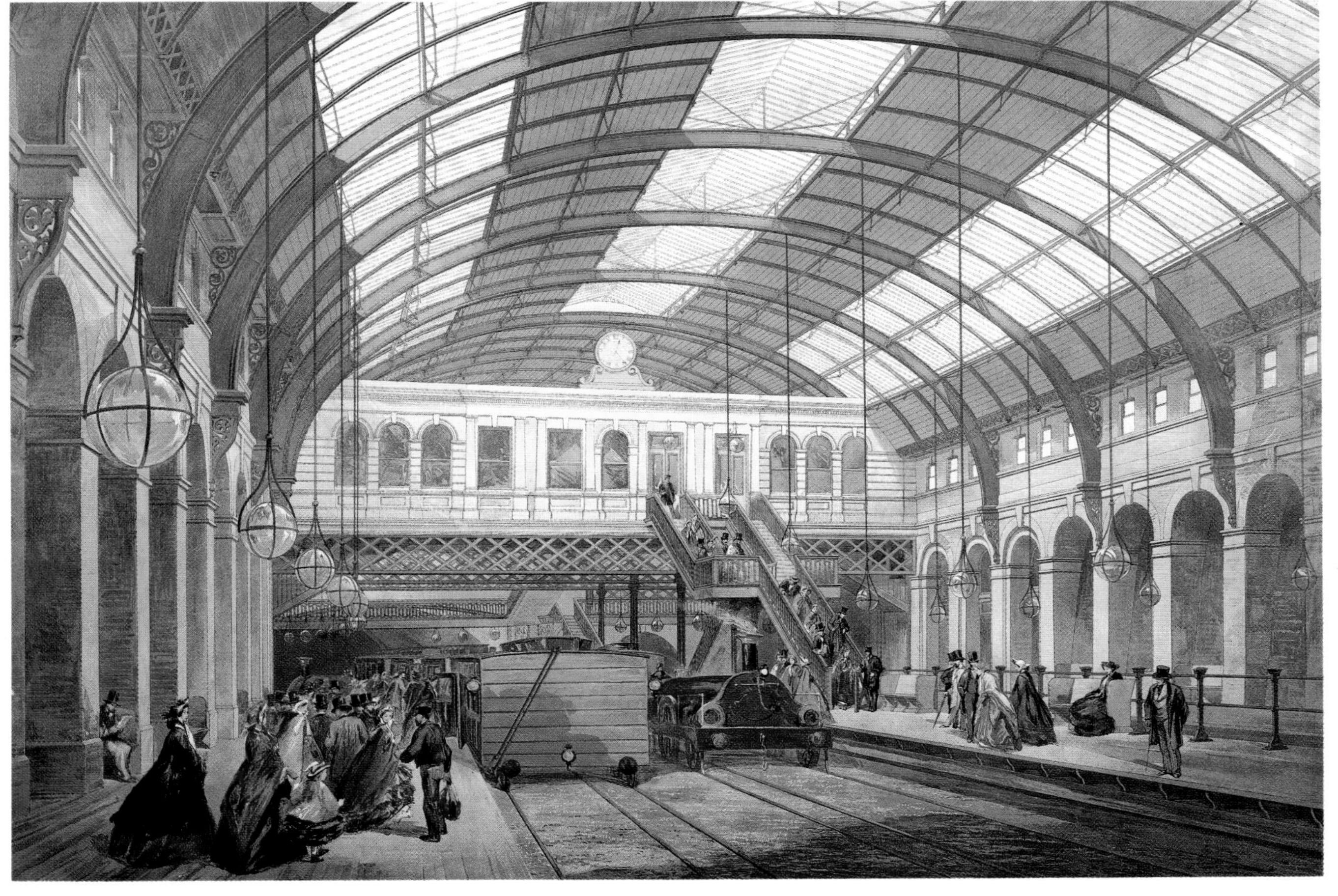

Left: Felix Samuely and Partners, Engineers; Powell and Moya, Architects; The Skylon, c. 1951, Festival of Britain. *Above:* Sir Benjamin Baker/Sir John Fowler, Engineers; 'The Forth Bridge under Construction', c. 1882-90
Opposite:
Ove Arup and Partners, Engineers; Foster Associates, Architects; Stansted Airport, UK, c. 1987: computer-generated drawing of structure and service integration (*top*), model of the terminal building (*bottom*)
Page 20:
Ove Arup and Partners, Engineers; Foster Associates, Architects; Hong Kong and Shanghai Bank, c. 1986, developing the structure
Page 21:
Ove Arup and Partners, Engineers; Richard Rogers Partnership, Architects; Lloyd's Building, London, c. 1987, facade detail

reminiscent of his Quaker forebears – those non-conformist heroes of Britain's 19th-century industrial success. In discussing engineering as a profession he mentions the division within 18th-century society between the aggressive merchant class and the industrialists served by the engineering profession. As Martin Wiener has shown, the merchant class consistently came out on top for a hundred years.

What is the solution? Not the proliferation of what Private Eye styles the bullshit middlemen industries of the 1980s. Happold's sombre view is that something is radically wrong when 'pop', public relations, video and a lot of non-jobs in the City are more financially rewarding and have higher status than engineering. More than that, it is the all-pervasive values of the successful merchant class which drives the wedge between art and technology. Anything old is respectable, while anything new is out; anything rural is respectable, while anything industrial is out. Perhaps Happold's fellow traveller, William Gosling, is likely to be more sanguine about the future of electronics – for one thing his field is hardly half-a-century old and so the British have not yet had time to institutionalise it or curb its consistent need for development and invention. It also has the richest assets – complexity and mystery – which mean that a much smaller group of experts and *afficionados* can offer gratuitous advice, making the building task that much easier.

What we have tried to accomplish in our joint narratives is a celebration of

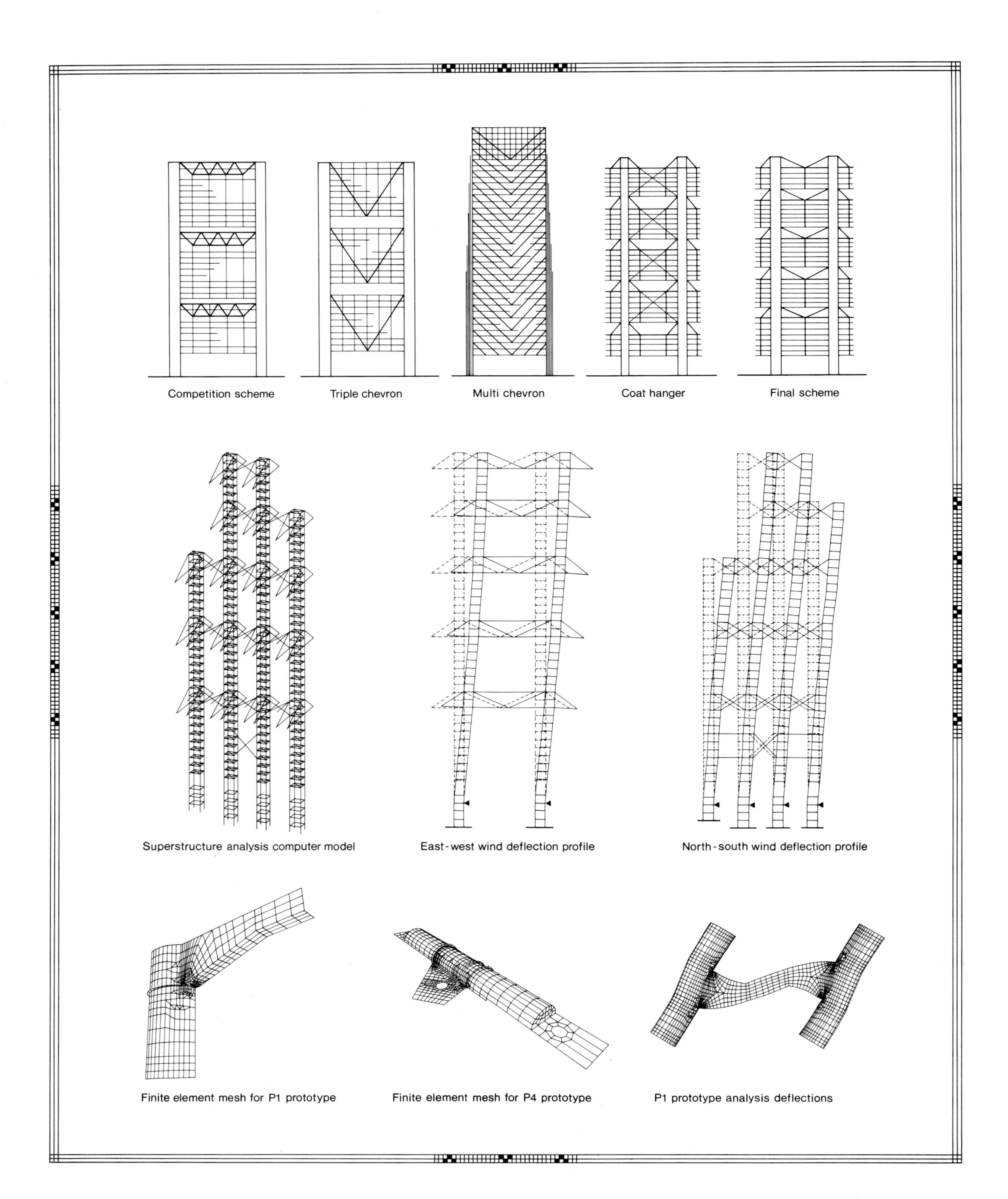
Competition scheme

Triple chevron

Multi chevron

Coat hanger

Final scheme

Superstructure analysis computer model

East-west wind deflection profile

North-south wind deflection profile

Finite element mesh for P1 prototype

Finite element mesh for P4 prototype

P1 prototype analysis deflections

the perpetual search for excellence that has dominated the lives of our greatest inventors and engineers. Visually and romantically the 19th century was extraordinary. What is equally revealing is that the 20th century does not pale in comparison. The status of Britain shorn of its Empire and industrial dominance is a fact – what must not be lost is the spirit and resources of the British technologist. The omens are optimistic; we have a generous base of talent in engineering, architecture and science – a new national awareness has started to preserve and extend our industrial base. Perhaps, as Sir Eric Ashby stated:

> Technological education is concerned with comprehending technology in its completeness, to teach students to state clearly to non-specialists what they do and why is a test of our education.

Perhaps Ashby should have been co-opted to the Rogers/Arup team to buttonhole Lloyds on the paradox of the Lutine Bell tradition and the arrival of electronic communications.

Peter Reyner Banham[2] was characteristically eloquent in describing the paradox last year when he surveyed the building in question.

> Surely in the high-tech world of electronic communications and personal computers most of these people could stay at home – stay in bed, even – and do their business without crowding on to this hopelessly overloaded site. However high the technology they employ the underwriters see no alternative to their common presence, blocking gangways and getting in one another's way generally, in this single room. Being there is what it's all about, the drama, the boredom, the irritation, the moments of high panic and all that ancient tradition, symbolised by the presence of the Lutine Bell to be rung to announce shipwrecks – under which sits a uniformed attendant calling for missing underwriters by means of a microphone and a current electronic paging system. Every item of high technology at Lloyd's – structural, environmental, mechanical or electrical – services a concept of business that still puts its ultimate faith in face-to-face contact and rumours whispered in passing ears.

If one finds this an irony – and it is extremely difficult not to – then it seems to be an irony proper to our late-Modern times. Architecture may have abandoned its utopian dreams of changing the world, because the world is perfectly capable of changing itself without architecture's aid, but the compulsion to try and make sense of the resulting human dilemmas is still the most essential quality of design.

Above:
Ove Arup and Partners, Engineers; Piano and Rogers, Architects; Centre Georges Pompidou, Paris, c. 1971-76: gerberette column and cross-bracing
Opposite:
Centre Georges Pompidou, Paris, c. 1971-76: services facade detail (*top left*), piazza facade detail (*top right*), cross-bracing (*bottom*)

2. Peter Reyner Banham, *Architectural Review*, 1986.

Above:
Structure of Graf Zeppelin Airship, c. 1928
Opposite:
Frei Otto/Büro Happold, Engineers; Munich Aviary, West Germany, c. 1980, detail

PROFESSOR EDMUND HAPPOLD

A Personal Perception of Engineering

Everyone knows the harshness of nature; how mistakes become extinct. Yet the unique characteristic of man is that his reason and imagination have enabled him to develop and adapt to his environment. And history shows how technological development has liberated him.

At first it was the discovery of a type of wheat which gave an abundant yield, achieved by ploughing, sowing and reaping, which led to the possibility of settlement and the building of permanent structures. Today there is the exploration of the deep – the conquest of space.

Yet in response there has been continuous concern at such advances. Fear of change. Fear of consequences. Technology, especially since the time of the industrial revolution, has been seen as a dehumanising force to be resisted.

Art has been seen as the civilising counterbalance to these advances. Art is seen as expressing the individuality of man and promoting cultural evolution. It is appropriate that this book has its roots in the Royal College of Art's 150th anniversary, since the college has its origin in a report on Arts and Manufactures which, in 1836, recommended the formation of a School of Design in which 'the direct application of the arts to manufactures should be deemed an essential element'.

But technology (defined by Galbraith as 'the systematic application of scientific or other organised knowledge to practical tasks') – or engineering if you prefer that word – obviously cannot in itself be a bad thing, it is how we use it. And a world which sees art and engineering as divided is not seeing the world as a whole. These days often only those people with an arts training are said to be creative. But, if the truth be told, it is technology that is creative because it gives new opportunities. Historic ideas of art and culture can entrap. It is technology that frees the scene.

Throughout history there has been a succession of turning points, achievements by engineers which represent a new conception of nature. This book is about these turning points, which express why engineering can be so intensely satisfying because it is, at its best, an art grounded in social responsiblity.

Industrial archaeology i.e. the history of engineering – is now culturally acceptable and certainly influences modern fashion. Yet as an engineer designer I represent an approach to design whose roots are not dependent on visual precedents. I am referring to engineering design as a technological idea, with its own aesthetic.

The engineering profession as we know it today developed to serve the non-conformist industrialists of the 18th century, who were the midwives to the industrial revolution. These men, predominantly Quakers, were barred by dint of their non-conformity from the established universities and professions. They found creative opportunities within their limited possibilities by turning to inventive industry. And because they believed in the equality of mankind guided by individual conscience, they backed humanistic management. They had very broad long-term aims. They coped with persecution by forming close bonds with their fellow industrialists and these inter relationships and the pooling of ideas and information facilitated the development of the industrial revolution.

The founders of our profession were creative mechanics. In 1712 Newcomen erected his first steam engine. In 1759 Smeaton carried out a classic study on water power. In the forty years that he worked as a consulting engineer, Smeaton regularly used the word 'professional' to describe himself. Each morning he was employed on a time basis to consider problems and design schemes. But the fact that he saw his scientific studies as the basis of his work is described by his daughter, Mary, who wrote 'his afternoons were regularly occupied by practical experiments, or some other branch of mechanics'. This interest in the scientific study of the sources of power in nature, together with the performance of materials, represents one aspect of a technologist's body of knowledge. The other aspect is in the development of construction methods; the organisation of work. Some of the achievements of the other accredited founder of the civil engineering profession in Britain, James Brindley, well illustrate this. In the construction of the Bridgewater Canal (completed 1769) he popularised the use of 'puddled clay': mixing together readily available sand and clay and getting his workmen to tramp it into the bottom of the canal with their boots to provide an impervious yet flexible lining.

Engineering technology really took off in 1760 when two foremen at a Quaker ironworks in Coalbrookdale produced cheap iron using coal, not wood, as fuel. It was the beginning of the era of making materials with creative possiblities using non-renewable resources, such as metals and fossil fuels.

In 1779 the world's first iron bridge at Coalbrookdale was complete. In 1801 the first steam carriage was built by Richard Trevithick. In 1826 Telford's great suspension bridge over the Menai was completed. George and Robert Stephenson built the first effective steam railway and I. K. Brunel's steam ships bridged the Atlantic. Technology transfer started early – every American knows the portrait of Whistler's mother – but his father came over to Britain to learn railway engineering from the Stephensons and not only returned to pioneer railways in America, but also went to Russia and started the construction of the trans-Siberian railway. These engineers started the staggering development of technologies which have in the last 130 years, so changed the world.

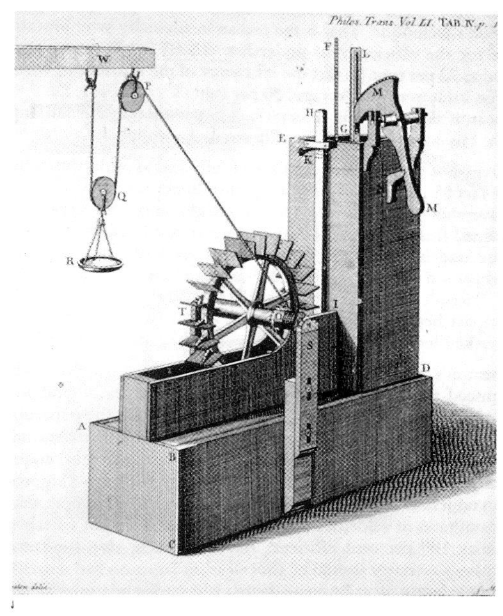

Above:
Top: John Smeaton, Engineer; 'An Experiment' ('his afternoons were occupied by experiments'), 1751. *Bottom:* Captain Savery and Thomas Newcomen; 'The Atmospheric Engine', 1712
Opposite:
'John Smeaton, the First Consulting Engineer' (engraving by W. Holl)

Above:
Top: James Brindley. *Bottom:* charcoal burning
Opposite:
Top: 'The World's First Iron Bridge', Coalbrookdale, Shropshire, c. 1779. *Bottom:* 'Ironworks at Coalbrookdale', c. 1775

As a structural engineer, or technologist, I acknowledge that my way of thinking about the world falls into what Persig designates the 'classical mode'. You may remember that in his book *Zen and the Art of Motor Cycle Maintenance* he describes those who see the world primarily in terms of immediate appearance as thinking in the 'romantic mode' and those who see the world primarily in terms of underlying form as thinking in the 'classical mode'.

His interesting example is of Mark Twain, who wrote in lyrical terms about the Mississippi River until he went to train as a river boat pilot. He gained a deeper understanding of the river through learning its science, but in the process it lost its original magic. Persig argues that the romantic mode tends only to develop existing forms, whereas the classical mode is capable of producing originality.

Most engineers would see themselves as falling into the latter category. Their craft is intensely creative; at its best it is art, in that it extends people's vision of what is possible and gives them new insights. But the aesthetic produced is 'bare', it may not have been seen before, and it is more likely to relate to a natural rather than historical, precedent. This is what I mean by engineering design as a technological idea as distinct from a visual style or fashion.

Perhaps the best known example of this is the Crystal Palace, produced not as an 'art object' but because it worked. Prefabrication and organisation of plant and labour, together with the significance of iron and glass as building materials, were demonstrated to the world. But to the art/architecture establishment of the time, notably Ruskin, the Crystal Palace was anathema. He advocated a return to the style of Gothic Mediaevalism. It is also interesting that although fire-proofed cast-iron structures were produced only two years after the erection of the Ironbridge at Coalbrookdale, it was a quarter-of-a-century before they were used in architect designed buildings.

Certainly in Britain there is a belief that technology must be tamed and controlled, largely by imposing standards of visual beauty which comply with criteria set up by those who have studied the arts.

Sixty-five years ago Roger Fry, amongst others, was complaining that the British were in love with ancient art and too little interested in the work of their contemporaries. I sometimes wonder whether we were not almost fatally damaged by the refugees from Hitler's Europe. Those who were gifted designers, frustrated by the reactionary environment here, went on to the USA and revolutionised design there. The art critics and historians, discovering our conservative, class-rich environment, stayed on to become famous and powerful in our establishment.

But technology is about change. It is concerned with the development of useful objects or processes which change our lives. It does this in response to people's aspirations or is restrained by people's fears; in this it relates to the arts. What it does must obey the laws of nature, which is why it uses science to examine behaviour. Technology is the making of things while science is the

explaining. So the roots of engineering are in nature.

Everything in the built environment has been achieved by technology. Every single man-made object in the world is the product of technology and traditionally the modern built environment divides into structures and machines.

The relationship between structures and machines is extremely interesting and there are many aspects they have in common. The most efficient use of materials, or perhaps energy, is a major one. Yet the energy in structures, which are seen as essentially permanent, is in the production of the materials and in the construction. Machines on the other hand, are designed to convert energy efficiently – into motion, heat, messages and the like. Structures are the steady parts of the system, machines the dynamic and the two are entirely interdependent. You cannot have an aeroplane without an airport, a generating turbine without foundations and a building, a chemical plant without a structure, cities without drains, power, etc., even a plane without engines and a frame. But in detail structures and machines are often completely different. You will find other essays in this book written by experts in the design of machines.

But does our society really understand and recognise the great contribution engineers have made to the modern world, and are still making today? Often, I think, the answer is 'No'.

Perhaps it is not surprising that the public, with its need to 'individualise' success, fixes its approbation on the package designer whose pencil co-ordinates and markets a design. In the design of machines, such as washing machines, computers or cars, the public often realises that the qualities which make the product excel are provided by engineers, but in my own field – the design of buildings – the engineer's role is less understood and therefore more undervalued. Nearly everyone in Britain, and perhaps elsewhere too, thinks that creative design in building is exclusively due to architects.

But in building, the product is usually a complex one, requiring many skills in order to put many values into it. I am a building or structural engineer working in partnership with architects and others, each group bringing a body of knowledge, experience and sensibility to a common problem. Today construction is about big money and to handle that successfully calls for toughness and rigour. Autocracy or selfishness are not called for, but a system of collective decision making is essential. For such a partnership means mutual authority and shared recognition amongst the members of the building team.

I hope this book and the exhibition it springs from will go some way towards improving public perception of the development of art in technology and their interrelation.

Engineers, almost by their nature, excel at group work and avoid extravagant claims. They are very conscious that design usually requires many specialists who are designers in their own right and who put different qualities into the product. Engineers are sensitive of claiming sole authorship.

Above:
Top: 'Whistler's Mother'. *Bottom:* 'Whistler's Father'
Opposite:
The Ironbridge, Coalbrookdale, c. 1779

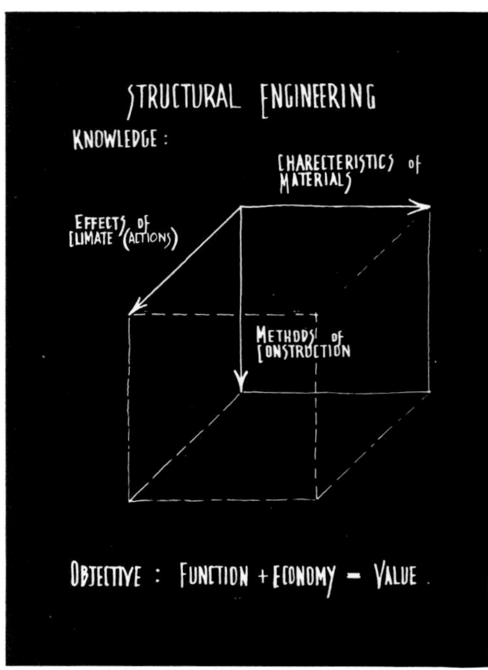

Above:
Top: Sir Joseph Paxton/Sir Charles Fox, Engineers; 'Building the Crystal Palace', c. 1851.
Bottom: Objective, Function + Economy = Value
Opposite:
Top and bottom: Richard Turner, Engineer; Decimus Burton, Architect; 'The Palm House at Kew', c. 1848

I am a structural engineer with an interest in building physics. Structural design is primarily concerned with the choice of form; the forces on that form and the analysis of its behaviour follow on from that choice. The whole process is influenced by the need for feasibility of execution, as success is proved by practicality.

Which qualities are essential to good structural engineering? Engineers should have an interest in the behaviour of materials and a knowledge of the physics of the environment. We need to give value for money. As Herbert Hoover said. 'An engineer is a man who can do for one dollar what any fool can do for two'. And I think our ambition is to achieve elegance as well as value; 'elegance' in the mathematical sense, meaning economy as well as appropriateness. Appropriateness (or function) + economy = value. As a French aircraft designer once said, 'When you cannot remove any element then you have the right design'. And here of course we can learn from nature in which structures have to be totally appropriate and mistakes become extinct.

The need for better living conditions for more people makes the engineer's struggle for efficiency worthwhile. I come from the generation which, largely due to accelerated technological changes in construction during the Second World War, came into engineering because of an interest in the efficient use of materials. This interest has pervaded my work as well as that of my colleagues and many others of my generation. We join a long line of engineers who have been working at long span, large space enclosures. The concrete shells of Nervi and Candela were modern versions of historic solutions; products of local materials and skills plus the advantages of mass in hot climates. Such structures are still relevant and we continue research into their construction methods. Buckminster Fuller emulated the Victorian engineers by following traditional forms, but he copied nature by reducing materials. The Pompidou Centre was not intended to be, but ended up as, a pastiche of the Crystal Palace –and what an expression of intermediate space that was. The energy input/output ratio of steel is high, so one becomes interested in timber. Thus the German architect, Frei Otto, using an equal mesh three-dimensional version of Robert Hooke's hanging chain model to define a tensile net form which, when reversed, provides a pure compression shell, then proposed that it be built with timber lathes. It would have collapsed. To act as a shell it needed diagonal cables to provide sheer stiffness but, because of its lightness, it was very economic to erect. It was test loaded with the town's dustbins and covered with a coated fabric.

To carry a force in tensions is, of course, in material terms the most effective, and a very old, building solution. Reinforced with steel cable, such roofs can safely and economically cover several acres. The actual skin can be concrete – as in Calgary, designed by Jan Bobrowski with our partnership as proof engineers – or it can be pvc, ptfe coated fibre glass, timber shingles, ceramic tiles or even, nowadays, stained glass. But most loading is from the wind and snow and thus cyclical. Steel cannot compare with the energy-

storing characteristics of timber – but, if the steel wires are crimped to act in the more lightly loaded condition like a spring, it can store energy mechanically. We have used crimped steel mesh as a fabric for aviaries in Munich, San Diego and Hong Kong. Compare the wires' similarity to a spider's web.

But, to return again to timber, recent comparative studies have shown us that it is the strain energy characteristic which also reduces the cost so radically. Timber, in proportion to its weight, is comparable in strength and stiffness to high strength steels. But the problems with timber have always been in achieving an effective tension connection between members, and it is only since the discovery of epoxy adhesives, impregnated with fibres developed for windmill blades in the USA, that 'high strength' collinear connections have been made possible. We can now achieve eighty-five per cent full strength.

For many years carpenters have resisted removing any more of a tree than the bark. This is because the tree, subjected to sudden gusts of wind and not wishing its outer capillaries to buckle, orientates some of its fibres diagonally around the capillaries so that in containing the sap an element of longitudinal pre-stressing is given to the outer fibres (to the order of 14 MN/m^2), while the centre of the tree is in compression. When the wind blows, the outer capillaries stay in tension, which they are well able to carry. Combining advances in timber connections with the tensile properties of raw timber has made possible the hanging forms in the School for New Woodland Industries at Hooke Park on which we are now working.

Using a structure not only to resist forces but also actively to amend the internal environment, becomes interesting. Light, of course, is the most powerful source of energy, and designing covered cities in the Arctic is simply an extreme version of this. In order to achieve a satisfactory all year round environment under such a cover, the quality of light is all important and that quality is dependent on as much of the spectrum as possible being transmitted through it. This is why glass is used for windows, even though it does reduce the ultra-violet part of the spectrum, thereby creating a 'greenhouse' effect. Studies have been carried out for an air-supported cover over a proposed thirty-six acre city in the Athabasca region of Alberta. These show how some of the new laminates transmit more of the light spectrum than glass – making possible even the growing of grass in the Arctic. Alas, the collapse of the oil market caused the project to be dropped. This development has moved slowly since then, but one of the German manufacturers has used this laminate as a cover in building a conservatory for tropical birds and alligators at Arnhem in Holland, and we used the lighting aspects of the concept for a roof in Brunei. It seems to perform very well.

Of course, our bubble idea is still relatively crude and needs further work. But one cannot help but be interested in further developing the possibilities of utilising energy from light. After all, this is only the old idea of the conservatory or greenhouse. When it is too hot we shade ourselves like plants

Above:
Top: 'Catch me who can', Trevithick's Locomotive in Euston Square, c. 1809. *Bottom*: 'These days people with an arts training are said to be creative. But if the truth be told, it is technology that is creative, because it gives us these new opportunities. Historic ideas of art and culture can entrap. It is technology.'
Opposite
Top: I.K. Brunel, Engineer; 'S.S. Great Britain', c. 1843. *Bottom:* Thomas Telford, Engineer; 'Bridge over the Menai Straits', c. 1818-26
Page 36:
Top and bottom right: Ove Arup and Partners, Engineers; Richard Rogers Partnership, Architects; Centre Georges Pompidou, Paris, c. 1971-76, final model and co-ordination drawing at street level of the services facade. *Bottom left:* Crystal Palace, interior of transept
Page 37:
Büro Happold, Engineers; Derek Walker Associates, Architects and Landscape Architects, Kowloon; Kowloon Park, Hong Kong, 1986-89, site model

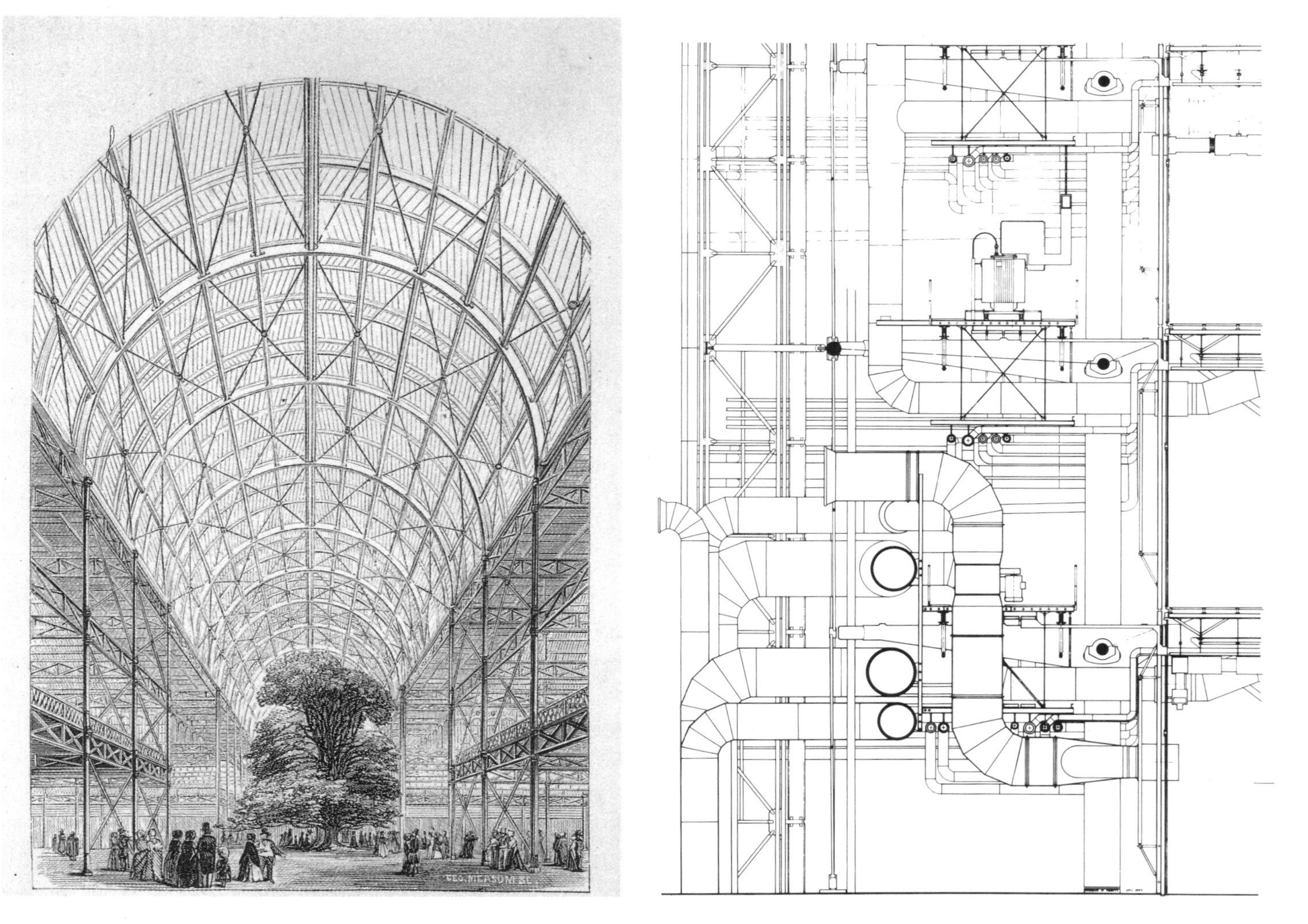

do with their leaves. But we still cannot build firm organic substances from carbon dioxide, water and light, as plants do. We must find out how, before the slight surplus store of carbon – coal, oil, gas and timber – which has been built up over millions of years and which at the moment provides most of our structural materials, is finally depleted.

Engineers are moving towards it. In fact organic materials are becoming more and more important. And just as 200 years ago those who resisted change thought iron and steel 'inhuman' materials, today people say the same about plastics. Yet such materials, in composites and laminates, are essential materials in aerospace and their use is growing.

All this is about change. Yet much engineering is at a simple level. The pace of advance in technology is generally set not by the most brilliant and able engineers but by the capacity of the individual – engineer or skilled mechanic – to master and use an improvement efficiently and harmoniously. My last example illustrates this: an exposed bridge over a busy road in a town. A simple steel trussed solution can be prefabricated and erected in a matter of hours. Then it can be clad like the building to reduce exposure and deflect the wind. A simple piece of street furniture.

You may have noticed that most of the ideas in these projects were first developed for overseas clients. Why is new technology not better utilised in Great Britain? The Prince of Wales is promoting community architecture in the cities. Yet there are few problems of the inner cities which could not be solved by the economic well-being which productive industry could provide. It is engineers who could bring this about. Our car industry is a failure, yet we have engineers who design the fastest, safest and most reliable cars in the world. Our car body stylists, trained at the Royal College of Arts, are in demand in Germany, the United States and Japan, where they work in Mercedes Benz and other successful car firms. If only these two groups could work together in Britain as equals, they could succeed.

Yet our intellectual and economic class systems are such that this seems impossible. We have the skills but not the business or social structure to enable our engineering expertise to revitalise our industry. Perhaps the failure is managerial. What is common to most successful overseas technical enterprises is the inevitability of collective decision making. Could it be that our national belief that someone should be in charge – preferably on an annual financial basis – inhibits this? Maybe those Quaker industrialists with their collective long-term ambitions were right and we should consider returning to those lost managerial values.

Above:
Top: Felix Candela, Engineer; La Jacaranda Night Club, Mexico, 1957. *Bottom:* Pier Luigi Nervi, The Palace of Sport, Rome, 1956-59
Opposite:
(*Top left to bottom right*) Ove Arup and Partners/Frei Otto, Engineers; testing a cable net structure. Büro Happold, Engineer; Diplomatic Club, Riyadh, Saudi Arabia, c. 1986. Buckminster Fuller, Engineer; American Pavilion, Montreal Exposition, c. 1967. Büro Happold, Engineer; 58° North Artic City Air Supported Structure, c. 1983. Ove Arup and Partners, Engineers; Frei Otto/Edmund Happold/ Ian Liddell; Mannheim Garden Centre, c. 1975. Le Mans. Büro Happold, Engineers; ABK Architects; Hock Park Scheme, c. 1987. Büro Happold, Engineer; Frei Otto/Rolf Gutbrod, Architects; Sports Stadium in Jeddah, Saudi Arabia, c. 1984

Above:
Top: SS Great Eastern Atlantic Cable paying out equipment, c. 1865. *Bottom*: I. K. Brunel, Engineer; 'Attempts to launch the SS Great Eastern', 1858 (photograph by Robert Howlett)
Opposite:
Top: Anthony Hunt Associates, Engineers; Richard Rogers Partnership, Architects; Inmos Microprocessor Factory, Newport, Wales, c. 1982 (painting by Ben Johnson). *Bottom*: a six-inch silicon wafer symbolises the latest microchip advance by Plessey

PROFESSOR WILLIAM GOSLING

A Short History of Electrical Communications

The understandable desire to be able to communicate at a distance is very old. Legends and stories come down to us from antiquity about magical scrying-stones, sympathetic needles and other such things supposed to convey this preternatural gift. The first historical reality was a visual signalling system invented at about the time of the Revolution by a Frenchman, Claude Chappe. His telegraph was really more of a semaphore: a high tower on top of which were fixed two jointed mechanical arms, controlled by rods from levers mounted inside the tower, where the operator could set the position of the arms to one of a series of patterns which signalled the letters of the alphabet.

Napoleon I very well understood the military utility of the system, and rapidly Chappe set up a network of such towers, all within telescope sight of the next in the chain, stretching right across French-occupied Europe. Completed in time of war, the building of the chain was an epic of human endurance – some of the construction crew actually died of starvation in the course of the work! Readers of Alexandre Dumas' famous novel will recall that, much later, it was by interfering with the transmission of stock exchange information along the Spanish part of the chain that the Count of Monte Christo succeeded in ruining the evil Baron Danglars.

For a time optical telegraphs of various kinds functioned in all the more advanced European countries. In England, Murray invented a simpler variant, and the Admiralty established a line from London to Chatham's Royal docks, but in England's imperfect weather the frequent mists and downpours must have made the system distinctly unreliable.

In 1753 a 'Mr. C M' – the initials have never been reliably deciphered – wrote to *The Scot's Magazine* with a proposal for an electric telegraph, which they published as a scientific curiosity. At that time the only electrical phenomena recognised were electrostatic but CM proposed to signal by means of the ability of an electrically charged wire to attract a small piece of paper.

The idea was to use twenty-six carefully insulated wires, one identified with each letter of the alphabet. At the transmitting end, the wire corresponding to the letter to be signalled was touched with an electric charge, for example from an electrostatic machine, whilst at the receiver each wire was located over one of twenty-six small pieces of paper, so arranged that when they were attracted by the charged wire above them they rose to reveal the letter written on them.

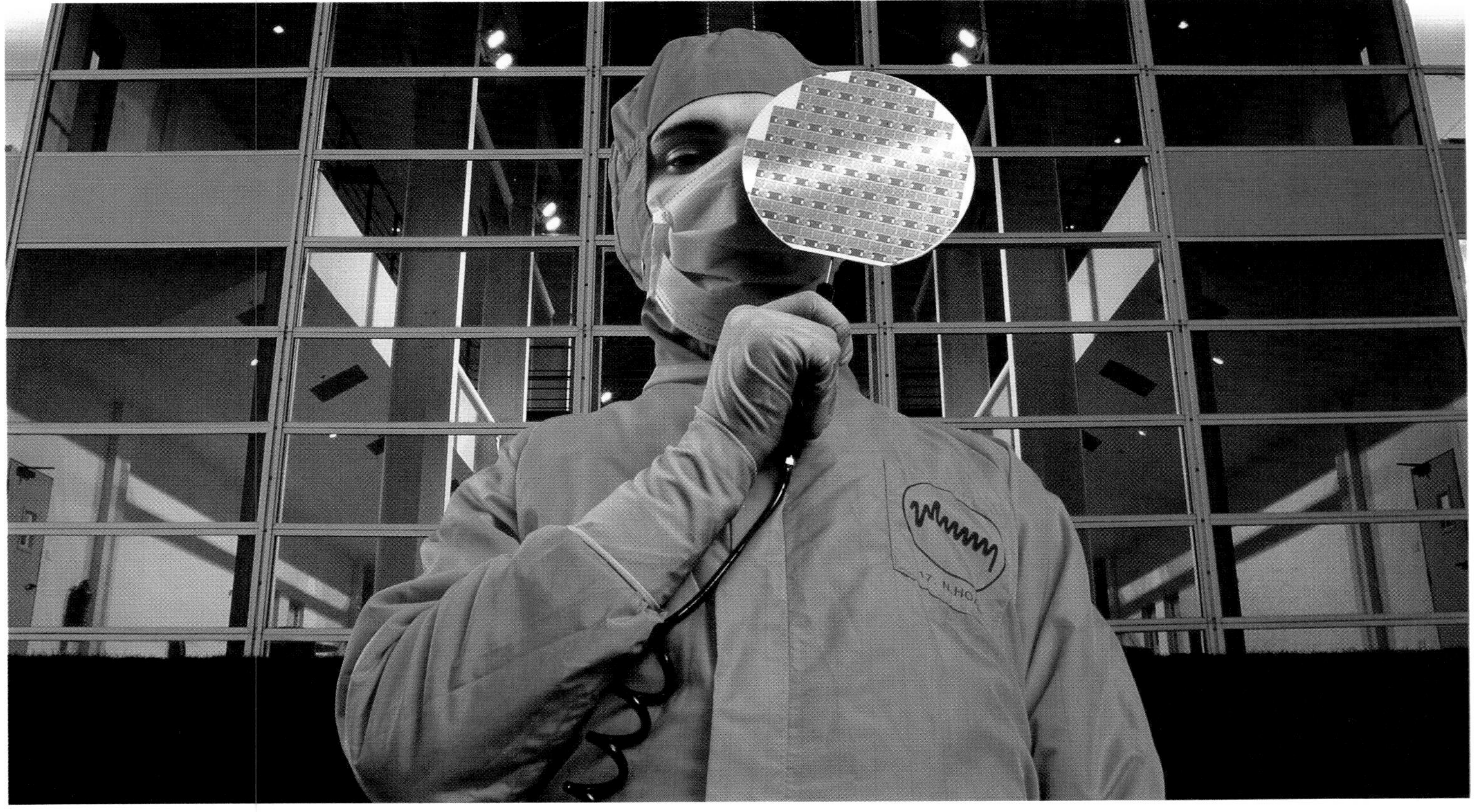

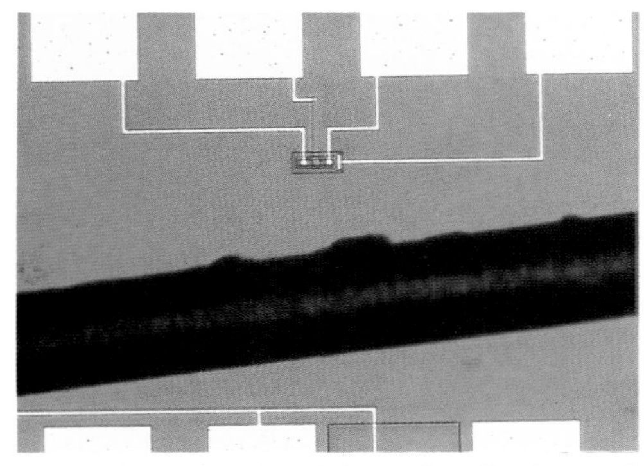

Top: The Electric Telegraph, Cooke and Wheatstone's instrument, c. 1845. *Bottom:* a human hair compared with a transistor and metal interconnection on a Plessey microchip

There is no evidence that CM's telegraph was ever built, although it could have worked if it had been. The problem, of course, was providing the twenty-six carefully insulated wires over any distance, particularly when one bears in mind the dearth of reliable insulating materials at the time – the most widely used then were glass and cast sulphur.

Later, many other multi-wire telegraphs of closely related kinds were proposed, and some even built. After 1800, when Volta had invented the electric pile – a kind of primitive battery – this device naturally replaced the electrostatic machine, easing the insulation problems, and, with the discovery in the same year of the electrical decomposition of water by Nicholson and Carlisle, the electrolytic effect was soon pushed into use as a means of detecting electric polarisation more senstive than electrostatic attraction.

In 1804 Francisco Salva Campilo built a multi-wire telegraph with complete success, using the Voltaic pile as the energy source and electrolytic effects as the means of detection. The signalling must have been very slow, because it takes a considerable time for a stream of gas to be evolved from an electrified wire dipped in water, and above all the invention was once again impracticable because of the multiplicity of wires.

Both CM's and Campilo's proposals were for alphabetic telegraphs, that is telegraphs which indicated directly the letter being signalled. Later, however, in the 19th century, commercial success was achieved exclusively by coded telegraphs, in which what was sent was a kind of coded sequence, meaningless directly but capable of being decoded by a skilled operator to give the intended alphabetic significance, Nevertheless, at a time when even literacy was very far from universal, it is not difficult to understand why so many early telegraph inventors strove so hard to produce alphabetic machines. Still later again, ABC telegraphs appeared, then subsequently teleprinters, both of which resurrected the principle of the alphabetic telegraph.

What was abundantly clear even in the early days, however, was that any system requiring twenty or more wires between transmitter and receiver was quite intolerable. Thus the next important alphabetic telegraph was an historic milestone because it succeeded with just one. Francis Ronalds, the son of a City of London merchant, offered an elegant solution to this problem in his telegraph of 1816, by combining two items of contemporary technology in an ingenious way. By the early 19th century accurate clocks were already easily obtainable and Ronalds realised that by combining a clock mechanism with a single-wire signal (electrostatic for speed of response) a practicable alphabetic telegraph could be constructed.

At both transmitter and receiver a clock was arranged with a disk geared up from the shaft which ordinarily would drive the second hand, thus rotating five times per minute. Around the edge of this disk were written the letters of the alphabet, and mounted in front of it was a second disk, normally fixed but capable of being rotated by hand, in which a small window had been cut to reveal one letter at a time on the rotating disk behind. Both clocks were carefully regulated to keep good time, with the result that if the windows were adjusted so that they were both showing the same letter together, they would stay in step for a considerable time.

This arrangement was combined with a single insulated wire between transmitter and receiver, having attached to it at each end a pair of very light pith balls hanging on strings from the wire, and designed to fly apart when the line was electrified. To send a message all that was necessary was to touch the output of an electrostatic generator onto the line at the moment that the required letter appeared on the rotating dial, whilst at the receiving end the letters were noted down whenever the pith balls flew apart. This simple but effective arrangement could signal over distances measured in miles at about two words per minute.

Ronalds gave great attention to the practical detail of the design, for example his line was insulated by means of glass tubing sealed with wax at the joints and contained within a wooden protective trough. Another detail: to synchronise the clocks it was understood that whenever an extra large charge was transmitted, giving a double deflection of the pith balls, the system was being synchronised, and the adjustable escutcheons at both ends would be set to reveal the letter 'A'.

It would have been expensive to build and slow to operate, yet even so the electrostatic telegraph might have marked the beginning of commercial telegraphy, but in the event it was not to be. Ronalds offered his invention to the Admiralty, who, after a very long delay, advised him that telegraphs of any kind were quite unnecessary. Taking the outcome quite philosophically, he did not seek other ways of exploiting his idea.

Primitive though it was, Ronalds' telegraph demonstrated many features of the most advanced data transmission systems of the present day. Firstly, like all telegraphs, it was digital, that is to say its signalling currents took up either one of two states (present or absent), and digital signalling is today characteristic of all forward-looking communications systems. Secondly, the telegraph depended on maintaining synchronicity between transmitter and receiver, and again the same is true in our most advanced systems today. The method of synchronisation, depending upon sending a labelled predetermined signal, also corresponds with contemporary practice. A principle difference is the rate of signalling: Ronalds could send one or two binary signals per second, but today fibre optic transmission can be at the rate of several thousands of millions per second!

The next major advance was the discovery of the deflection of a compass needle by an electric current flowing in a wire nearby. The year was 1820, the month July, when the Danish experimenter Oersted published this startling result. By September, Johann Schweigger in Germany announced his 'multiplier' – later to be called a galvanometer – consisting of one hundred turns of wire in the form of a coil, with the magnetic needle at its centre. A phenomenally sensitive detector of electric currents resulted, and it was this sensitivity which gave the electromagnetic telegraph the ability to signal over

Sebastian Ferranti, Engineer; Deptford Power Station, c. 1889, the first major electric generating station in the world: (*top*) the 200 KW Alternator at Deptford, (*centre and bottom*) views of the interior, c. 1889

Circuit probe test of a 2 micron integrated circuit

extreme ranges. Indeed, for sending single timing signals and similar 'one off' messages, telegraphs were consructed almost at once, for example by Gauss and Weber. However, obsessed as most experimenters still were by the need for an alphabetic telegraph, true long-distance communications remained elusive.

One experimenter who did not follow the popular view in this respect was Schilling von Canstatt, who (despite his fashionably Germanic name) was a Russian in the diplomatic service of the Czar. No doubt he would have been quite used to the idea of cipher clerks in the embassy reading coded messages, and perhaps for that reason he was unusual for his time in being prepared to consider a coded telegraph.

His instrument consisted of five galvanometers of the Schweigger type, to the moving needle of each of which was attached a disk of card painted white on one side and black on the other. Thus it was easy to arrange that when a galvanometer was passing no current a white disk was was seen by the operator, but when the coil was energised the disk would turn to show its black face.

Since there were five such disks, there are evidently thirty-two distinct patterns of black and white possible, enough for the alphabet and a few punctuation marks. What is more, with a simple printed form and a pencil, it would have been quick and easy to note down the code groups for subsequent interpretation. Probably the speed of signalling would have been a few words per minute.

The system was entirely practicable and could in principle have given a service over very long ranges, bearing in mind that by that time the Voltaic pile and been replaced by much more reliable primary batteries, notably the Grove cell. It had, however, one major snag. Because of the use of five galvanometers, at least six connections were needed between transmitters and receivers (assuming that the return line was common to all five), and this would necessarily be a costly business. In modern terminology one could say that the system used parallel signalling. The concept of serial signalling – sending the five signal code group in sequence over a single line and galvanometer – seems never to have been considered.

Great plans were approved by the Czar for a long telegraph line to be built in Russia, from Kronstadt to St. Petersburg, perhaps even to be carried under the sea in the Gulf of Finland, but Schilling died suddenly and it all came to nothing.

The next, and in some ways most incongruous, telegraph pioneer, coming on to the scene in the early 1830s, was an Englishman, William Fothergill Cooke. A man of negligible scientific education, he had been invalided out of the Indian Army and made a precarious living selling wax anatomical models to medical schools. In the course of his business he visited Germany and there saw Steinheil's telegraph – a simple 'two state' device similar to Gauss and Weber's. His imagination was fired, and like all the experimenters of the day, he went away to try to convert what he had seen into a true alphabetic telegraph.

Quality control in gallium arsenide integrated circuit production at Plessey

Over the details of his resulting 'musical box telegraph' it is better to draw a veil. It worked just a little, enough to encourage him to go on, but could signal over only a pathetically short range. Realising that he needed scientific help to overcome his problems Cooke applied in the first instance to Michael Faraday, who, distinctly unenthusiastic, directed him on to Charles Wheatstone, then Professor of Physics at King's College London.

What followed has passed into legend. It was the subject of a great and litigious subsequent dispute between the two men, and the truth of it will never be known for certain. The facts are simply that Wheatstone gave Cooke a polite and considerate hearing, examined the musical box telegraph, and then announced that he was too busy to do anything about it at the time, but if Mr. Cooke would care to return in two weeks he would be happy to discuss the matter further with him then. After the fortnight had passed the unsuspecting Cooke came back, only to be told by Wheatstone that he himself had long had a far superior telegraph, which overcame the disadvantages that beset Cooke's effort.

Although Wheatstone firmly claimed to have been working on his 'five needle telegraph' long before Cooke appeared on his doorstep, one can well understand the other man's scepticism on the point. In truth, though, if Wheatstone took anything at all from Cooke it cannot have been much more than the mere idea of an alphabetic telegraph, for his instrument bore not the slightest resemblance to the other man's. Indeed, if it had antecedents they were much more plausibly in Schilling's work, who also used five galvanometers, it will be recalled.

Truth to tell, the only significant difference between Wheatstone's and Schilling's proposals was that the English experimenter turned the galvanometers on their sides, so that the axis of rotation of the magnetic needles was horizontal. Pointers attached to the needle shafts were normally vertical but could be deflected to left or right depending on the sense of the current sent through the galvanometer coil. Mounting the pointers side by side on a vertical 'hatchment board', marked out with a diamond pattern of lines having letters marked at the intersections, twenty distinct letters could be unequivocally signalled, a reduced but adequate alphabet. (The number is twenty and not thirty-two because only two needles may be deflected at one time in Wheatstone's system, and also deflection of the two needles in the same sense does not convey a message. Schilling did not labour under these constraints.)

Cooke and Wheatstone formed a partnership and patented the electromagnetic telegraph in 1837. They realised that railways, which were developing vigorously at that time, needed the telegraph for control of their activities, and also had a valuable way-leave over which the telegraph lines could be established. In 1837 the telegraph was demonstrated to the Directors of the London and Birmingham Railway, but they declined to take it up.

However, in 1839 a five-needle telegraph service was established from Paddington to West Drayton, on Brunel's Great Western Railway, and later

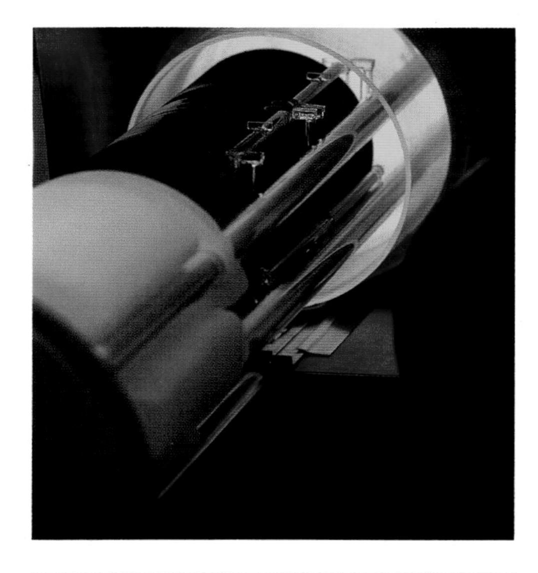

Top: a high temperature furnace where oxide layers are deposited and grown on a silicon wafer. *Bottom*: testing materials in simulated sunlight conditions at the Plessey Research Centre

was extended to Slough. This was the first commercial public telegraph service to be established anywhere in the world. It could hardly be described as an immediate success for in the first year more money was made from the admission charges for showing people around 'the telegraph cottage' than from charges for sending telegrams. Nevertheless, as the geographical range of the telegraph extended the volume of traffic grew, and before long highly profitable telegraph services were established over the length and breadth of Britain, and indeed of Continental Europe.

In a sense the early Cooke and Wheatstone telegraph was not technically innovative, being little if at all in advance of Schilling's ideas. However, in the course of designing the system as a whole one technical milestone was achieved which proved to be of epoch-making significance.

A trivial-seeming but none the less difficult problem of the early telegraph was the operation of a call bell to draw the operator's attention to the fact that his presence was needed at the equipment. The currents over the line were too weak for this purpose, so Wheatstone invented a device to overcome the problem. It is still kept in the collection of Wheatstone memorabilia at King's College.

The device consisted of the usual magnetic needle mounted on a horizontal axis and inside a coil, but now attached to the spindle was a counterbalanced arm, just a few centimetres long, on the end of which was attached a wire in the form of a horseshoe. When a faint current passed through the coil, the needle was deflected and the arm dropped, so that the 'horseshoe' bridged two small cups of mercury, thus switching on the vastly stronger current from a local battery circuit.

This 'relay', as it came to be called, is the first historical example of an electrical amplifier, or, from another point of view, of a digital 'gate' circuit. Horribly slow and impractical though they were for such a purpose, it would have been possible, in principle, to have built a digital computer or an automatic telephone exchange from a few thousand devices like this!

In fact all such developments lay unimagined far in the future, but much more immediately the use of relays made possible extension without limit of the geographic range of the electric telegraph. Whenever the signals began to be unduly weak, as the lines were further and further lengthened, a relay would be used to return them to full strength once more. In this invention of Wheatstone's – the first electrical 'active device' – we undoubtedly see the origins of electronics technology as we know it today.

However, the Cooke and Wheatstone telegraph, in its initial form, still used the parallel signalling that was inherited from Schilling, and the first to break this barrier with an electromagnetic telegraph were Morse and Vaile, in the United States, who received much scientific advice from Joseph Henry.

Perhaps with an eye to the long distances over which their lines would have to be constructed, Morse rejected all multi-wire telegraphs, and tried to achieve an alphabetic system in quite a different way. Accepting the need to use a coded mode of transmission, and basing it on the 'dots and dashes' code which is now so familiar, he sought to solve the drawbacks by developing a simple automatic code group sender and also a printing receiver, which produced a permanent ink-on-paper record of the received code groups. Thus the telegraphist was not obliged to memorise the code at all, being able to send at will and, using a book, decode at leisure the coded message received in 'written' form.

Curiously, both in Europe and America, the alphabetic telegraphs which had cost so much effort to devise, did not last long. Morse found to his chagrin that his telegraphists did not bother to load the paper rolls into his printing receiver, because they could 'read directly' the incoming signals simply by listening to the characteristic sound that the receiving equipment made.

In England, Cooke, the practical man trying to make a viable business out of the telegraph, soon abandoned the economic burden of five-wire transmission, and his coded two-needle and one-needle telegraphs became supreme in commercial service. These too were adapted, by fitting distinctively tuned small metal 'sounders' where the tips of the deflected needles would strike them, so that the telegraphist could 'read' them by ear, rather than eye, and so be free to write the messages down directly at the same time.

The one-needle telegraph had two distinct musical notes, and the two-needle version, used for circuits on which the traffic was very heavy because it could achieve almost double the sending speed, had four. In the hands of a skilled telegraphist, Cooke's two-needle telegraph could send and receive at about forty words per minute.

The unforgettable, almost continuous, musical tinkling noise of these Victorian 'speaking telegraphs' was, in its day, the characteristic sound of telecommunications, carrying news of great affairs of state and of private triumph and disaster alike, making a modern-style press possible for the first time, and serving to create the nervous system which co-ordinated the far-flung British Empire.

The alphabetic telegraphs did not die, however, and Charles Wheatstone perfected an ABC telegraph which depended on sending trains of mechanically generated pulses, like the finger dial on later automatic telephones. At the receiver a dial and pointer mechanism unequivocally indicated the letter being sent. It was ideal for the amateur user, and in the third quarter of the 19th century a considerable ABC telegraph network was established in London for communication between private houses and places of business. This system was the fore-runner of the private telephone, and taught those who set up and operated it many lessons which were to prove invaluable later on when the telephone came. The ABC telegraph, excluded by Wheatstone from his eventual sale of his interest in the commercial telegraph to Cooke's company, also brought its inventor a considerable fortune.

However, the 19th-century telegraph could not transmit the human voice. To be sure, that feat can indeed be accomplished by transmitting code groups, for that is how the modern digital telephone systems work, but even to communicate a barely recognisable voice requires the transmission of the

Top: Marconi with his transmitter and receiver, c. 1888. *Bottom*: Marconi's receiver and transmitter during wireless transmission across the Channel

Top: the photophone of Bell and Tainter: the transmitter, 1881. *Centre*: Boston audience listening to a speech by Graham Bell fifteen miles away, c. 1877. *Bottom*: printing press driven by solar energy, c. 1882

equivalent of several thousand binary digits – 'bits' – per second, and at its fastest the Victorian telegraph achieved only a hundredth of this.

Thus the way was open for the introduction of the telephone, in which the voice signals are passed not as a sequence of code groups but by means of a continuously varying electrical signal which faithfully represents the variations of acoustic pressure on the diaphragm of a microphone, and which can readily be reconverted into sound in the receiving telephone earpiece. Thus signalling by code group – what today would be called digital transmission – is replaced by a system in which the electric current is an analogue of the sound wave.

Unhappily the telephone did not spread as rapidly in Britain as in Continental Europe, in part because the authorities of the day were short-sightedly unwilling to allow the telephone services to use the existing network of telegraph lines, when in fact the experience of other countries was that the two could coexist quite well. As a result, Britain lagged in the commercial exploitation of the telephone, a handicap which was many decades in being recovered, and it was therefore between Paris and Brussels that the first international telephone line opened in 1887.

Although it lasted the major part of a century, Alexander Graham Bell's invention of the analogue telephone will be regarded historically as a technological blind alley, creating a great industry and having profound social consequences in its day, but in the end conceding to the technology which preceded it – digital communication. Despite that, however, in the meantime many important milestones were passed which ought to be recorded.

Perhaps the most significant of the technical advances was the invention by Strowger of the automatic telephone exchange, first demonstrated in public at the Paris Exhibition of 1927, which made it possible for the telephone to become the all-pervasive method of communication that we know today. The early realisation of this system was based on electromechanical devices so crude that a normal part of the routine maintenance was sweeping up the brass dust which fell out of them in the course of their operation, nevertheless it was the recognisable starting point for a revival of interest in digital technology which was, in due course, to revolutionise the whole communications scene.

For that to come to fruition needed electronics technology, and the late 19th-century transition from the telephone to radio as the focus of interest, which required operation with currents of far too high a frequency for electromechanical devices to be of any value, forced this development.

James Clerk Maxwell, a Scot and one of the greatest theoretical physicists of all time, published his *Treatise on Electricity and Magnetism* in 1873, in which he predicted the existence of electromagnetic waves in what we would now call the radio bands. By 1887 Heinrich Hertz had demonstrated the reality of the radio waves and sent them over a few metres of space. By 1890 Branley had invented the 'coherer' – the first reasonably sensitive radio receiver.

Marconi drove this scientific curiosity forward, making it into a practical and commercial reality, rather as Cooke had done for the telegraph two generations before, and by 1901 he succeeded in demonstrating transatlantic radio communication. At the same period Fleming was conducting the first experiments on the use of a diode valve in radio. By 1906 this had evolved into Lee de Forest's 'audion' or triode, the first electronic amplifier, and the immediate origin of all our electronics technology.

Like the first communications by wire, the first radio systems were telegraphic: digital, in fact. However, the drive to analogue telephony was naturally irresistible, and by 1912 de Forest had succeeded in transmitting between adjacent building using his triode tube. The step to broadcasting was not long coming, and the first station to transmit entertainment programmes on a regular basis was on the air by 1920 from Pittsburg (KDKA).

The British Broadcasting Company was established in 1922, the same year that the Plessey Company began the first UK volume production of valve and crystal radio receivers, under contract to the Marconi Company.

Electronics technology developed apace, among the most important milestones being the invention of the pentode valve by Bernard Tellegen in 1925, and the evolution of the Theory of Feedback and Systems Stability by Nyquist in 1932 – one of the seminal achievements of the 20th century, and, along with Hartley's studies on modulation, the beginning of a theoretical basis for information technology which had no pre-existent intellectual origins in earlier technology or physical science.

After sound broadcasting came television, also using analogue transmission. Early systems, pioneered by John L. Baird in London, were electromechanical. The first television receiver in volume production anywhere in the world was made by Plessey in the late 1920s, and exported as far afield as Japan, which it reached in 1931, to be used in television research.

However, it was all-electronic television, pioneered by Isaac Schönberg and his team at EMI – most notable among them the brilliant Alan Blumlein – who created the high-definition electronic system with which the BBC began the world's first full television service in 1936.

World War II resulted in great advances in digital electronics, both from the development of effective radar systems and from the evolution of the digital computer. The fundamental theoretical concepts on which the electronic digital computer is based were evolved by Alan Turing, starting with his famous paper in 1936 'On Computable Numbers'. Colossus, the world's first stored-programme digital computer was built at Bletchley Park, England, in 1943 by T. H. Flowers to Turing's design concepts. The time was ripe for a return to digital communications in place of the analogue techniques which were up to that time almost universal.

In the late 1940s, proposals for voice communication by digital means were investigated, including one (pulse time modulation) which was remotely descended from Ronalds' ideas. Ultimately it was PCM – pulse code modulation – invented as early as 1938 by Alec Reeves, an Englishman then working for ITT in France, which achieved general acceptance. This was in part because of the firm theoretical backing lent it by Claud Shannon's

Top: a Plessey three-dimensional, long-range defence system. *Bottom*: Oakhanger Satellite Ground Station in the UK

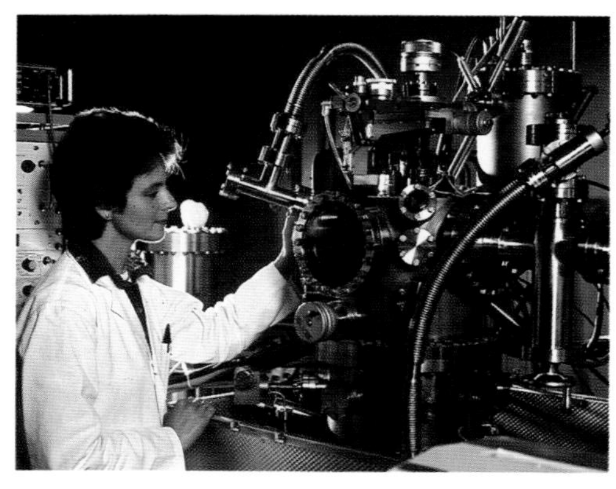

Top: working with lasers. *Centre*: Auger and X-ray photoelectron spectrometer for chemical analysis. *Bottom*: Plessey air defence facility

mathematical Information Theory of 1948, another crucial intellectual advance of the new information science.

With the rediscovery of the transistor by Shockley, Bardeen and Brattain in 1947, and the subsequent development of solid state electronics, the transition of all forms of information transmission, processing and storage to digital form began. Even now that change is not yet complete, but the final triumph of digital techniques in every aspect of information technology is today quite obvious and inevitable.

A necessary prerequisite was the evolution of integrated microelectronic circuits, the 'silicon chip', for which the first patent was claimed by Jack Kilby in 1959. Over the last thirty years the consequence of this micro-electronics revolution has been that many electronic functions can now be carried out in one millionth of the space, ten thousand times as fast, and at one hundred thousandth of the cost than was possible in the 1950s.

Not only has this modified techniques of signal processing, making the digital epoch a practicality, but it has also made possible many subsidiary inventions of the greatest significance. The most important of these are the geostationary satellite and the optical fibre.

Although terrestrial radio transmissions are exremely valuable for communications over ranges of up to a hundred kilometres or so, for intercontinental contacts they are limited to relatively low signalling rates. Thus the idea of putting a station in the sky which could receive transmissions from stations within its 'line of sight' on the earth's surface and re-transmit them to quite another surface site, out of contact with the first, and perhaps a continent away, was proposed by Arthur C. Clark in 1946. It was not practicable until advances in space travel technology had made possible the launching of man made satellites in geostationary orbits, 36,000 kilometres out from earth, and over the equator. Here their motion exactly matches the rotation of our planet, so that they seem to the observer on the ground to hang motionless in the sky.

Sputnik I, the world's first artificial satellite, was launched by the USSR in 1957, and SCORE, the first communications satellite, by the US in 1958. As early as 1962 the geostationary Telstar was carrying international television traffic.

By contrast, cables had been used for short and long distance communications from the very beginning, but always in the form of insulated wires carrying electrical signals. This has many disadvantages, not least that such signals are easily degraded, and that electrical cables pick up interference from other sources.

In 1966 Charles Kao and George Hockham, working in the STC laboratories at Harlow, England, demonstrated that it would be possible to communicate by means of pulses of light carried in an hair-thin optically transparent fibre. Using semi-conductor light sources and detectors, this technology has already swept out the electrical cable and microwave links as the optimal technique for fixed point-point transmission on land, and the

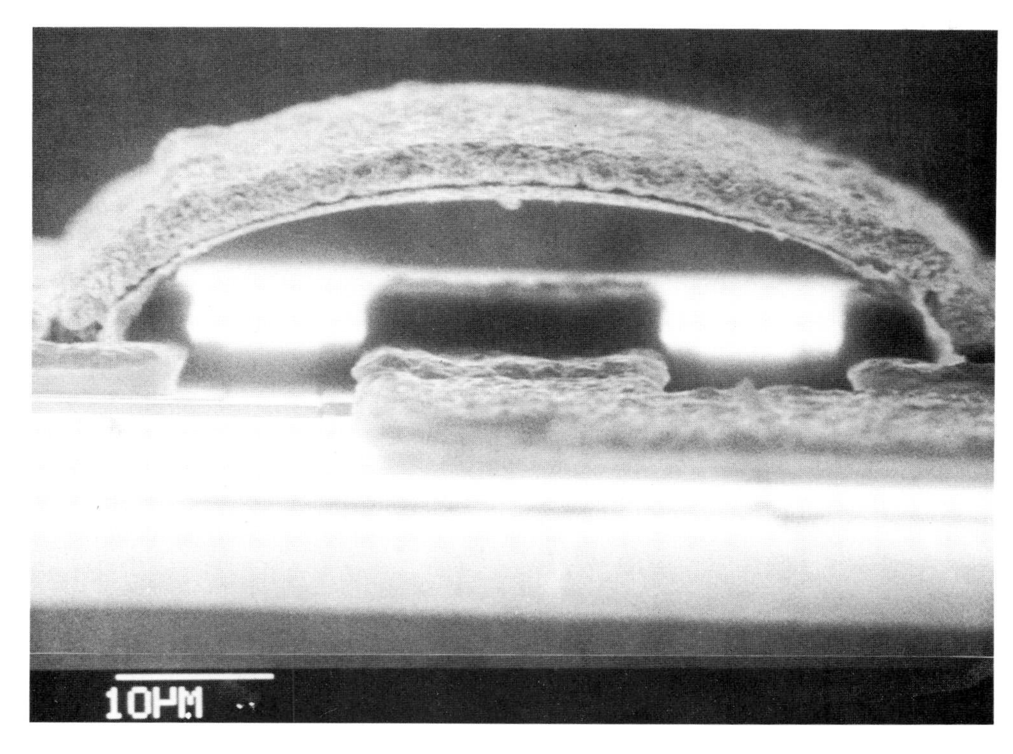

Bridge on microchip, dramatically enlarged by an electro magnetic microscope

undersea domain is now being successfully attacked.

Soon even local telephone connections are likely to be increasingly made by optical fibre, using digital transmission, of course. A 'spin-off' from optical transmission is optical storage of data, the most familiar example of which is the Compact Disc high-fidelity audio recording system introduced in 1983, which combines digital and optical technologies to give hitherto unheard of quality and longevity in audio recordings at modest price.

To conclude: electrical communications has grown up alongside electronics engineering, developing its own scientific base in the meantime. Its relationship with computer technology has always been close, too, and now grows even closer.

From the beginning, British pioneers played a more than proportionate role in the headlong advance of communications technology, and they do so still. It is a just cause for national pride.

But it is an historic irony that in the century and a half that it has been with us it has in a sense come full circle. It began with electrical telegraphy, using on-off signals in an essentially digital technology. For many decades it was deflected into the analogue domain, to meet the needs first of voice and later of vision transmissions.

Communications engineering is returning once more to explore the digital domain in which it came to birth, but now strengthened by an electronics engineering virtuosity that those who began it all could not have dreamed of. The limit to what can yet be attained, or the direction in which this exciting technology will go, is hardly possible to conjecture. What we can say with assurance is that its capacity to transform the world in which we live is by no means yet exhausted.

Above:
Structural details
Opposite:
'Waiting for the Queen' (lithograph after Joseph Nash)

ROBERT THORNE

Paxton and Prefabrication

In the history of prefabrication the Crystal Palace has always been awarded legendary status. The story of its design and erection, which never seems to wear thin however often repeated, fully justifies that reputation, but even the most compulsive accounts of the building works in Hyde Park and later at Sydenham cannot conceal that it is a story with a slightly unsatisfactory ending. Within a matter of years, if not months, an ideal method of component construction was developed to a peak of efficiency in preparation for the most celebrated public event of the 19th century; and then almost as fast it faded from view, confined to marginal projects none of which matched the success of the prototype. Descriptions of the Crystal Palace generally reach a crescendo at the opening of the Great Exhibition, with a less hearty burst of enthusiasm at the opening of the Sydenham building, and then leaves the rest to silence. The hiatus which followed, which is in its way just as instructive to any discussion of prefabrication, has gone largely unconsidered, perhaps above all because lessons based on non-events can never hope to match those that deal with things that really happened. But it is also the case that the failure to examine the aftermath of the erection of the Crystal Palace has helped keep from view some of the more problematic aspects of the building's history.

The conventional account of the preparations for the Great Exhibition is by now so familiar as to hardly need reiteration.[1] It assigns an heroic role to Sir Joseph Paxton, once an humble gardener's boy, who saved the Commissioners of the exhibition from the apparently impossible quandary of how to house the event. In alliance with the contractors, Fox Henderson, he was able to exploit a loop-hole in the document inviting tenders for the Commissioners' own design so as to secure acceptance of his alternative scheme, while at the same time he published his design in order to rally public support behind it. The same publicity machine which helped win victory for his proposal stayed in action to celebrate its miraculous realisation in Hyde Park during the winter months of 1850-1. Led by the *Illustrated London News*, press reports of what was happening behind the site hoardings provided a more detailed coverage of the construction process than any other building has been privileged to receive: the raising of the columns and girders, the cutting of sash-bars, the manufacture and fitting of the glazing, the debates about paint colours – nothing was thought too esoteric or technical for inclusion. Through the accumulated effect of such reports the

1. Christopher Hobhouse, *1851 and the Crystal Palace*, 1937; Yvonne French, *The Great Exhibition: 1851*, 1950; Ralph Lieberman, 'The Crystal Palace', *AA Files* 12 summer 1986, pp. 46-58.

INDIA

building was acclaimed for the speed and ingenuity of its creation well before any reasonable assessment could be made of its final effect.

In a sense the Crystal Palace was a gift to the press because its process of assembly was so easy to grasp. The wet and messy business of conventional masonry construction, which left the spectator at a loss to understand the progress and importance of what was happening, was supplanted by an easily understood sequence of construction. In design, the key to its modular system was in its roof. The largest feasible panes of glass for use in Paxton's ridge-and-furrow roofing system were forty-nine by ten inches. Two such panes set at the correct angle constituted a roofing segment eight feet wide: from that dimension stemmed the standard lengths of roof and gallery girders (twenty-four, forty-eight, and seventy-two feet) and thus the dimensions of the building as a whole. Strictly speaking, therefore, the entire building, 1,848 feet in length with its galleries, aisles and transept, obeyed a structural logic based on just one of its features.

As with any prefabricated structure, the principal virtue of this standardisation was in permitting most of the work to be done off-site, in this case at the Fox Henderson factory in Smethwick and at other firms in the West Midlands and London. But much as that helped speed the process, it depended on sophisticated timing to organise the arrival of components in the right order for erection. The chief attraction of the site in Hyde Park while work was in progress was the spectacle of the Fox Henderson team orchestrating the receipt of the different parts, testing them where necessary, and then dispatching them to the right spot for assembly. With the system working at full tilt, castings could be ready for use in Hyde Park within twenty-four hours of leaving Smethwick foundry, and Paxton claimed to have seen two columns and three girders erected in just sixteen minutes.

As if the story of its erection was not enough, the Crystal Palace could claim to be a landmark of prefabrication in two other senses. First, its removal from central London and re-erection in an adapted form at Sydenham demonstrated that a building made up from a kit of industrial parts could be dismantled as easily as it had been put up, with the capability of being reused in a variety of other forms. In that respect it had more in common with the machines which helped bring it into being than with buildings of traditional construction. And secondly, an account of its design was published that was sufficiently exact for it to be a template for other buildings following the same system. The prime intention of Charles Downes and Charles Cowper in issuing their volume of the Fox Henderson working drawings was to round off the technical record of the Hyde Park building, but as they were quick to point out, the book which resulted was 'a work so correct and complete as to enable an architect or engineer to erect a similar building if necessary'.[2]

So at first glance the Crystal Palace seems to have fulfilled more than a fair share of the requirements expected of a prefabricated building – in its mode of construction, its adaptability, and its potential as the basis for a repeatable

Above:
Top: drilling and punching machines. *Bottom:* the power plant
Opposite:
Top: engraving from the *Illustrated London News*. *Bottom:* 'Opening Ceremony' (lithograph after Joseph Nash)

2. Charles Downes and Charles Cowper, *The Building Erected in Hyde Park for the Great Exhibition of the Works of Industry of All Nations*, 1852, reprinted 1971.

Above:
Top: view of the glass roof. *Centre:* testing the structure. *Bottom:* hydraulic pressure apparatus
Opposite:
Top: United States Court. *Bottom:* View down the Nave, 1850 (lithograph after Owen Jones)
Page 58:
Top: British Court, agricultural machinery. *Bottom:* British Court, De la Rue's patent envelope folding machine
Page 59:
Hoisting the Nave Arches into Place, c. 1851

architectural system. In the opinion of many 20th-century commentators its catalogue of virtues was such that any discussion of why its promised progeny failed to appear is bound to look to causes beyond its immediate sphere. Among such explanations the most readily cited has been the professional jealousy that the Hyde Park project engendered; the fact that architects (and consulting engineers for that matter) were pushed aside by an outsider in a project of such significance continued to rankle for many years after. *The Builder*, in its obituary of Paxton, emphasised that 'the building of 1851 was accomplished only by a course of proceeding that would not be recommended for imitation, and by omission of what would be in every other case an architect'. And whenever the subject of iron came up in debates at the RIBA the example of the Crystal Palace was taken as testimony that the brutish qualities of such a material needed a proper architectural sensibility to tame and develop them. 'In my opinion', announced the sententious Robert Kerr, 'the more the system of iron and glass is amplified, the more it will be seen that no really permanent scientific work can be produced by that means; and, though some excellent effects may be accomplished for a moment, yet architecture at large is not so greatly advanced constructively, as we might expect by that innovation'.[3]

The territorial wranglings in the building world that helped engender such remarks may also be understood as part of a wider reaction in which the reputation of the Crystal Palace was bound to suffer. Architectural historicism of the kind which Kerr and his contemporaries took as their starting-point was one manifestation of the rejection of utilitarian rationalism which took hold in the 1850s. The Crystal Palace was caught up in that reversal because it was a symbol of a way of thought that was running out of favour and it suited many people to single it out as a victim.

Such explanations, pragmatic or philosophical, for the failure of the Crystal Palace to initiate a general architectural transformation, presuppose that the building itself was almost faultless and that the methods which brought it into being could have readily been applied elsewhere. In the 1930s, when the building was canonised as a classic of modular construction, its lessons seemed so urgently obvious that the reluctance of previous generations to heed them could only be attributed to jealousy and short-sightedness. 'To my knowledge', said Siegfried Giedion, 'the possibilities dormant in the modern civilisation we have created have never been so clearly expressed'.[4] All that was required was to pick up where Paxton left off.

Half-a-century later, the projects of those who were nurtured on that view of the 19th century have their own tale to tell, and in the knowledge of what they achieved the building which they looked to as an exemplar takes on a different hue. In the emergencies of the post-war period a Paxtonian solution was more than simply a way of closing an historical gap: for architects designing schools and housing in the public sector it seemed the only way to meet the building crisis. By background and necessity they were attracted to

3. *Builder*, June 24 1865, p. 443; *RIBA Transactions*, 1st. ser. Vol. XXV, 1874-5, p. 214.
4. Quoted in *Architectural Review*, Vol. LXXXI, Feb. 1937, p. 66.

ZOLLVEREIN
UNITED ST
OFFICIAL CATALOGUES
SPICER BROTHERS, WILLIAM CLOWES &
JOINT CONTRACTORS
UNITED STATES
UNITED STATES
UNITED STATES

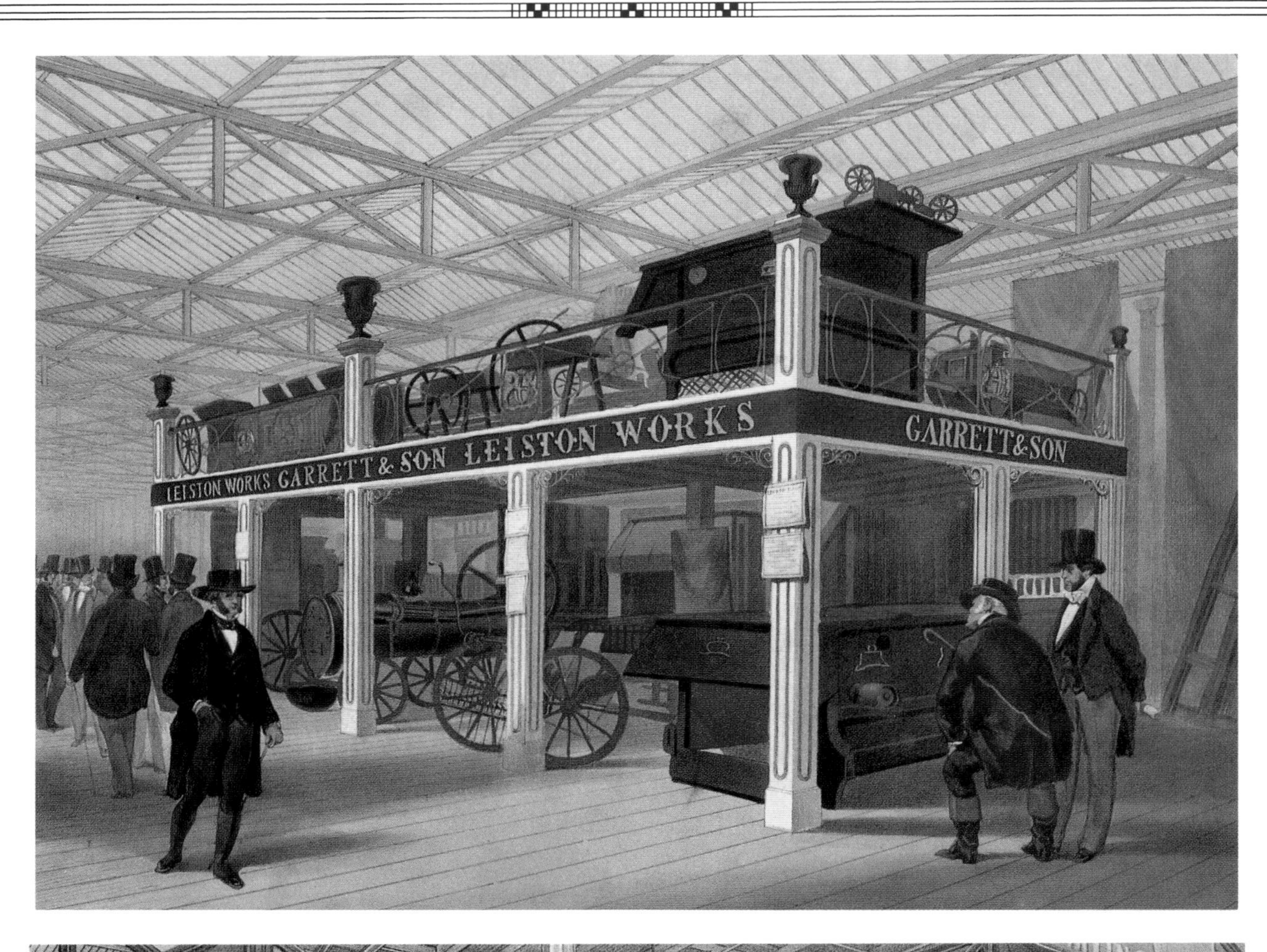
LEISTON WORKS GARRETT & SON LEISTON WORKS
GARRETT & SON

NOT TO TOUCH
ANY ARTICLE

Above:
Top: Erecting the Columns. *Bottom:* Enclosing the Site
Opposite:
British Court – Statue of Shakespeare

industrialised building techniques; to the Hills system, the Derwent timber frame, Intergrid, and most famously, to CLASP. How these systems were developed, in an almost Victorian atmosphere of energy and expediency, has been told with proper sympathy by Andrew Saint in his book, *Towards a Social Architecture* (1987). As with so many episodes it turns out to have been a more complicated and uneven matter than is generally supposed. By abstracting the lessons of his account, and applying them to what happened at the time of the Great Exhibition that story in its turn loses some of its heroic simplicity.

As post-war architects readily appreciated, putting buildings on the production line requires a different approach to every aspect of the building process. For a start, as Paxton demonstrated when he delivered his preliminary drawings to Fox Henderson, the component manufacturer is bound to have thrust upon him many aspects of design, analysis and research which traditionally are the architect's prerogative. The question of what is feasible in conditions of industrial production is as critical as conventional aspects of design, so responsibility for shaping a project inevitably shifts towards the manufacturer, although whether credit for that responsibility shifts as well is another matter. At the other end of the process, prefabrication presents another kind of challenge to customary architectural practice. By definition, a modular system, however flexible, is not tailored to a single site and can (as critics of post-war schools never tired of pointing out) seem clumsy in many locations. If it works well, and stands the test of other expectations, that may be counted only a slight drawback, but to many eyes the individuality of a building in a particular setting is a first consideration, not an added luxury. As Andrew Saint concedes in his assessment of system-built schools: 'There is no getting away from appearances and aesthetics'.[5] If that is so for buildings of one type, most of them modest and unobtrusive, it becomes doubly problematic when the application of a prefabricated system to a variety of buildings is proposed.

Returning to the Crystal Palace, how do the lessons of subsequent prefabricated projects alter the popularly accepted view of the role played by the manufacturer? As long as Paxton is cast as a Napoleonic figure armed with a visionary scheme and the ability to see it realised, the ancestry of the building is bound to be discussed primarily in terms of his career. The Crystal Palace was a greenhouse writ large, and his previous experience was in the design of gardens structures, so the links can easily be made. In its most horticultural qualities it was indeed a metropolitan version of what he had accomplished at Chatsworth, particularly in it glazing, its ridge-and-furrow roofing (a system devised to improve plant propagation) and its use of standard, machine-cut wooden parts.[6] But nothing in Paxton's previous career had prepared him for a project of such a scale, nor for one requiring the extensive use of cast and wrought iron, materials which be never felt at home with. Hence his dependence on the firm of Fox Henderson, not just for submitting a tender on his behalf but for taking the risk to, as Sir Charles Fox

5. *Towards a Social Architecture*, 1987, p. 235.
6. On Paxton at Chatsworth see George F. Chadwick, *The Works of Sir Joseph Paxton 1803-1865*, 1961, pp. 72-103, and articles by Chadwick on the Great Stove in *Architectural History*, Vol. IV, 1961, pp. 77-91, and Vol. VI 1963, pp. 106-9.

DAY & SON. LITHRS TO THE QUEEN.
LONDON. PUBLISHED BY LLOYD BROTHERS & Co 22, LUDGATE HILL, & SIMPKIN MARSHALL & Co STATIONERS HALL COURT.
SEPTEMBER 1ST 1851.

WAX FLOWERS
PERSIA

put it, 'mature and realise' his scheme.[7]

The Fox Henderson contribution, though less romantic, puts the Crystal Palace in a more realistic context. The company had started life in 1839 as Bramah, Fox and Co., a partnership of John Joseph Bramah and Charles Fox. Its works in Smethwick, designed by Fox, was completed in 1841, in time to catch the engineering contracts of the railway boom in the mid-1840s. By 1845 Bramah had retired, and the firm changed its name to Fox Henderson and Co. with the arrival of the Scottish engineer, John Henderson. As far as is known, Fox was more of a designer while Henderson looked after running the works. Though the company fabricated iron products of every description (including prefabricated houses for California) its attraction for Paxton was its reputation for the design and manufacture of iron roofs. Two of its station roofs have long been known, though the structures themselves have been demolished: Tithebarn Street, Liverpool (1849-50), designed in conjunction with John Hawkshaw, and New Street, Birmingham (completed 1854), designed by one of the firm's staff at Smethwick.[8] Other roofs, constructed to cover ship-building slips in naval dockyards, have recently won comparable attention, thanks to the detective work of James Sutherland. Fox Henderson's first naval contract seems to have been for two roofs at Pembroke Dock, begun in 1844, both of them composite trusses spanning eighty feet and eight inches, with twenty foot overhanging eaves.[9] They have not survived, but a similar roof of about the same date was erected by the company at Woolwich and subsequently, after that dockyard closed, reappeared in a transmogrified state as No. Eight Machine Shop at Chatham, where it can still be seen. Nearby at Chatham is another migrant Fox Henderson roof: the plaque on the Boiler Shop, dated 1876, marks its re-erection there after a previous life as the roof of No. Four Slip at Woolwich.

True to Paxton's predilections, the Hyde Park Crystal Palace had a great deal of wood in it, but, because of the iron structure on which it depended, Fox Henderson were a natural choice as contractors. Their role provides a more convincing connection between the project (plus the Exhibition it housed) and the industrial society which spawned it than an ancestry derived from hothouses and gardening. Their commitment made them, as was remarked at the time, 'masters of the situation', from the first stage of preparing working drawings and making structural calculations, through the fabrication of components and their assembly on site, to the eventual dismantling of the structure and its transfer elsewhere.[10] On the apparent evidence of such a chronicle of success the project seems to vindicate the prominent role which prefabrication gives to component manufacturers. Yet, as post-war experience has demonstrated, such involvement can also be a costly trap for a company which is lured into taking a greater responsibility for design and performance than it had allowed for. Fox Henderson, despite their track record on previous engineering projects, do not seem to have wholly escaped such risks. In November 1851, after the Exhibiton had closed, the firm was granted an additional £35,000 on top of its original

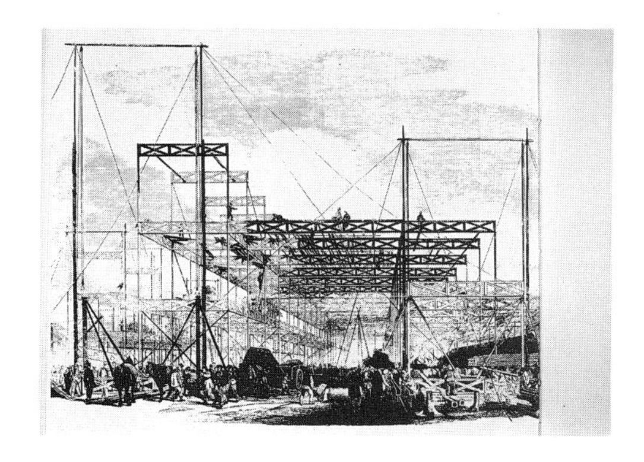

Above:
Top: Raising the Trusses. *Bottom*: Fox-Talbot calotype photograph, 1851
Opposite:
Top: British Exhibit. *Bottom*: The Crystal Fountain

7. *Illustrated London News*, July 5 1851, p. 21.
8. 'Description of a Large Roof Recently Erected at the Liverpool Terminus of the Lancashire and Yorkshire Railway', *Transactions of the Royal Scottish Society of Arts*, Vol. IV, 1856, pp. 94-7; Edwin A. Cowper, 'Description of the Wrought-Iron Roof over the Central Railway Station at Birmingham', *Proceedings of the Institution of Mechanical Engineers*, 1854, pp. 79-87.
9. Capt. M. Williams, 'Description of Wrought Iron Roofs Erected Over Two Building Slips in the Royal Dockyard at Pembroke, South Wales', *Papers of the Corps of Royal Engineers*, Vol. IX, 1847, pp. 50-58.
10. 'Helix', (pseud., W. Bridges Adams), 'The Industrial Exhibiton', *Quarterly Review*, Vol. LV, July 1851, p. 354.

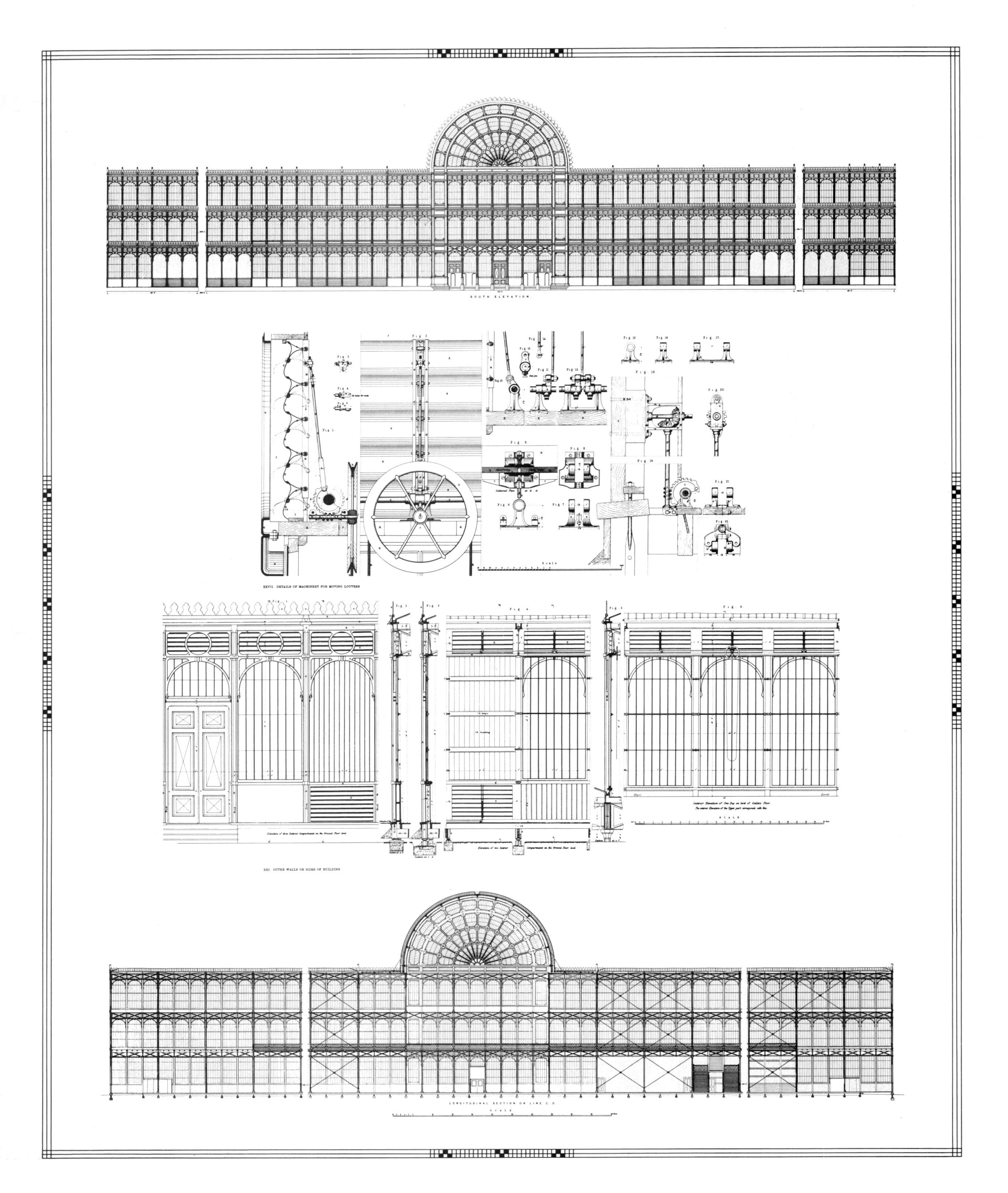
SOUTH ELEVATION
XXVII. DETAILS OF MACHINERY FOR MOVING LOUVRES
Elevation of three Exterior Compartments on the Ground Floor level.
XXI. OUTER WALLS OR SIDES OF BUILDING
LONGITUDINAL SECTION ON LINE C.D.
SCALE

contract of £79,800 plus later additions. This was necessary, said the Commissioners, because 'Fox Henderson discovered that their net liabilities were far beyond those to which they were entitled in the contract'.[11] Even taking account of the fact that the contract was a complicated one in the first place, the Commissioners' generosity, in the flush of the Exhibition's success, conceals a significant drawback in the method used to get the event housed on time. In less sympathetic circumstances Fox Henderson might never have seen the year out.

In the other principal debate which prefabrication has engendered the question whether an industrial method can produce an acceptable architectural aesthetic – the Crystal Palace again has a less secure position as the indisputable ideal of its kind. Some of the resentment that the architects harboured towards Paxton derived from the belief that one of their number, Sir Charles Barry, had suggested the round-arched transept, the most highly appreciated feature of the building. Regardless of the truth of that claim, or the professional interest which fuelled it, the transept certainly was the idol of those who doubted whether a modular structure, unrelieved by such a feature, could achieve a pleasing effect. As a contributor to the *Architectural Quarterly Review* put it: 'The interior of the transept of the Exhibition Building, with its arched roof – due to Mr Barry – has . . . a noble effect; but this stamps only more completely the inferiority of ART in the general building'.[12] None doubted that the perspective down the nave, heightened by Owen Jones' clever colour scheme, had a dazzling impact on first viewing: 'fairy-like' were the words on everyone's lips. But translucent length on its own was not enough. Apart from the transept, nothing else interrupted the rhythmic monotony of columns and girders, which in repetition were eventually confusing and tiring.

As for the exterior, Henry Mayhew described the 'disappointment on the countenances of the newcomers on first beholding the building', as they realised how far the magazine illustrations were from reality. Except from certain angles, the magnificent scale of the building was hard to grasp:

> Look along it, and the trees soon block out the view; look up at it from either side, and the extreme width of the building prevents you seeing the upper storey, so that it has a most unseemly 'squat' or 'dumpy' appearance'.

And approaching closer, the building belied its name by presenting a ground floor sheathed in wooden hoardings.[13]

If the critics of the Crystal Palace had been simply a few disparate voices – disgruntled architects, militant Ruskinians, and those who find something to complain about in everything – their opinions might reasonably be discounted in any final judgement of the project. What gives weight to their views is that those most responsible for the building recognised their validity and endeavoured, in the years immediately after the Exhibition closed, to take account of them. Paxton, Fox Henderson and other members of the building team went on, singly or together, to other projects which were

Above:
Top: Erection of Crystal Palace, 1851. *Bottom*: Routing the Sash Bars
Opposite:
Elevations and details from Downes Cowper's book of 1852

11. First Report of the Commissioners for the Exhibition of 1851, *Parliamentary Papers*, 1852, XXVI, p. 30.
12. *Architectural Quarterly Review*, June 1851, Vol. 1, p. 26.
13. *Edinburgh News*, May 3 1851, p. 4. For similar remarks by a visiting Dutch architect see *Builder*, Oct. 18 1851, p. 654.

regarded as the direct successors to the Hyde Park one, notably the re-erection of the Crystal Palace at Sydenham and the construction of the new G.W.R. terminus at Paddington. As they did so they fell over themselves to show that they too were eager to advance iron and glass architecture beyond the 'scaffold pattern' of the prototype.

Paxton had always hoped that the success of the Crystal Palace would ensure its retention in Hyde Park as a winter garden in which 'the great truths of Nature and Art would be constantly exemplified'.[14] But having insisted in almost the same breath that it could easily be dismantled, his campaign to keep it standing had a difficult case to make: defeat was conceded with the government's decision in March 1852 that the park should be cleared. Two months later the Crystal Palace Company, established to organise the building's resurrection as a temple of national recreation, had issued its first prospectus: by August the same year the first column had been raised at Sydenham. Paxton was the moving force in the company; Fox Henderson were again the contractors, with John Cochrane and John Henderson in charge; and many others associated with the Great Exhibition had a part in fitting up the interior.

The progress of reconstruction never attracted the same attention as the original project, except on the melancholy occasion in 1853 when a portion of the timber scaffold used to erect the ribs of the central transept collapsed, killing thirteen men. But even relying on just the inquest reports from that event it is easy to appreciate that the Sydenham structure was much more than simply a rerun of its predecessor.[15] The ribs of the celebrated Hyde Park transept were of laminated timber, assembled on the ground before being raised into position in pairs. The new central transept was higher, wider (one-hundred-and-twenty as against seventy-two feet) and built of wrought iron; its principle ribs, set in pairs at seventy-two foot intervals, were of uniform depth with their top and bottom flanges braced by a lattice of flat diagonal bars. Each pair of ribs was carried, via connecting frames, on four columns, two of which were set forward from the main line of the gallery. The scaffold collapse related to the problem of fabricating a roof for which the Hyde Park building provided no precedent. On the same count, although the transept (like the rest of the building) incorporated columns and girders which had already done one term of service, it also called for hundreds of new castings fresh from the Fox Henderson foundry.

Changes of the kind to be seen in the central transept were repeated in almost every other aspect of the rebuilt structure. At Sydenham two end transepts were added, built on the same principle as their larger brother, and the nave was raised to their status by being given an arched roof. This remodelling took place within a length ten bays less than the original, and appeared yet shorter because the Jones colour scheme had disappeared beneath dignified coats of dark red. The team responsible for the redesign were not as prone to make public statements about their intentions as they had been first time round, but the effect of what they did was to produce a

Above:
Dismantling and re-erecting the Crystal Palace at Sydenham
Opposite:
Contemporary photographs by Delamotte showing the dismantling and re-erecting of the Crystal Palace, 1852

14. Joseph Paxton, *What is to Become of the Crystal Palace*? 1851, p. 13.
15. *Builder*, Aug. 20 1853, pp. 529-30, Sep. 17 1853, pp. 590-1, Sep. 24 1853, pp. 602-4; *Civil Engineer and Architect's Journal*, Sep. 1853, Vol. XVI, p. 355-8. The collapse of the scaffold seems to have been caused by the failure of the upper part of a truss (principally a jointed beam ninety foot long) while it was being moved into position.

Top: Owen Jones' design for Saint-Cloud Glass House. *Bottom*: Paddington Station, late 19th-century photograph

DIMENSIONS

Hyde Park
Length of main building 1848 ft
Breadth of main building 408 ft
Height of nave 64 ft
Height of transept 108 ft
Weight of iron used 4500 tons (700 wrought, 3800 cast)
Timber 600,000 cu. ft
Panes of glass 293,655 (9000,000 sq. ft)
Guttering 24 miles
Space occupied by exhibits 991,857 sq. ft
Number of exhibits over 100,000
Number of exhibitors
British 7381
Foreign 6556
TOTAL 13,937

Number of visitors
Season ticket holders 773,776
£1 days (2 days) 1042
5s days (28 days) 245,389
2s 6d days (30 days) 579,579
1s days (80 days) 4,439,419
TOTAL 6,039,205

Area of floor 772,824 sq. ft
Area of galleries 217,100 sq. ft
TOTAL 989,924 sq. ft (this figure only includes the exhibition area)

tighter, more architectural treatment, creating as much of an effect of permanence as its prefabricated structure allowed. The next possible concession to critics might have been to produce a yet more articulated profile by adding the three domes that Barry had suggested, one at each of the transept intersections. But even that might not have been enough. Barry's son, speaking in 1871, kept up the chorus that his father's friends had started:

> I cannot conceive that anybody can look upon that structure as either artistic or of scientific construction. It is a huge glass case, situated in a position where the eye wanders from the building to the beautiful surroundings of Nature; therefore it gets a great amount of adventitious praise of which, architecturally, it is undeserving.[17]

Paddington Station, completed the same year as the Sydenham Crystal Palace, offers a yet more instructive example of how the lessons of the Exhibition building were absorbed almost before they had happened. I.K. Brunel, hurt that his own proposal for accommodating the Exhibition had been so mauled by other members of the Building Committee, was ready to demonstrate the virtues of what he dubbed the 'railway shed style'. Early in 1851 he enlisted Matthew Digby Wyatt, known to have the same views, and Fox Henderson, who were than so delirious with the progress of their Hyde Park contract that they felt they could tackle anything. Independent minded though Brunel was, he and his team could not avoid conceiving of the station as anything other than a mature variant of the Crystal Palace.

Today, looking along the lines of columns which support the triple-arched roof, the parentage is unmistakable, even though their spacing is wider than in the Hyde Park building and the column to girder connections are differently handled. Also, the sheds are twice interruped by fifty foot wide transepts, introduced to modulate and relieve their length just like the one which Paxton had been persuaded to adopt. But a description of the station cannot be founded on such analogies alone, leaving unremarked the parts of its structure which stretch the Paxtonian model into a more consciously architectural form. The wrought iron arched ribs used at Paddington had no precedent in the original Crystal Palace, nor had the idea of piercing such ironwork with a pattern of stars and planets to relieve its tunnel-like effect. In other kinds of decoration, as well, Brunel and Digby Wyatt were innovators: the swirling iron tracery of the end gables, the column capitals, and the foliage pattern which sprouts in the haunches of the ribs. 'A station after my own fancy', as Brunel called it, was not just a workable terminus but a lesson in how the temporary qualities of iron and glass could be reinterpreted for enduring use.[18]

Paddington received high praise ('by far the finest work of its kind in Europe', pronounced *The Builder*) but on its own was not enough.[19] The new order of architecture, as dreamt of by Paxton, had to extend further than obvious kin such as railway stations to make his case proven. He was ready to talk in terms of housing, churches and factories, not to mention his old

16. C. R. Von Wessely, 'On Arched Roofs', *Civil Engineer and Architect's Journal*, Apr. 1866, Vol. XXIX, pp. 107-8.
17. RIBA Proceedings, 1871-2, p 84.
18. Robert Thorne, 'Masters of Building: Paddington Station', *Architects' Journal*, Nov. 13 1985, pp. 44-58.
19. *Builder*, Dec. 24 1870, p. 1020.

stamping ground of horticultural buildings[21], but in practice his movement (if it can be so called) ended where it had started, with exhibition buildings. Such structures, by their very nature exotic and ephemeral, were unhelpful in establishing a continuity of experience in modular construction, though they might advance debate and practice in certain ways. But as it turned out, those responsible were so conscious of the assumed faults of the Crystal Palace that their contribution was to rein in the tendencies which it had started, without providing a strong alternative. Owen Jones, an old Crystal Palace associate, was almost alone in advancing the cause of iron and glass for exhibitions, but none of his designs were built.

Of the exhibition buildings that were completed in the 1850s, the New York Crystal Palace of 1853 shows the often bizarre way that the lessons of Hyde Park, valid and otherwise, were digested. An iron and glass building was stipulated by the New York City Council and of such designs submitted in competition (including an entry by Paxton himself) one by the partnership of Carstensen and Gildermeister was chosen: a Greek cross surmounted by a dome, with lean-tos marked by towers in the angles of the cross. 'Columns, girders, and arches', it was said, 'will be connected according to the system of Messrs. Paxton, and Fox and Henderson'. Whatever form of borrowing that constituted, 'the monotony of Mr. Paxton's design' was eschewed in the adoption of a plan and system of embellishment which consciously suppressed memories of the Hyde Park model. This was a project completed before the lessons of the Sydenham Crystal Palace, or indeed Brunel's Paddington, were available, but it illustrates a movement which their influence could do nothing to deflect.[20]

If any one event marked the termination of the expectations first engendered by the Crystal Palace it was not the death of Paxton in 1865 but the collapse of Fox Henderson nine years before. The immediate cause of the failure was a loss of about £70,000 on a Danish railway contract but the indications were clear, from their abysmally slow completion of their two Paddington contracts if nothing else, that the reputation they had gained at Hyde Park had got the better of them.[21] Even there, as already suggested, their conjunction of skills was not as wholeheartedly satisfactory as is generally thought. Other firms took their place, but no other engineering contractors applied their own skills to the cause of structural innovation to quite the same degree as Fox Henderson; and perhaps for that reason none could claim a roll-call of projects to compare with those that passed through the Smethwick works in the decade 1844-54. But even if Fox Henderson had survived, that would not have saved the day for prefabrication. They brought into being a model system and helped mature it, but nothing they could do would have ensured its perpetuation.

Top: Boiler Room, Chatham Dockyard. *Bottom*: floor to slip No. 4, Woolwich, re-erected Chatham, 1876

Sydenham
Length of nave 1608 ft
Width of nave 312 ft
Height of nave 68 ft
Length of each wing 574 ft
Width of transept (central) 387 ft
Height of transept 175 ft
Height of towers 284 ft
Weight of iron 9642 tons
Area of glass used 1,650,000 sq. ft superficial
Weight of glass used 500 tons
Amount of brickwork 15,391 cu. yd
Hot water piping (12 in.) 50 miles
Total area of main building 603,072 sq. ft

20. George Carstensen and Charles Goldemeister, *New York Crystal Illustrated Description of the Building*, N.Y. 1854, p.48.
21. *Illustrated London News*, Nov. 1 1856, p. 441; Thorne, op. cit., p. 51.

FRANCIS PUGH

Design, Engineering and the Art of Drawing

A sophisticated computer graphics programme can produce multiple views of an object rapidly and cheaply, a task which not so long ago required the time-consuming labour of a large drawing office. Yet the sheer facility of this process can easily make us forget the conceptual power of the basic language of drawing. One has only to imagine the process of design, manufacture or construction without such a system to realise that in common with other tools of prediction and analysis, it greatly expands our capacity for sustained innovation. But how did this language evolve and what were the circumstances that brought its various elements into being?

Architectural Drawing

Even in the distant past any major work which required the deployment of large numbers of craftsmen, also needed a means for organising and directing their labour. Plans drawn by master masons served this purpose in Mediaeval Europe and had done so in some form since Graeco-Roman times, but for the most part traditional methods of full-scale drawings were sufficient for even the largest undertaking. The shapes of vaults and arches were inscribed on the plaster floors of mould lofts, while stonemasons and carpenters used templates as well as marking directly onto materials. Drawings of elevations seem first to have appeared towards the end of the Middle Ages when the complexity of surface design had reached a point where the work of the stone carvers needed more precise direction. But to our eyes Mediaeval drawings lack a coherent system for relating the parts of a structure to the whole. The system which seems so familiar today is based on the concept of scientific perspective, a discovery usually attributed to the Florentine sculptor and architect Brunelleschi. To Brunelleschi, and later 15th-century humanist philosophers, perspective represented much more than a technical device: it symbolised a new objective view of man's place in a divinely ordered universe. They sought to understand this idea further through the study of Euclidean geometry, from which was gained a greater awareness of pattern, order, proportion and relationship; principles which also underlay the desire to create a new harmonious architecture based on classical ideals.

Architects of the generation after Brunelleschi began the process of inventing a new basis for the language of drawing when their attempts to communicate novel architectural forms led them to the notion, implicit in scientific perspective, that three related views were sufficient to define any

Above:
Elevation of one bay of Bishop Fox's chantry, Winchester Cathedral, attributed to William Vertue, early 16th century
Opposite:
Detail of longitudinal section of the SS Great Eastern (lithographic print hand-coloured from John Scott Russell's shipyard drawing office, 1860)

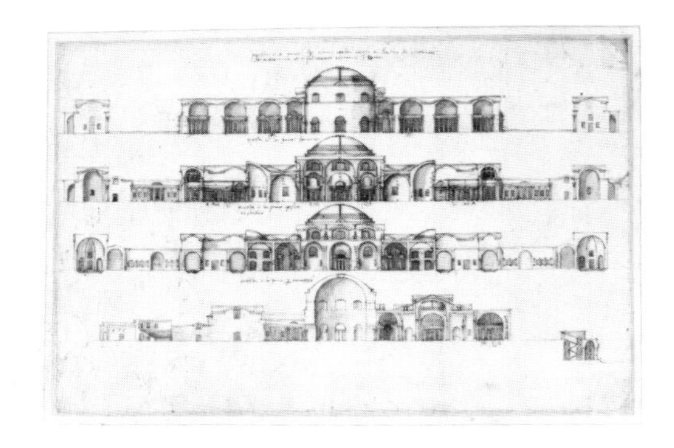

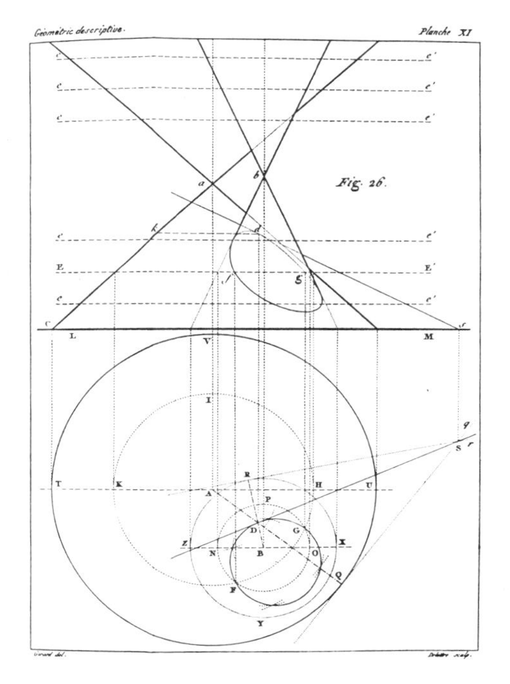

Above:
Top: elevations and sections of Baths of Caracalla, Rome, Andrea Palladio, c. 1450. *Bottom*: method of determining the curve of intersection of two cones (Plate XI from Gaspard Monge, *Geometrie Descriptive*, 1795)
Opposite:
Top: Craigellachie Bridge over the River Spey (from *Atlas to the Life of Thomas Telford*). *Bottom left*: transformation of alternating to continuous motion, Leonardo da Vinci (*Codex Atlanticus*, folio 8 V-b). *Bottom right*: Robert Stephenson, Engineer; Newcastle upon Tyne High Level Bridge drawing

object or structure. Artists and architects including Leonardo da Vinci and Bramante contributed to this development but Raphael was probably the first to make use of plans, sections and elevations related in logical sequence and to a consistent scale. Whatever his underlying philosophical motive, the immediate cause seems to have been the need to oversee his many workshop commissions while at the same time fulfilling the terms of his appointment as architect of St. Peter's after 1515. Drawings supplied in advance would allow building work to continue during his frequent absences from Rome. This business-like approach is equally evident in his proposal that the ruins of ancient Rome should be measured and drawn to give architects and scholars a more thorough understanding of the scale, appearance and construction of Roman buildings. In both instances, it is apparent that Raphael, in common with many of his contemporaries, held a comprehensive view of the function of drawing. Not only could it be used to implement a design, it could also serve as part of a business transaction, a record, a form of instruction and of objective analysis. Later generations of architects used drawing for all these purposes so that knowledge of current designs and knowledge of the craft of building were no longer regarded as 'mysteries' communicated only to initiates in workshops and guilds, but could be freely disseminated in sketchbooks and technical treatises.

The key figure in this transition to something akin to modern practice is Andrea Palladio who began his career as an apprentice stonemason and only later became acquainted with the philosophy of humanist architecture through his patron and teacher Giangiorgio Trissino. Like Raphael, Palladio regarded measurement and drawing as more effective ways of gaining accurate knowledge about the remains of the ancient past than the study of humanist theory or the writings of Vitruvius. His own architectural treatise, the *Four Books on Architecture* of 1570, while modelled on Vitruvius, is essentially a practical work. Its wood block illustrations, though less than perfect for conveying subtleties of line and tone, were to have an influence far beyond the confines of the society and period in which they were conceived, because they conveyed, as no drawings had done before, a sense of the scale and proportion of classical architecture in a simplified and readily understandable form. Palladio's use of a lucid and orderly sequence of carefully chosen cross-sections is apparent in his drawings of the Baths of Caracalla. It has been suggested that this orderliness was imposed by Palladio and is not evident in the buildings themselves.[1] If so, it further emphasises the way in which our present language derives from a system of drawing that sought to organise experience for philosophical and spiritual ends while never losing sight of the practical advantages that had been gained in consequence.

Engineers and the Art of Drawing

Palladio's drawings were influential in many ways, not least in providing exemplary models for the illustrators of scientific or technical works and for the working drawings of engineers and architects. It is worth noting that John

1. James S. Ackermann, *Palladio*, 1966, pp 171-172.

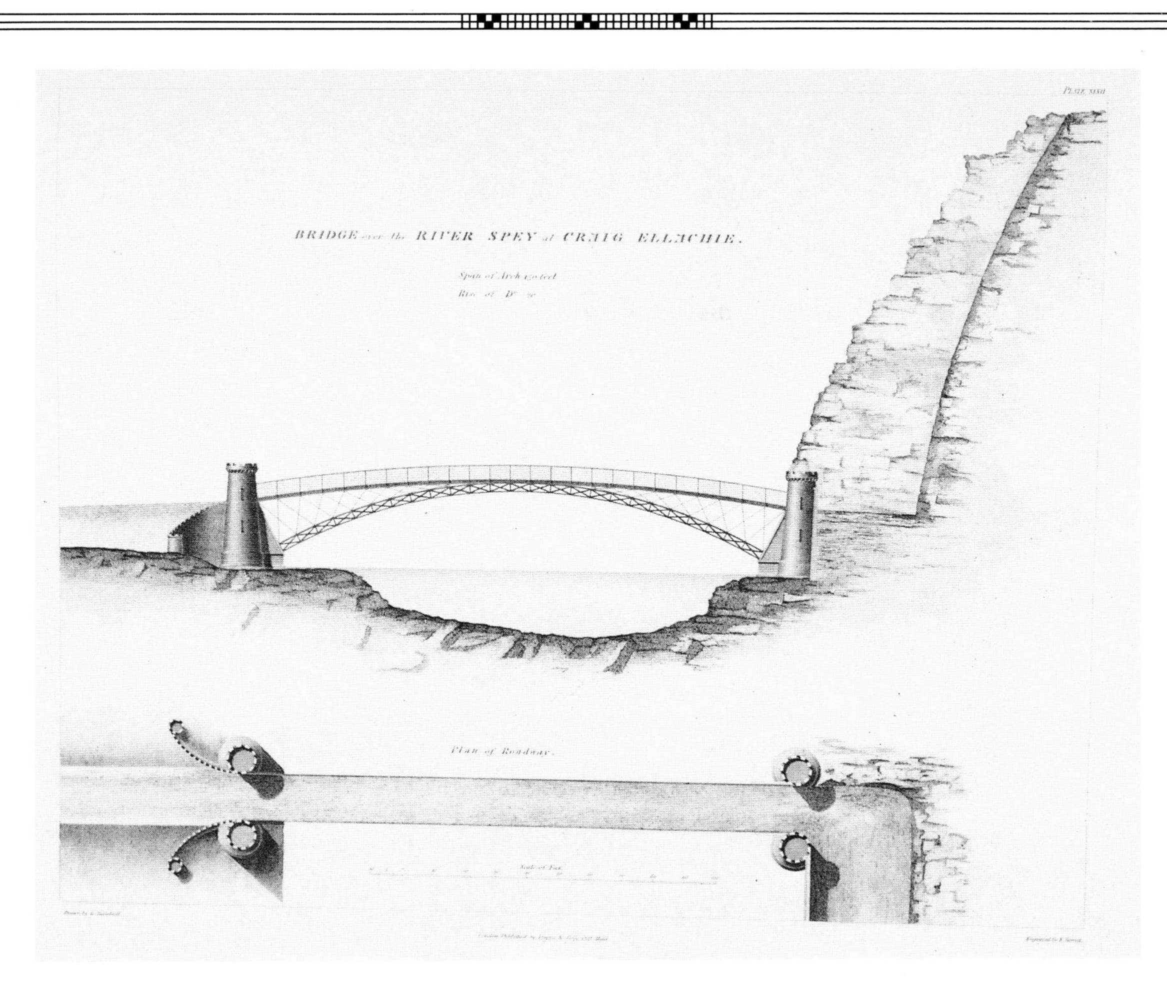
BRIDGE over the RIVER SPEY at CRAIG ELLACHIE.
Span of Arch 150 feet
Rise of D° 20
Plan of Roadway.

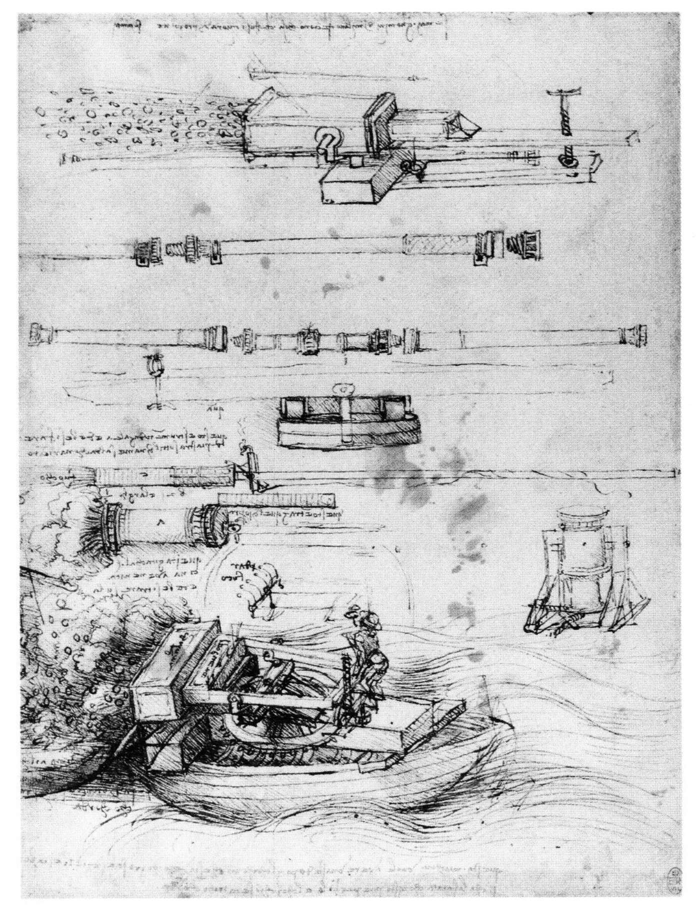

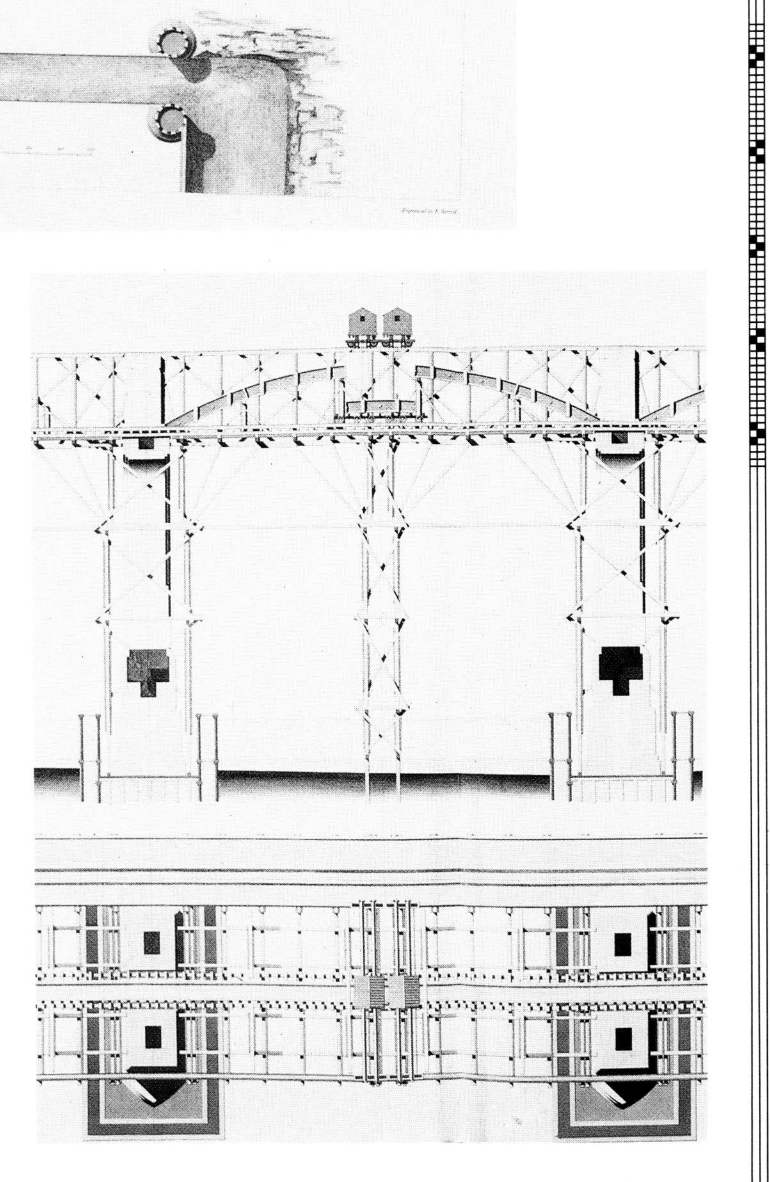

EXAMPLES OF FINISHED SHADING.
Fig.1.
Fig.2.
Fig.3.
Fig.4.
Fig.5.
Fig.6.
Fig.7.
Fig.8.
BLACKIE & SON, GLASGOW, EDINBURGH & LONDON.

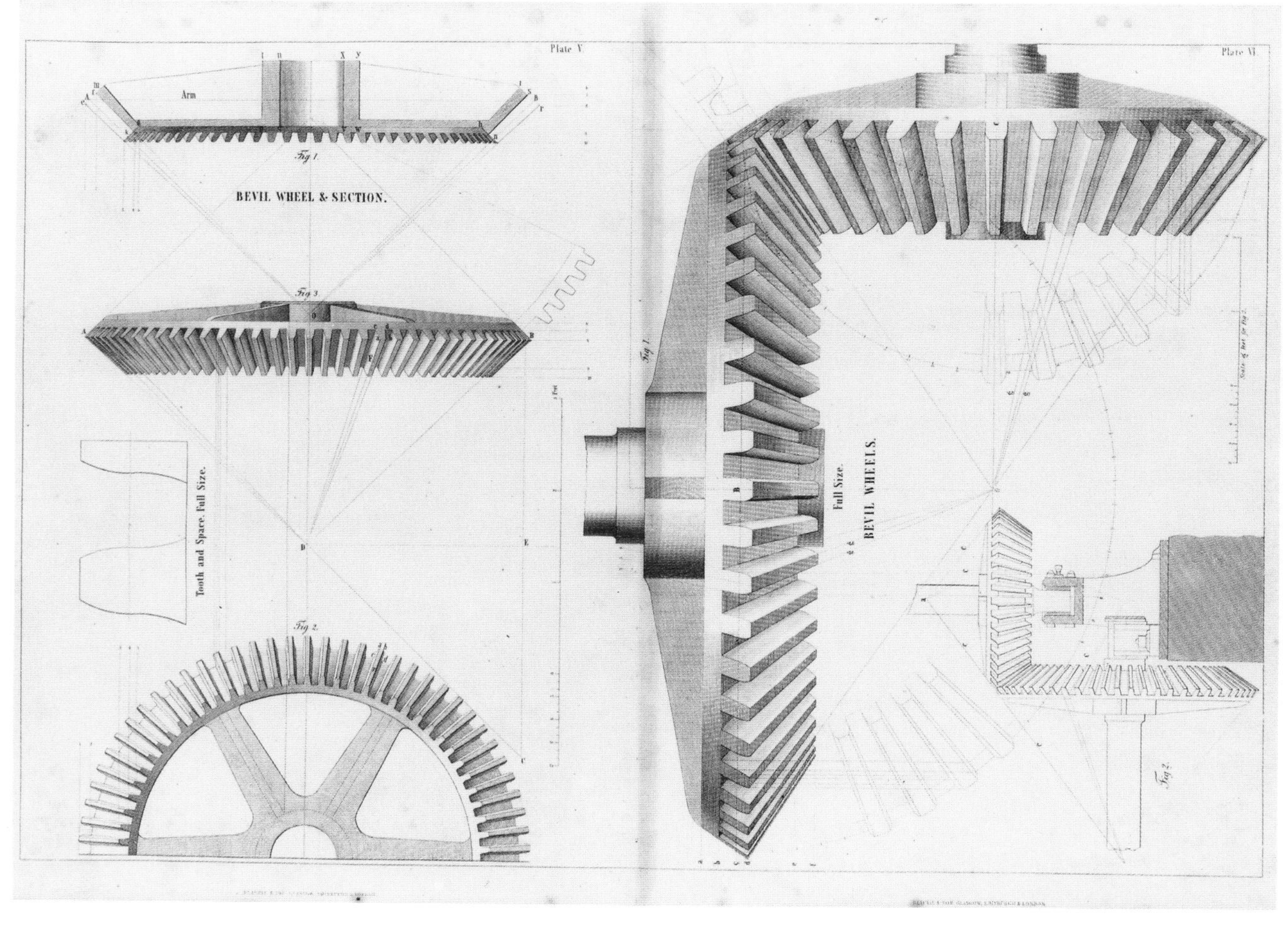
Plate V
Plate VI
Arm
Fig 1
BEVIL WHEEL & SECTION.
Fig 3
Tooth and Space. Full Size.
Fig 2
Full Size.
BEVIL WHEELS.

Smeaton, one of the founders of an independent engineering profession in Britain, owned a number of books, either edited or written by architects in the Palladian tradition; Vitruvius in Claude Perrault's edition of 1684, *The Architecture of L. B. Alberti* by Giacomo Leoni (1755) and William Chambers' *Treatise on Civil Architecture* (1759)[2]. Their illustrations would have exposed him to a methodical and sober classicism both in their architectural style and the way in which the structure of buildings was explained through sequences of precisely engraved sections and elevations. That an engineer found such sources useful is hardly surprising. Architecture and what later came to be known as civil engineering were, until the late 18th century, essentially the same discipline. Together they formed part of what were regarded as the 'designing arts' – disciplines characterised by their use of drawings, models and mathematical calculation as means for constructive forethought.

What I shall call the Palladian tradition of drawing emphasised pictorial means, especially the use of case shadows to give a greater illusion of reality. It was particularly suited to explanatory drawings and to a derivative, the presentation drawing; a work intended to impress clients and patrons, and record a project in its finished state. In this form it continued to be used by engineers, with the addition of colour and an increasing number of standardised symbols, until the end of the 19th century, and in more traditional industries until the beginning of the First World War. Some of the finest examples of engineering drawing belong to this category. The sequence of hand-coloured lithographs of the 'Great Eastern' and illustrations to the *Atlas to the Life of Thomas Telford*, a companion volume to his autobiography – were made expressly for the purpose of promoting the reputation of the engineer. In the eyes of the draughtsmen who produced them they were identical in every respect to similar works produced by architects and shared a similar taste for decorative colour combined with immaculate finish. For this reason they are the best known and most frequently preserved of the many types of industrial drawing.

Above:
'Rubanier', ribbon-making machine (plates I, IV and V from *Dictionnaire des Sciences*, Tome XXXVIII, 1772)
Opposite:
Top: plate showing Examples of Finished Shading. *Bottom*: plate illustrating method of drawing bevil wheels (both from Le Blanc and Armengauld, *The Engineers and Machinists Book*, 1847)

Drawing the Machine

Architectural drawing provided engineers with a substantial part of their graphic language but other sources were also important. The depiction of machines is an area where the conventions of architectural drawing could provide only partial solutions. From the late 15th century, attempts to describe the interrelation of moving parts led artists to develop a variety of graphic innovations. Leonardo's sketchbooks contain numerous descriptions of both practical and fanciful machines drawn with an extraordinary grasp of explanatory technique. Among his drawings which seek to elucidate the elements of mechanics is one which uses an exploded view to describe each part and then an assembled view to show the complete mechanism. But these drawings made little impact in Leonardo's lifetime and their dispersal ensured that they remained largely unknown until the 19th century.

2. A. W. Skempton, (ed.), *John Smeaton, FRS*, Appendix I, 'Smeaton's Library and Instruments', 1981.

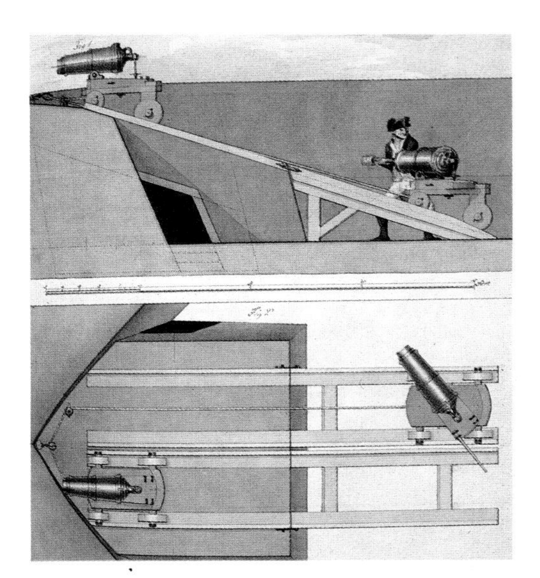

Garrison gun carriages for sea lines and barbet batteries etc, Brigadier General Robert Lawson, c. 1790

When the artist Rudolf Manuel Deutsch used an exploded view in Georgius Agricola's treatise on mining and the metal industries, *De re metallica* (1556), it can probably be assumed that the idea had evolved as part of the general response of artists and engineers to the desire for more accurate representation. Thus it is unlikely to have been solely Leonardo's invention. Agricola's treatise is also notable for some of the earliest uses of the cutaway to reveal areas which in reality would be hidden and the numeric or alphabetic key linking descriptive text to the parts of an illustration. Both were devices which soon became part of the standard repertoire of the illustrator of technical treatises and from this source were eventually incorporated in everyday engineering practice.

From the 16th century onwards the growing sophistication of graphic technique can be traced in a succession of works on the mechanical sciences, especially those describing machines driven by wind and water power. The earliest drawings incorporate all the information in one picture while later examples often present different aspects of a machine or mechanical process in two or three separate pictures. The illustrators of Diderot and d'Alembert's *Dictionnaire des Sciences*, one of the most accomplished of the great 18th-century works on arts and manufactures,[3] frequently adopt a sequence of views including an orthographic projection, and horizontal and longitudinal sections, followed by details of parts and illustrations of workers engaged in manufacture. Smeaton however reverts to the earlier type of single view in his illustration of the process of constructing the Eddystone lighthouse, so that the various methods for landing and hoisting stones during the six stages of construction can be readily comprehended in one illustration. Books of this type also reproduced the style of working drawing employed by specific crafts or trades. Possibly the earliest published working drawings of machines were those contained in the early 18th-century Dutch treatises on mill design which like those in architects' and builders' pattern books were intended to give practical guidance on matters of design and construction.[4]

Despite their increasingly sophisticated presentation, illustrative techniques had many drawbacks when it came to depicting machinery. To some engineers it was now apparent that a more mathematically precise and consistent drawing system was needed to cope with the particular problem of describing the complex geometry of machine forms. The theoretical foundation was established by the French military engineer Gaspard Monge whose *Descriptive Geometry* published in 1795 uses a theoretical structure derived from the mathematics of Descartes, Pascal, and others to bring clarity and logic to a variety of *ad hoc* techniques customarily used by stonemasons and carpenters. The technique was first applied to the design of fortifications and was consequently regarded as a state secret for the decade or so prior to its publication. However, Monge eventually persuaded the French government that his system could provide much needed stimulus to manufactures and this led to its incorporation in the curriculum of the Ecole Polytechnique, an institution which Monge himself had founded. After 1815 it was taught in the

3. *The Dictionnaire des Sciences* was published by Diderot and d'Alembert for the Academie des Sciences at Paris. It consists of thirty-five volumes issued between 1752 and 1780 and includes twelve volumes of plates.

4. Works on mill design were published in Holland in 1727, 1734 and 1736 and were given the collective title *Groot Volkommen Moolenbock*, a complete book on mills. Batty Langley, the author of the most influential early 18th-century architect and builders' pattern books also produced one of the earliest English works on the practical use of drawings: *Practical Geometry, Applied to the Useful Arts of Building, Surveying, Gardening and Mensuration*, 1726.

5. The teaching of descriptive geometry at West Point was the responsibility of Claude Crozer, a graduate of the Ecole Polytechnique and an artillery officer under Napoleon. From 1816 to 1817 he was assistant professor of engineering and a full professor from 1817 to 1823. In 1821 Crozer published his *Treatise on Descriptive Geometry*. The cadets were also trained to draw topography and the human figure by the French miniature painter Thomas Gimbrede. Further information is contained in: M. Welch, 'Early West Point, French Teachers and Influences'; *The American Society Legion of Honor Magazine*, spring 1955, Vol. 26, No. 1, pp 27-43.

new technical high schools established in the German states, while *émigré* French instructors at the West Point military academy introduced the theory to America.[5] More general acceptance, in Britain especially, came only with the publication of the first textbooks on machine drawing in the 1840s. Many of these were written by French authors. The sole aspect of Monge's theory to be widely adopted in his lifetime was the standardised relationship of plan, side and front views in the form known as 'first angle projection' – a system which rapidly replaced older forms of oblique projection and other casually related views. It soon became the standard method for organising general arrangements, the type of working drawing that serves as the master to which all others relate and thus a key constituent of the engineer's design vocabulary.

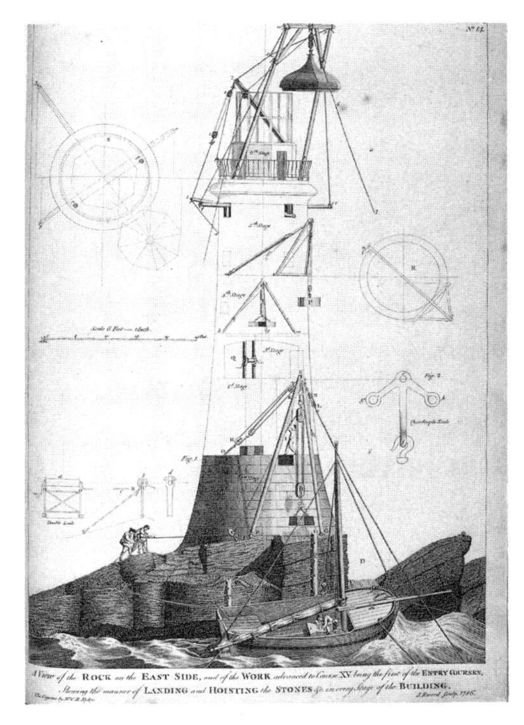

John Smeaton, Engineer; Eddystone Lighthouse, 1856 (Engravings by James Record from drawings by John Smeaton, c. 1791)

Ships, Maps, Surveying and Instrument Making

This investigation would be incomplete if it concentrated on theoretical developments to the detriment of those aspects of drawing, like the majority of British examples, which evolved more or less empirically from the practice of a variety of existing trades. Those who were most influential in this respect, other than architects and builders, were shipwrights, surveyors, map makers and instrument makers.

The basic form of ships' draught had first appeared in the late 16th century in response to the shipwrights' need to predict the shape of new types of ocean-going, gun-carrying vessel. By the mid 18th century the evolution of ships' draughts, and of the mathematical calculations associated with them, had arrived at a description of the irregular solid of a ship's hull which dispensed with oblique projection in favour of a series of vertical and horizontal sections superimposed in line only on the three principal views. In British naval draughts this resulted in partly schematic, partly pictorial representation, which left much to the discretion of the shipyard foreman. A typical lines plan employs a simple colour code; black for the outline, red for the decks and fittings that formed part of the internal structure of the ship, and green for those lines like the hull contours which describe form rather than physical structure. Only works on naval architecture like Hendrik Chapman's *Architectural Navalis Mercatoria* (1768) which were intended principally for wealthy subscribers, would include projected drawings in addition to a simple lines plan; the process being too laborious and time-consuming to be part of everyday shipyard practice.

The technique of land surveying by triangulation began to develop in the 15th century but accurate map-making was not possible until the invention of the first surveying instruments with telescopic sights in the early 17th century. Even then instruments were hardly precise and it is not until the mid 18th century that accurate surveying became a reality. It is no accident that the makers of the finest surveying instruments were also innovators in the field of drawing. The trade card of the London instrument maker George Adams lists instruments for copying drawings, and camera obscuras for making

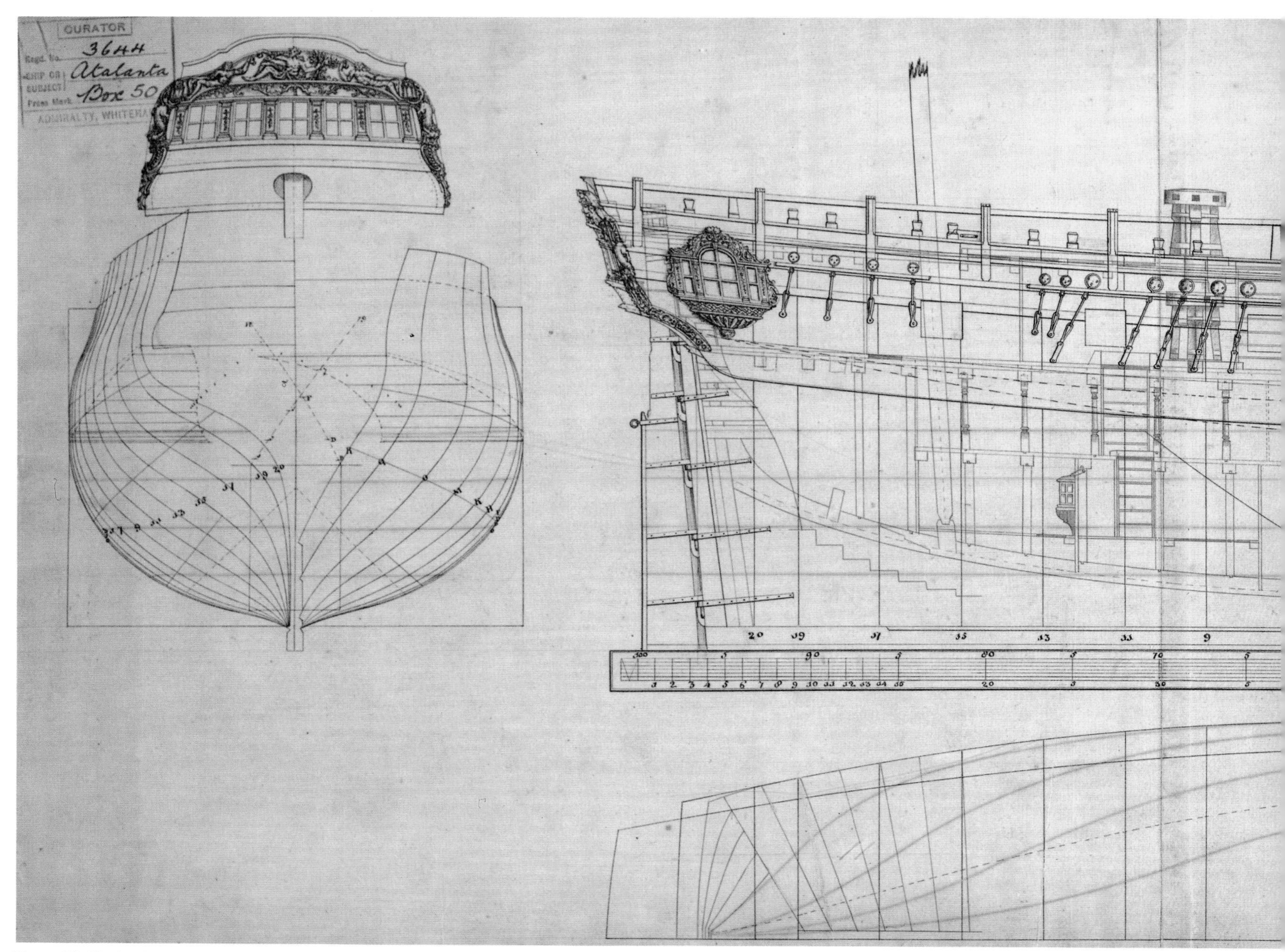

Top: HMS Atalanta showing ship as built before fitting out, September 1775 (Admiralty Collection). *Opposite*: ship's draught with four sections, attributed to Mathew Baker, c. 1600 (from *Fragments of Ancient English Shipwrighting*)

perspectives in addition to the more standard surveyors' equipment of theodolites, water levels, and measuring wheels. Map-making in consequence began to abandon earlier pictorial styles for a more precise plan view often with fine hatching to indicate hill shapes and a range of conventional symbols. Engineers needed a thorough knowledge of map making and surveying, thus a familiarity with the equipment and methods seems to have played an important part in the careers of many engineers. Smeaton and Marc Brunel were trained as instrument makers, and both retained habits of precision and accuracy derived from their early experience. Drawing, surveying, map making and engineering are linked even more closely in the career of the topographical artist Paul Sandby, who trained as a draughtsman in the Board of Ordnance Drawing Room at the Tower of London before

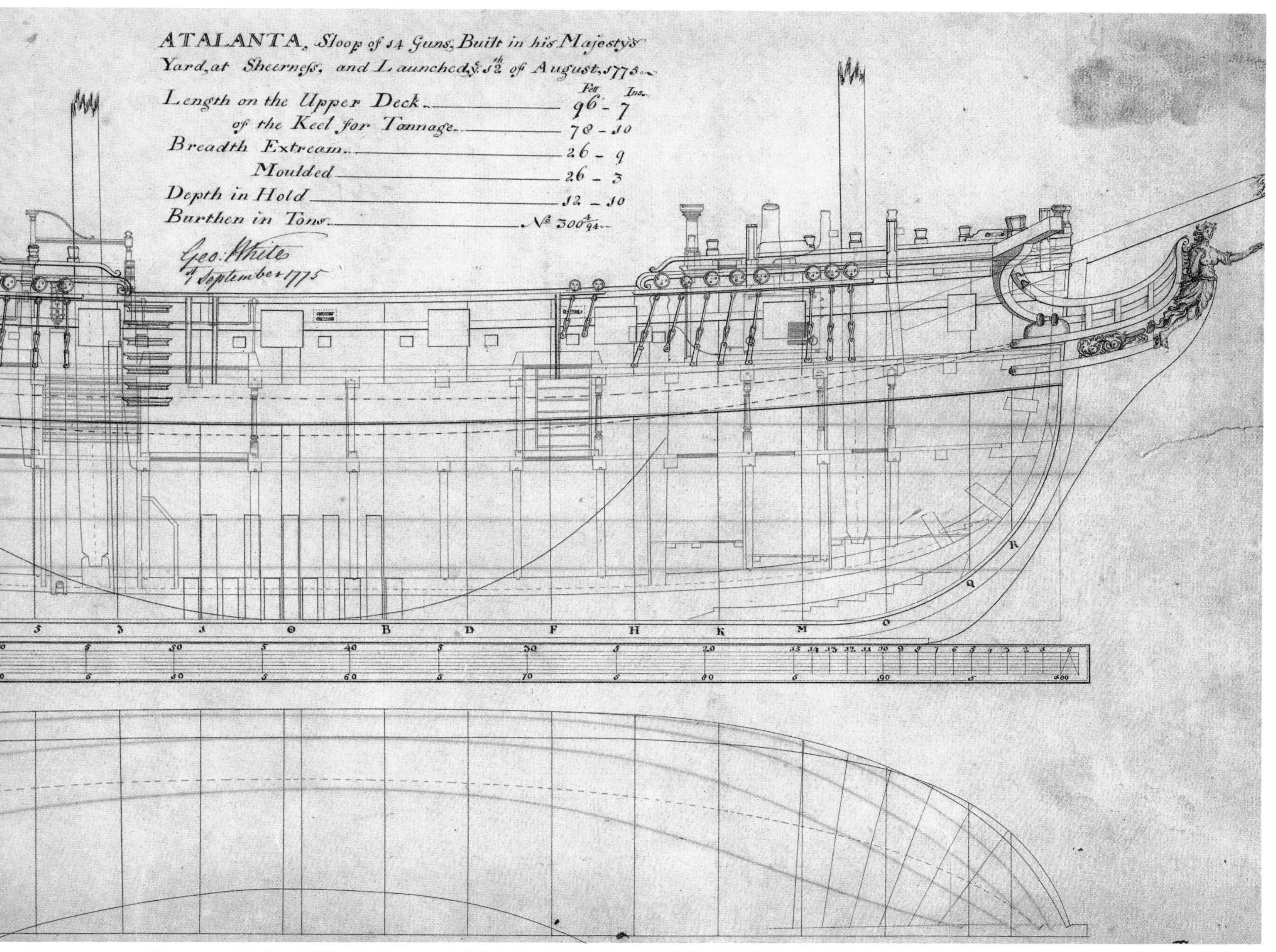

taking part in the Military Survey of Scotland between 1746 and 1751, and was later engaged one day a week as drawing master at the Royal Military Academy, Woolwich – the first such government appointment in Britain. The influence of his methods and style of drawing are apparent in artillery manuals, reports and other military documents of the period. The importance attached to surveying as a training for the engineer is clearly evident in this passage from L.T.C. Rolt's biography of I.K. Brunel:

> Young Isambard began to display his talent for drawing when he was only four years old, and by the time he was six he had mastered Euclid. Such a precocious display of inherited talent obviously delighted Marc (Brunel) so that he determined to foster it to the limit of his means. He first sent him to Dr. Morell's boarding school at Hove where the boy

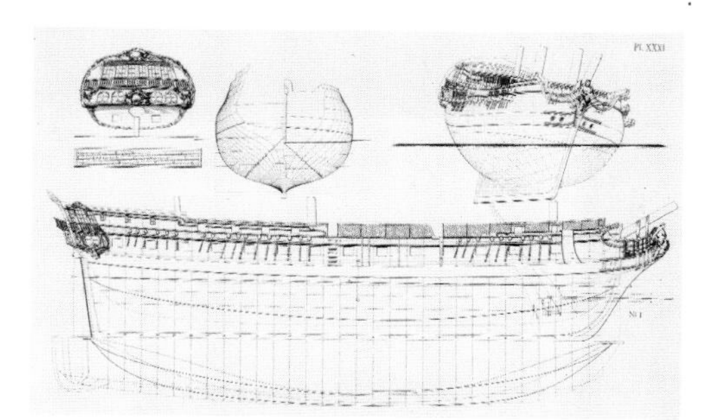

Above:
One of the first technological books: Georgius Agricola's *De Re Metallica*, c. 1556
Opposite:
Top: John Wilkinson's forge engines (drawings by James Watt and John Southern, c. 1782).
Bottom: I.K. Brunel, Engineer; general arrangement drawing for proposed Paddle Engine SS Great Britain, by Maudslay Sons and Field, c. 1839

> amused himself in his spare time by making a survey . . . and sketching its buildings, just as his father had done in his youth at Rouen. Marc always insisted that this drawing habit was as important to an engineer as a knowledge of the alphabet, and it was undoubtedly in this way that both father and son developed such extraordinarily acute powers of observation.[6]

Engineering Drawing

It would be foolish to suggest that drawing and theoretical knowledge at any stage replaced the craftsman's skill at working with materials. Throughout the 19th century much emphasis was placed on the need for all engineers to acquire practical experience in addition to learning how to draw. But the evidence of collections like the Boulton and Watt archive suggests that substantial changes in working methods took place in the crucial decades between 1760 and 1790, particularly in industries associated with the steam engine and iron replaced wood as the principal construction material. The pace of technological change was leading to a new form of industrial organisation in which older craft diciplines were gradually subordinated to a controlling hierarchy of engineers and entrepreneurs. James Watt and his assistant John Southern were in the forefront of these developments when they produced drawings to control and record each aspect of the engine building activities of the Boulton and Watt manufactory; a situation which in scope and ambition anticipates the central importance attached to the drawing office in British industry after 1830.

The growing specialisation that brought an independent engineering profession into being was also responsible for giving drawing its new industrial prominence; and in turn the chief draughtsman, a figure who assumes growing importance as the 19th century progresses, became indistinguishable from the industrial designer of today. It was to facilitate these changes that drawing itself began to change. Borrowing extensively from each of the sources I have described it gradually became more standardised until by the close of the 19th century it had largely abandoned pictorialism for more schematic forms of representation which were less ambiguous and capable of more precise interpretation. From 1870 onwards the descriptive function of drawing was challenged with increasing success by photography. In contrast, its importance as a tool for prediction and analysis continued to grow. Perhaps one of the more thought-provoking aspects of the computer drawing systems is the possiblility they offer for reintegrating pictorial depiction back into the design language. The gain in perceptual clarity would be especially beneficial to clients and users as well as to designers themselves. Then, a fuller understanding of the origins and developments of the language of drawing becomes more than a matter of passing interest. A closer study of the experience of earlier generations of artists, architects, engineers and designers is essential in attempting to define new forms of visual language appropriate to our immediate needs.

6. L. T. C. Rolt, *Isambard Kingdom Brunel, a Biography*, 1857, p. 16.

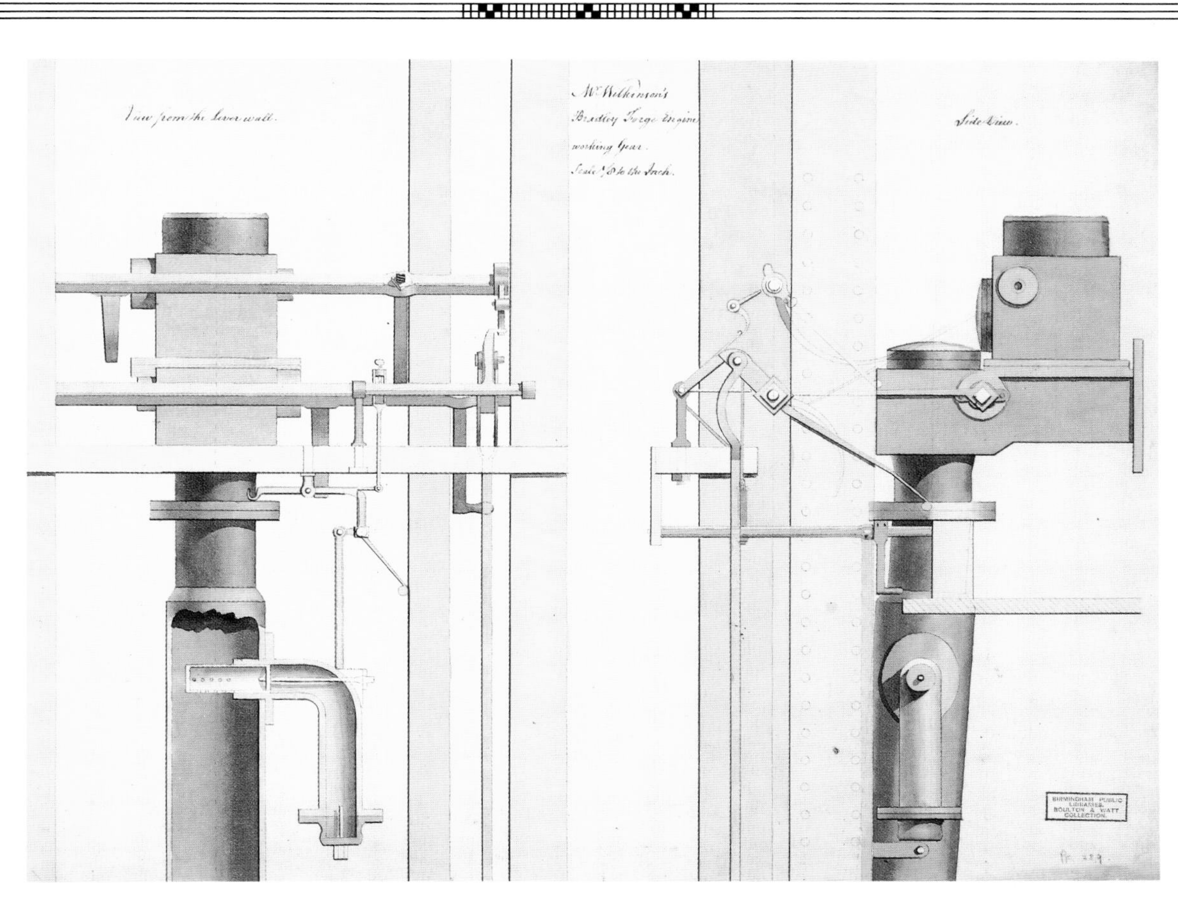
View from the Lever wall.
Mr Wilkinson's
Bradley Forge Engine
working Gear.
Side View.
BIRMINGHAM PUBLIC LIBRARIES. BOULTON & WATT COLLECTION.

DR. R. ANGUS BUCHANAN

The Life-Style of the Victorian Engineers

Above:
Top: Joseph Locke. *Bottom*: William Kennard, Engineer; Crumlin Viaduct, c. 1857
Opposite:
Top: Robert Stephenson in committee: 'Celebration of the Britannia Bridge', contemporary painting, c. 1850. *Bottom*: the competition committee for the Crystal Palace, c. 1850

British engineers of the Victorian period were a heroic generation of men engaged in massive constructional operations which transformed the landscape and environment of 19th-century society. Their works are still apparent in railways and waterworks, ports and engine houses, and in the design of our towns and cities. But the life-styles of these engineers remain comparatively obscure, because all aspects of the ways in which professional engineers have gone about their routine activities, organising their business, undertaking appropriate training, and planning such leisure and other activities as they allowed themselves, can only be reconstructed by a close study of the lives of individual engineers. Such an exercise is impossible to achieve in total for the simple reason that so many engineers have been virtually anonymous, and even when their names and dates are known any detailed knowledge of their working lives has gone unrecorded and is, for the most part, completely beyond recall. It is possible, however, to make some generalisations on the basis of the information which is available about the relatively small number of engineers about whose careers we possess adequate documentary evidence. Fortunately, the engineers about whose working practices we do have some knowledge include some of the most distinguished of the 19th-century leaders of the profession, and, while it must be admitted that these figures cannot be regarded as typical, it is certain that any generalisations based upon them will be representative of some of the dominant trends in the profession.

The most striking characteristic about the work of the 19th-century engineers in Britain was its quantity: as a group and as individuals they tended to work exceedingly hard. Conder records railway engineers working for days on end in order to meet the Parliamentary timetable for submitting schemes for legislative approval,[1] and there is plenty of reason to think that this was a regular and even an habitual occurence. I.K. Brunel drove himself and his staff through long periods of almost unremitting labour, and other leading engineers like Robert Stephenson, C. B. Vignoles, and G.P. Bidder behaved in the same way.[2] There were several reasons for this commitment to what would now be regarded as an excessively heavy programme of work, and one which was responsible in part for the premature death of some of these men including Brunel and Stephenson. For one thing, it was made necessary by the extraordinary volume of work which fell upon the shoulders of a relatively small number of men able to cope with it when the first railway

1. F. R. Conder, *The Men who Built Railways*, Jack Simmons (ed.), 1983, especially Chap. V, 'Work in 1835 – Preparation for Parliament'. See also Thomas Mackay, *The Life of Sir John Fowler*, 1900, pp. 28-30.
2. See L.T.C. Rolt, *I. K. Brunel*, 1957, and *George and Robert Stephenson*, 1960; K. H. Vignoles, *Charles Blacker Vignoles – Romantic Engineer*, Cambridge, 1982; and E. F. Clark, *George Parker Bidder – The Calculating Boy*, Bedford, 1983.

Top: James Watt. *Bottom*: Robert Stephenson

mania created an unprecedented demand for competent engineers in the 1830s. But there was more to it than that, for even before the advent of the railways, and for long after they had ceased to be the leading factors in the economic expansion of the nation, the habit of hard work was apparent amongst the engineering fraternity. There was an element of inevitability about it. Many of the engineers came from social and family backgrounds in which hard work was a prerequisite of survival, and the habit, once formed, lasted a professional lifetime. And many of them had imbibed the ethic of hard work from religious sources and from the expectations of parents and masters during apprenticeship and other forms of training. Most important of all, the reason for the chronic hard work of the engineers, however, was the attitude towards their jobs generated by the successful members of the profession: stated simply, the engineers enjoyed their work and preferred it to most other activities. It was a matter of the engineering psyche, a point it will be necessary to return to later.

Although engineers worked hard, they developed a strongly gregarious spirit in some aspects of their professional associations. In particular, they demonstrated a readiness to form professional associations. Almost as soon as the profession of civil engineering had begun to take shape, Smeaton and a few of his colleagues formed the Society of Civil Engineers in 1771. Some modern scholars have dismissed this body as being nothing more than a dining club.[3] While it could not be denied that the Society had a strong element of conviviality, it should be remembered that the meetings were held reglularly during the Parliamentary session and that they gave those members who were in London an opportunity to meet and discuss problems and matters of common concern. Such discussions amounted to a preliminary approach to control over the profession, and even though the controls were rudimentary they deserve to be recognized as a move in this direction. The formation of the Institution of Civil Engineers in 1818 took the business of professional control a stage further, especially under the assiduous presidency of Thomas Telford, by formulating strict entry requirements and a system of nomination, as well as grades of membership determined according to professional credentials. The model devised by the civils was so successful that it was replicated many times in the course of the 19th century, first by the Institution of Engineers of Ireland and then by the Institution of Mechanical Engineers, and thereafter by more than a dozen different specialised groups of engineers who sought autonomy while preserving all the distinctive features of the civil's organisation. It is doubtful whether this proliferation of institutions served the best interests of the engineering profession, because it deprived the engineers of the advantage of possessing a single undivided voice in matters of public debate and thus weakened their collective strength and professional coherence. But there can be no doubt that the engineers greatly appreciated the benefits of institutional affiliation, because they joined them in large numbers. Between 1850 and 1914 the number of people registered as members of national professional associations increased from

3. Göran Ahlström, *Engineers and Industrial Growth*, 1982; also Eric Ashby, *Technology and the Academics*, 1959. For the Smeatonians, see A.W. Skempton (ed.), *John Smeaton FRS*, 1981.
4. See R. A. Buchanan, 'Institutional Proliferation in the British Engineering Profession, 1847-1914' in

about one thousand to forty thousand.[4]

Among the important services performed for the burgeoning engineering profession by the institutions which it created was that of education and training. It is necessary to emphasise this because it is usually ignored. British engineering in the 19th century has frequently been criticised on account of its failure to encourage the introduction of theoretical training and its unwillingness to recognize the competence of young men seeking to enter the profession with academic qualifications obtained from universities and colleges. The critics span the period from the 1850s, when Lyon Playfair and John Scott Russell were trying to establish technical education as a basis of engineering instruction, to the 1980s, when the members of the Finniston Committee drew attention to the need for more theoretical training in the 'formation' of engineers.[5] There can be no doubt that the rapid development of thermodynamics, electrical science and organic chemistry in the second half of the 19th century created a pressure for greater theoretical knowledge on the part of engineers operating within these fields. But the critics have been unwilling to accept the fact that British engineers had already made substantial provision for the education and training of their members. In the first place, this had been achieved by the traditional device of apprenticeship or pupilage, by which young men hoping to enter the profession were required to serve for three or four years in the office of a practising engineer of standing. Admittedly, this form of instruction had serious limitations and was particularly ill-equipped to convey theoretical instruction in fields with which the master engineer was not himself familiar, but the outstanding success of the apprenticeship system in training up a couple of generations of British engineers made it impossible to abandon it immediately in favour of new methods. For, engineers who had themselves enjoyed the benefits of the old system were understandably reluctant to acknowledge the advantages of a system which placed more emphasis on theory than on practice.

Secondly, however, the critics have also failed to perceive the genuinely educational role of the professional institutions. From the very early days, Telford had insisted on members presenting papers on their engineering work, which were then subjected to close discussion by the community. This process was gradually formalised into the presentation of official papers at the regular meetings of the institutions, for discussion and subsequent publication, together with a full report of the discussion in the published transactions of the institution. This pattern, established by the civils, was adopted by all the other British engineering institutions, which thereby acquired sizable reference libaries of great educational value. While this mass of readily available experience could not, any more than the traditional techniques of apprenticeship, impart new theoretical information, it did provide a very substantial foundation for the continuing education of British engineers, and made it relatively easy for those who applied themselves to the study of the new sciences to assimilate the material. It was certainly a distinctive quality of the British engineering training that engineers were able

Top: G. P. Bidder. *Bottom*: George Stephenson

Economic History Review, 2nd Series, Vol. 38, No. 1, Feb. 1985, pp. 42-60.
5. See Finniston Committee Report, *Engineering our Future*, HMSO, 1980.

Top: I. K. Brunel. *Bottom*: Great Western Railway (J. C. Bourne)

to draw on this body of common experience and to add to it according to their ability, and it was a quality which deserves to be taken into account in any discussion of engineering training and education.

Another aspect of the gregariousness of engineering work was the tendency of practitioners to operate in teams. This derived in part from the large scale of most engineering tasks, requiring the combined efforts of several skilled people, often in charge of large bodies of labourers and sub-contractors. In part, also, it stemmed from the success of a few outstanding engineers who were in such heavy demand that they needed to organize many assistants in order to perform all their commitments. Thus, the leading consultant engineers were able to establish themselves, from the days of Smeaton onwards, as men whose professional services were available for advice and direction in return for a fee, but without a commitment to specific hours of work provided the job was done to professional standards. This meant that the consultant was free to undertake any number of engagements as long as he could arrange the labour force to perfom the work involved in his absence, and many of those engineers whose services were highly sought were able to build up formidable teams of men to assist them. It was obviously important, in these circumstances, to design a clear line of command, and Smeaton pioneered such a pattern which was then adopted by John Rennie and Thomas Telford and every other top rank engineer. The usual form of this pattern was one of a number of 'assistant engineers', working directly under the chief and dividing their time between office work and field work on particular projects as allocated by him, and then a number of 'resident engineers' engaged to oversee the actual construction work on the project, usually with the responsibility of dealing with contractors and directing labour. There would then also be the pupils to fit into this pattern, placed to gain experience of work in the office and the field, and providing a useful supplementary source of labour in the process. Moreover, there would additionally be secretarial staff and draughtsmen needed to service the team. Most engineering practices seem to have managed with one head secretary, like the admirable Mr. Bennet who managed Brunel's manifold activities for him, and a few clerks to transcribe letters, or type them when this innovation in office practice became available. Altogether, a substantial office could consist of a team of twenty or thirty people, responsible for a dozen or more different projects in widely scattered parts of the country. It was clearly important that this team should work smoothly, and it was up to the chief to make sure that it did so. His powers of delegation were crucial to the morale of the team, and while some, like I.K. Brunel, displayed considerable authoritarianism, others, like Robert Stephenson, were prepared to give their subordinates more scope for initiative. Most teams regarded their head with considerable respect, and sometimes with awe. Robert Stephenson is said to have kindled deep affection among the people who worked for him, while Brunel was regarded with rather more respect than affection. But all consultant engineers knew that their ability to function on a large scale depended upon harmony among their team, and they were usually at some pains to ensure such conditions.[6]

The communal character of engineering does not appear to have carried over into camaraderie beyond the workplace and the professional institutions. It is well known, of course, that Robert Stephenson and I. K. Brunel, despite disagreement about the broad gauge and the pneumatic railway, enjoyed each others' company, and were always glad to take an opportunity of going off together in order to inspect a waterworks scheme or some other matter of professional concern. George Stephenson, moreover, even when of mature years, was not above wrestling a fall with the young G. P. Bidder in Robert's office, and Robert occasionally entertained fellow engineers on board his yacht 'Titania'. But there is surprisingly little evidence of social or convivial activity beyond the professional commitments of engineers. To some extent, this was the result of the great pressure of work under which most 19th-century engineers operated, but it could also be attributed to the quality of the engineering 'psyche', as already mentioned. The fact of the matter seems to be that engineers generally did not care much for socialising.

They rarely took any part in political activities, and when they did it was not with much distinction. Both Joseph Locke and Robert Stephenson became Members of Parliament and sat for Honiton and Whitby respectively, representing conflicting Liberal and Conservative interests. But neither took much part in debates, and when they did it was usually to speak on professional matters such as railway finance and the Suez Canal project rather than on any subject of partisan feeling.[7] Daniel Gooch subsequently sat for twenty years as M. P. for Cricklade and prided himself on never having spoken in a debate, although he enjoyed the 'club' facilities of the House of Commons.[8] John Fowler sought a parliamentay nomination rather half-heartedly, but did not achieve it, and no other 19th-century engineer of any standing expressed any political ambitions.

In religious associations, also, the engineers were not notable for any distinctive commitments in the 19th century. Several of them held strong religious convictions which they would express privately, as did Francis Fox who regularly sought to ensure that workmen engaged on his projects should have a rest-day on Sunday in order to allow them to go to church.[9] But most, like Fox, rode light to denominational commitments even though their views tended to a high degree to be conventional and conformist. Unlike contemporary scientists, who included sectarians like Michael Faraday, fundamentalists like Philip Gosse and spiritualists like William Crooke, it cannot be said that any engineer of the period had really interesting religous views. Similarly with other social relationships, engineers appear to have been reserved and non-commital. Certainly, the Brunels gave some minor patronage to the arts and Scott Russell maintained a circle of friends with musical inclinations, but these activities could not be regarded as typical. It was as much as most leading engineers could do to have their portraits painted and a Landseer hung in their sitting room.[10] For the most part, 19th-century

I.K. Brunel, Engineer: building of the SS Great Eastern (construction photographs by Robert Howlett, c. 1857-58)

6. For Brunel's office, see R. A. Buchanan, 'I. K. Brunel: Engineer' in Sir Alfred Pugsley (ed.), *The Works of Isambard Kingdom Brunel*, 1976.
7. See L. T. C. Rolt, *Stephensons*. Robert Stephenson spoke strongly against British involvement in de Lesseps' scheme for a Suez Canal.
8. See Daniel Gooch, *Memoirs & Diary*, R. B. Wilson (ed.), 1972, p. 345.
9. See Francis Fox, *Sixty-Three Years of Engineering*, 1924, pp. 242-3.
10. See Rolt, Brunel, and George S. Emmerson, *John Scott Russell*, 1977.

Above:
Sir Benjamin Baker/Sir John Fowler, Engineers; The Forth Bridge, c. 1882-1890: (*top*) constructing the central girder, (*bottom*) looking through the cantilever below rail level
Opposite:
I. K. Brunel, Engineer; (*top*) Ivy Bridge Viaduct, c. 1846, (*bottom*) Gover Viaduct, Cornwall Railway, c. 1858

engineers were non-radical conformists in all matters unrelated to their professional concerns. They eschewed controversy and preferred the quiet life. Again, the exception tends to confirm the rule, for although Fleeming Jenkin, the talented Professor of Engineering who did much to build up the School of Engineering at Edinburgh, was an enthusiastic supporter of amateur dramatics and a controversialist whose views were taken seriously by economists and by scientists such as Charles Darwin, there were few other engineers who could be compared with him for his breadth of interest and influence.[11]

Work, then, played a paramount role in the lives of the 19th-century engineers, and little was allowed to compete for their attention with professional commitments. Some began to think of building country houses for themselves, and a few, like Armstrong at Cragside and Fowler at Braemore, succeeded. Armstrong commissioned Norman Scott to construct a mansion on which no expense was to be spared, but, even though top quality materials were used throughout, Cragside survives as a monument to bizarre taste rather than as a building of architectural excellence.[12] Some, as they grew older – if they survived the rigours of hyper-active working years and many did not – began to think of spending more time with wives and families and enjoying country pursuits. To some extent, the lives of these engineers confirm the findings of the Wiener Thesis that the leaders of British society turned their attention too readily from industry to the pleasures of the countryside, but it is important to remember that such diversionary activity only affected engineers, when it affected them at all, in their declining years, so that it makes no substantial contribution to Wiener's argument.[13]

Thus, while the works of the 19th-century British engineers survive in large numbers, distinguishing our urban and industrial landscape and the pattern of our transport systems, the engineers themselves remain largely shadowy figures except for the brilliance of a handful of outstanding luminaries. Their life-styles and patterns of work can only be reconstructed in outline, but the main features are clear. They lived very busy lives, working excessively hard to fulfil the mass of commissions which came to them, having little time or inclination to participate in anything beyond the immediate demands of their professional engagements. Within these limits, they devised an apparatus of institutional organization which provided them with a basic training and a continuing education as well as a professional image and the facilities of a club in which they could meet and share experiences with fellow engineers. They also established a system of consultancy with its attendant hierarchy of assistants which worked outstandingly well in the rapidly expanding field of engineering activity througout the 19th century. The life-style of the engineers, in short, was important in determining the extent and shape of their works. For that reason it deserves serious attention.

11. For Fleeming Jenkin, see R. L. Stevenson, *Memoir of Fleeming Jenkin,* 1888; see also D. R. Oldroyd, *Darwinian Impacts*, 1980, pp. 135-7; J. A. Schumpeter, *History of Economic Analysis*, 1955; and Mark Blaug, *Economic Theory in Retrospect,* 1962, 3rd. edition 1978, p. 341 etc.
12. For Armstrong, see David Dougan, *The Great Gun-maker – The Story of Lord Armstrong*, 1970, especially pp. 118-122. Cragside is now owned by the National Trust.
13. Martin J.. Wiener, *English Culture and the Decline of the Industrial Spirit, 1850-1980,* Cambridge, 1981.

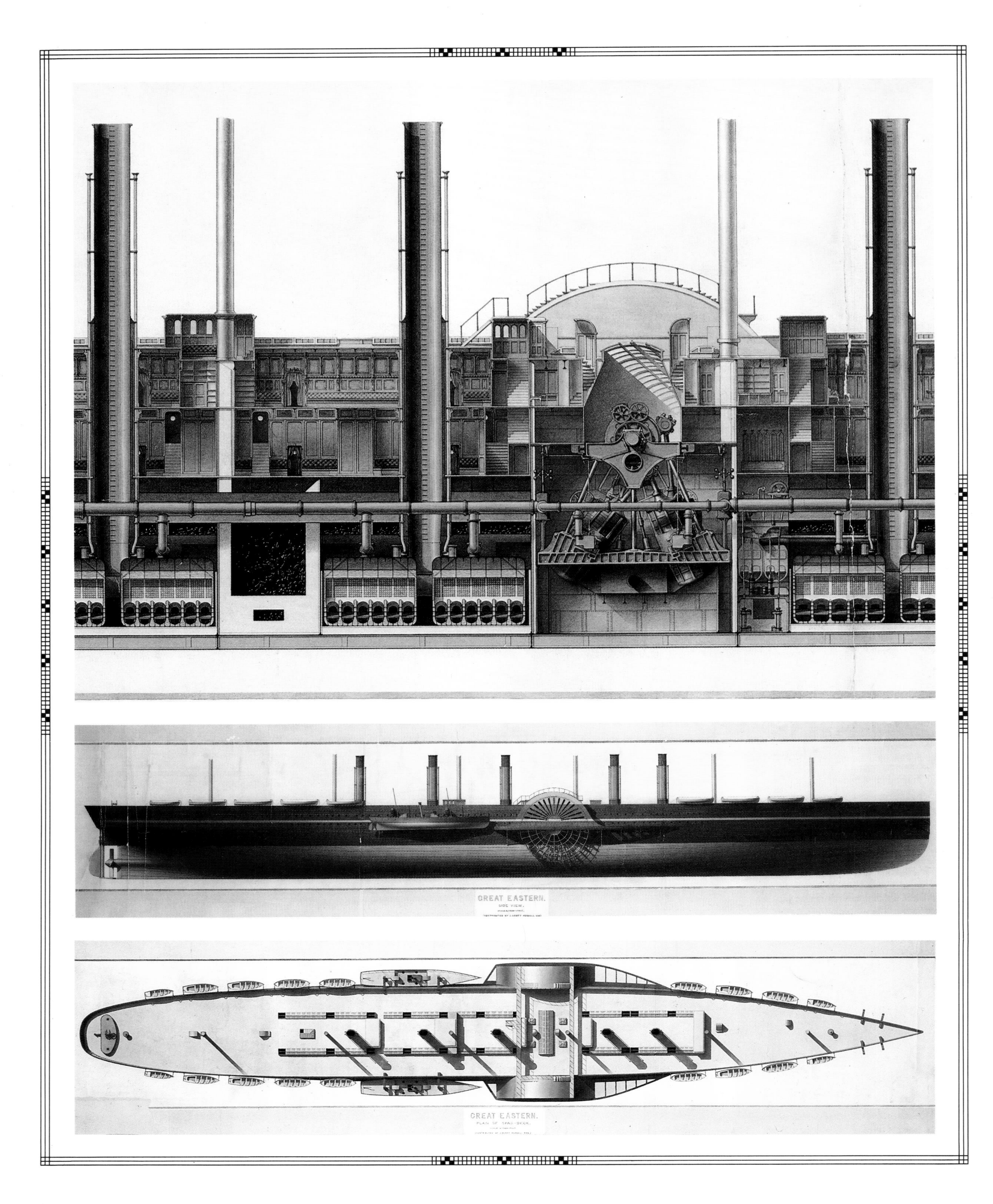
GREAT EASTERN.
SIDE VIEW.
GREAT EASTERN.
PLAN OF SPAR-DECK.

PROFESSOR DEREK WALKER

The Railway Engineers

Engineering by inference is concerned with getting things built or made. In the recent history of British engineering, no two men have dominated a profession with more skill or energy than Robert Stephenson and Isambard Kingdom Brunel. Railway Mania offered the platform to parade their individual skills.

When I first read the biography of Brunel, written by his son eleven years after his death, I realised what ambition was all about. For here was a man – brash, confident, self-seeking, ruthless, with little time for relaxation – who was drawn to perform deeds of endurance which are inconceivable today. A perfectionist who found it difficult to delegate, Brunel produced in a short working life a body of work that would not be thought credible today.

Asa Briggs draws attention to an anonymous autobiography dated 1868 in which a civil engineer published 'Personal Recollections of English Engineers and of the Introduction of the Railway System into the United Kingdom'. The book is a fascinating study of individual engineers and what went on behind the scenes. He was particularly enlightening in the comparison he drew between Brunel and Stephenson.

> The imperfections of the character of Mr. Brunel were of the heroic order. He always saw clearly before him the thing to be done and the way to do it, although he might be deficient in the choice of the human agency which was necessary to effect his design; Mr. Stephenson on the other hand knew how to derive from his staff and his friends a support and aid that carried at times over real engineering difficulties with a flowing sheet.

Nevertheless, Stephenson, like Brunel, exhibited a similar shrewdness and professionalism in scrupulously avoiding the many 'bubble' schemes floated by speculators at that time. Both men would only undertake projects that would be of lasting benefit.

The astonishing variety, scale and location of their projects excite the imagination even today: the Britannia Bridge, the Royal Albert Bridge at Saltash, the High Level Bridge at Newcastle, the London to Birmingham Railway, the Great Western Railway, the SS Great Britain and the SS Great Eastern to name but a few. A visual audit is perhaps the best way of bringing home the quality of their work and how it was achieved by perpetual striving in search of the innovative and elegant solution. That Robert Stephenson was the last engineer to be honoured with a state funeral is a fitting memorial to the great contribution made by engineers of this period to British prosperity.

Opposite:
I.K. Brunel, Engineer; SS Great Eastern, c. 1857-60, section, elevation and plan (Scott Russell drawings)
Page 92:
Top: SS Great Eastern laying the Atlantic cable, c. 1865. *Bottom, left to right*: Joseph Locke, I.K. Brunel standing in front of the chains of the SS Great Eastern, and Robert Stephenson
Page 93:
Laying the Atlantic cable, c. 1865

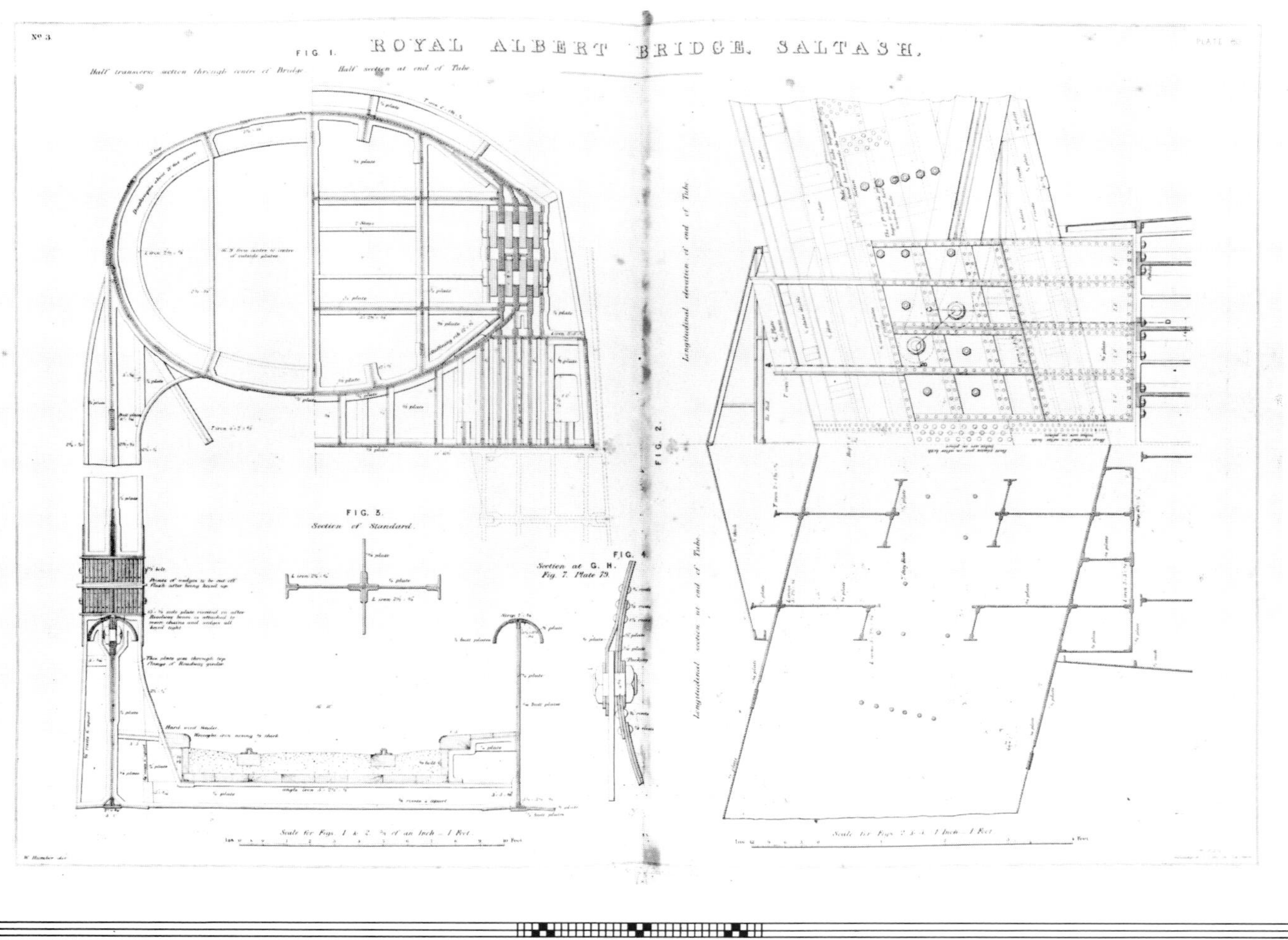

Nº 3
FIG. 1.
ROYAL ALBERT BRIDGE, SALTASH.
FIG. 5.
Section of Standard
Section at G. H.
Fig. 7. Plate 79.
FIG. 2.

Opposite:
I.K. Brunel, Engineer; Saltash Bridge, c. 1858, raising the second tube, and structural drawing
Above:
Saltash Bridge, c. 1857, floating the first tube, and the ceremonial opening by Prince Albert (painting by T.V. Robbins)
Page 96:
I.K. Brunel, Engineer; The Box Tunnel, c. 1841, and Taff Vale Railway Bridge, 1845
Page 97:
I.K. Brunel, Engineer; Blatchford Viaduct, South Devon Railway, 1849 (drawing by J.C. Bourne), and Robert Stephenson, Engineer; Newcastle upon Tyne High Level Bridge, 1849
Page 98:
Robert Stephenson, Engineer; London-Birmingham Railway, entrance to tunnel at Watford, c. 1837
Page 99:
Robert Stephenson, Engineer; Watford tunnel under construction, and Blissworth cutting, c. 1836 (J.C. Bourne drawings)
Page 100:
Robert Stephenson, Engineer; Britannia Tubular Bridge, c. 1850, Anglesey entrance, isometric projection, and construction junction
Page 101:
Brittania Bridge under construction, c. 1847, and London-Birmingham iron bridge under construction, c. 1838

BRITANNIA BRIDGE
ANGLESEY ENTRANCE.

PLATE 13
ISOMETRICAL PROJECTION
OF A
PORTION OF ONE OF THE TUBES
OF THE
BRITANNIA BRIDGE.
Scale of an Inch = 1 Foot.

PERSPECTIVE VIEW OF A PORTION OF THE BRITANNIA TUBES
RESTING UPON THE CENTRE TOWER IN THE MIDDLE OF THE MENAI STRAIT.
ROBT. STEPHENSON AND WM. FAIRBAIRN ENGINEERS.
Published by John Weale, 59 High Holborn, London, 1849.

Page 103
Opposite:
Robert Stephenson, Engineer; Primrose Hill Tunnel under construction, c. 1836, and William Kennard, Engineer; Crumlin Viaduct, c. 1857
Above:
Hanwell Viaduct proposal, c. 1830 for Brunel's Great Western Railway (painting by John Belcher the Elder), and Robert Stephenson, Engineer; Camden Town stationary engine house, c. 1836
Page 104:
Navvies constructing the London-Birmingham Railway (drawings by J.C. Bourne, 1830s)
Page 105:
John Hawkshaw, Engineer; Old Charing Cross Station, c. 1864, and W.H. Barlow, Engineer, St. Pancras engine shed with Owen Jones' hotel competition design on the right, 1865
Page 106:
Hampstead Road cutting and Old Euston Station on the London-Birmingham Railway (aquatints by T.T. Bury)
Page 107:
I.K. Brunel, Engineer; Great Western Railway, first terminus at Bristol, c. 1841, and Slade Viaduct, South Devon Railway, c. 1844

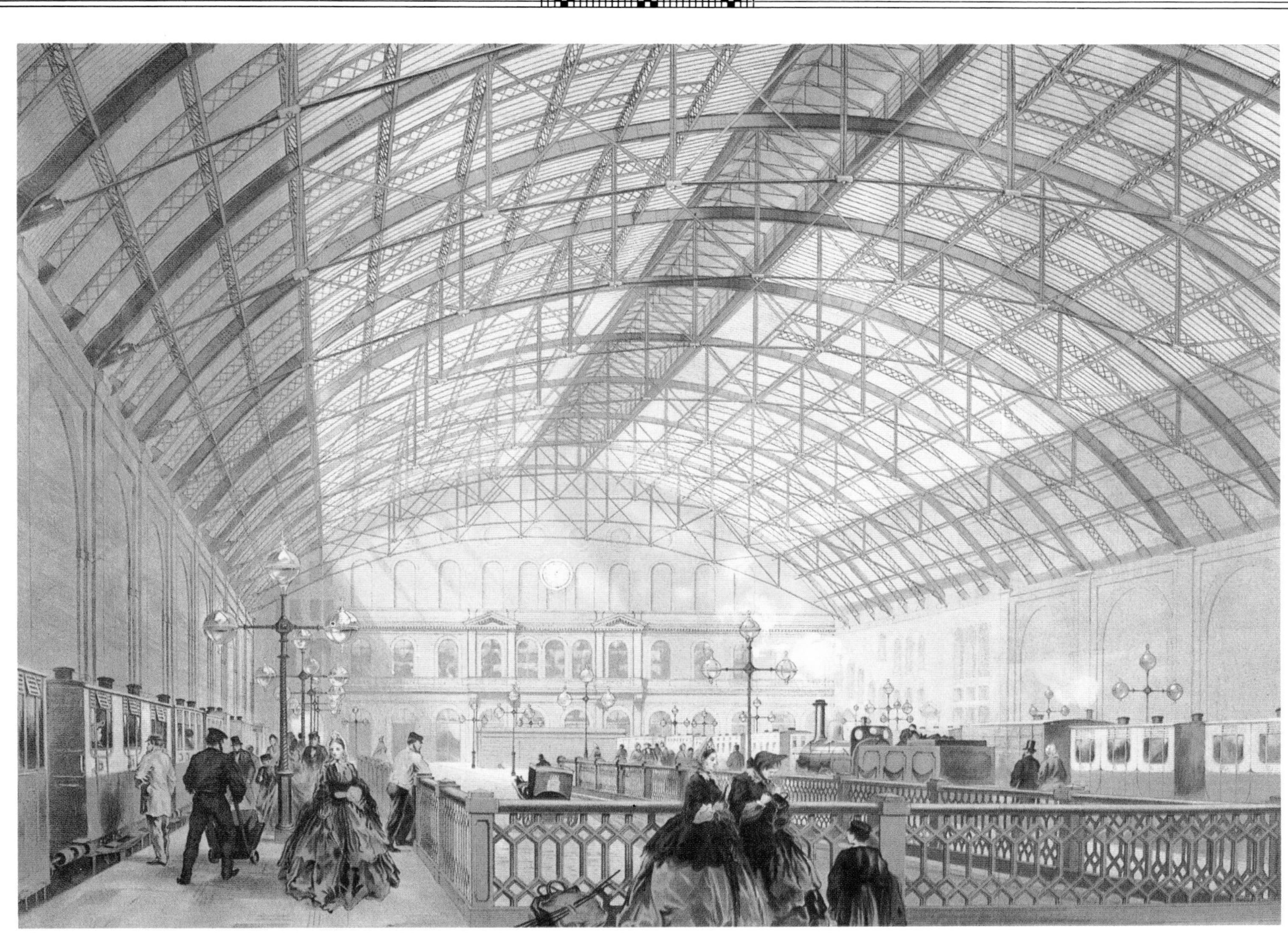

LIVERPOOL
DUBLIN
BELFAST
DERBY
WORCESTER
CHELTENHAM
GLOUCESTER
BATH
BRISTOL
EXETER

Above:
Structural wrought-iron beams, cast-iron columns, monastery of Alcobaça, Portugal, c. 1752
Opposite:
Thomas Telford, Engineer; Tewkesbury Bridge, c. 1823-6 – the climax of elegant cast-iron design

JAMES SUTHERLAND

Developments of the Use of Materials in Structures

Today all major building and civil engineering construction depends on metals and, to a lesser extent, cement, yet up to two hundred years ago these hardly figured at all. The materials then were stone, brick, lime and timber. Iron and bronze were used sparingly in the form of nails, cramps or tie rods and only rarely on a large scale, as in the tie rods at St. Sofia, Constantinople of the 6th and 10th centuries,[1] or the later iron chains to the domes of St. Peter's in Rome and St. Paul's in London.

Evidence of really early applications of iron on a major scale tends to be conjectural or at best indirect. This is certainly the case with the six-metre long iron lintels in the Temple of Olympian Zeus at Agrigento (c. 470 BC) and the pre-Christian structural iron in China.

There is a tendency to see structural achievements as stemming from specific 'inventions' or as the products of individual inspiration. In practice inventions are usually only the culmination of long periods of improvement, often by several generations of thinkers and experimenters. Also the diverse paths of structural development have always depended on the gradual growth in our knowledge of the properties of different materials. It is thus essential to understand these properties to some extent in order to appreciate the work of the engineers who used them.[2]

The first metal used by man is thought to have been copper. In spite of high resistance to corrosion and a useful level of strength, copper and its alloys by whatever name – bronze, gun-metal, latten or brass – never became major structural materials. Likewise lead, tin and latterly zinc were only used in sheet form or, in the case of lead, for pipes.

Iron was the first metal to be used extensively for structures and for centuries it has symbolised strength, even supernatural strength.

Of the ferrous metals, wrought iron was the earliest available in Britain. It was smelted in simple charcoal furnaces or hearths, from about 500 BC onwards. This early wrought iron, which may be called 'blacksmith's iron' to distinguish it from the later industrialised product, was both purified and shaped by hammering in a pasty, but never quite molten, state. It was almost pure iron, virtually free of carbon.

Cast iron first became available in Europe when the blast furnace was perfected around 1,500 AD, thus making truly molten iron a possibility. Iron could now be cast as a liquid directly into moulds of the required shape or stored as 'pigs' for remelting and casting at a later stage. Cast iron has a much

1. R. Mainstone as quoted by J. G. James in note 3 below.
2. For fuller descriptions of the process see *A History of Technology*, Singer, Holmyard, Hall and Williams (eds.), Vols. III-V, 1958.

higher carbon content than wrought iron (2.25-4.0 per cent) and, as a result, a lower melting point.

For a long time the smelting of cast iron depended on charcoal, which tended to crush in the blast furnace thus limiting the scale of the operation and the size of individual castings. This limitation was overcome when Abraham Darby developed the first successful coke smelting process in 1709. Curiously it was not until sixty to eighty years later that full advantage was taken of this process, but, once established, it soon became clear that the production of cast iron could be highly industrialised.

Essentially, cast iron is very strong in compression but much less so in tension. Also it is brittle and subject to hidden flaws.

Physically and chemically there is no difference between industrially rolled wrought iron and the blacksmith's hand-wrought iron. Industrialised production became possible with the invention by Henry Cort of the reverbatory or puddling furnace in 1793. In this, carbon was removed from molten pig iron, thus leaving the same pure, semi-molten material as in the case of the blacksmith's iron. A related invention by Cort in 1794 of grooved rollers enabled the semi-molten iron to be formed into continuous sections – bars, angles, tees and ultimately channels and I-beams.

With these two inventions the way was open for the production of wrought iron on a large scale but, as in the case of coke-smelted cast iron, there was a gap, this time of some fifty years, between the means of production and large-scale application in the 1840s.

The great advantages of wrought iron are its strength – equally high in tension and compression – and its ductility; unlike cast iron it is not brittle. The disadvantage of wrought iron when first introduced was its greater cost and the fact that it could not be moulded into precisely tailored shapes.

Steel, with an intermediate carbon content between that of wrought and cast iron (0.2 – 1.0 per cent), has been said to combine the virtues of both. It is produced industrially in a molten form and then rolled into long uniform sections, visually indistinguishable from rolled wrought iron. However, steel is stronger than wrought iron, both in tension and compression, and, produced as a liquid, it can be handled and formed into much larger single elements than the semi-molten wrought iron.

In most history books steel is said to date from Bessemer's converter of 1856, although, strictly speaking, Kelly got there a year earlier in the USA. The truth is that steel is much older than either, being produced on a limited scale in the early 19th century by the controlled impregnation of wrought iron with carbon (cementation) or by the crucible process (by Huntsman in the 1740s and others later).

The Bessemer converter certainly pointed the way to the large-scale production of steel, but the material's acceptance for structures was largely due to the later perfection of the Siemens-Martin open-hearth process, which was slower but more controllable than the Bessemer one.

There is so much confusion over the recognition and physical properties of

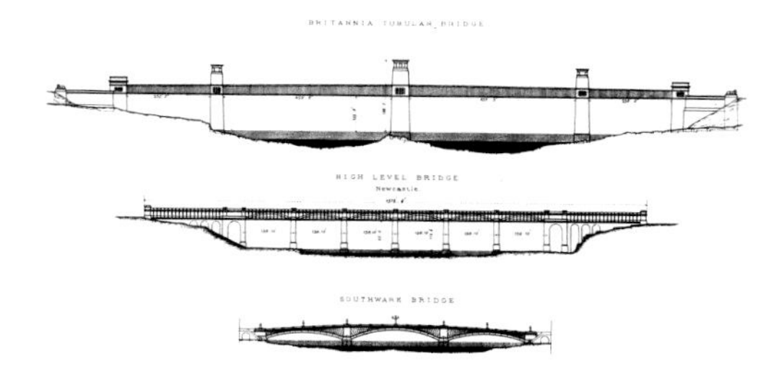

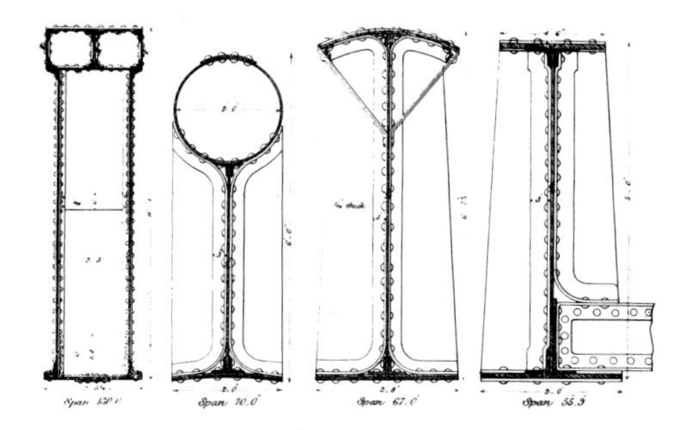

Above:
Top: the Britannia Bridge compared with the Newcastle upon Tyne High Level Bridge and the Southwark Bridge. *Bottom*: sections through riveted wrought-iron beams of c. 1850-60 showing transition from box form (left) to the I beam familiar today (from *Humber's Bridge Construction)*
Opposite:
I. K. Brunel, Engineer; The Engine House, Swindon, timber and cast iron, c. 1846. *Bottom*: Robert Stephenson, Engineer; The Britannia Bridge, c. 1845-50

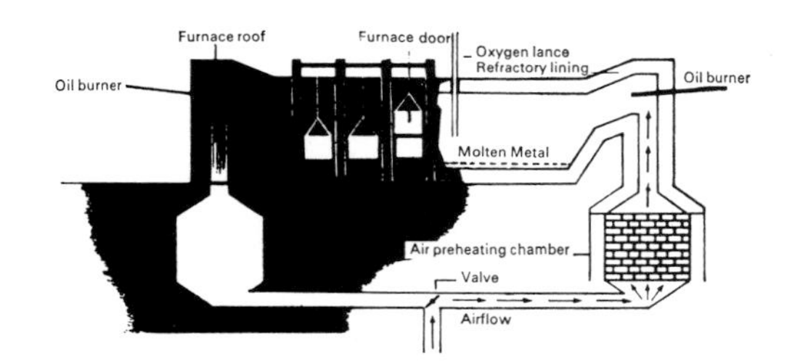

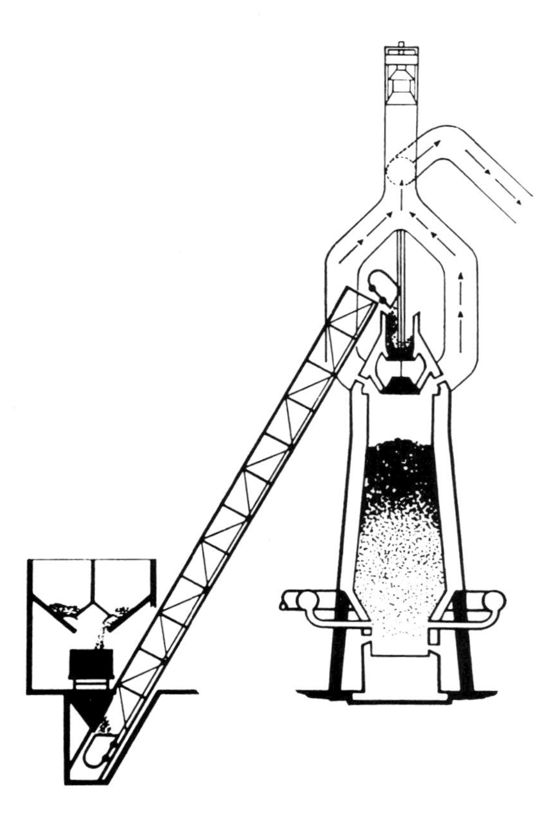

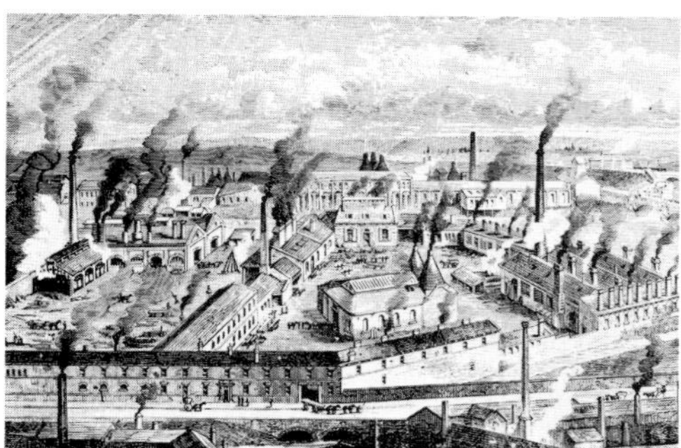

Top: Open hearth furnace. *Centre:* Blast furnace. *Bottom:* Scotia Steelworks, 1864

the three ferrous metals (wrought iron, cast iron and steel) that it may be worth summarising these before considering structural uses.

Put at its simplest, wrought iron is strong and ductile, with a fibrous texture like timber when fractural, while cast iron is coarsely granular, extremely strong when compressed but brittle like chalk. Steel is finely granular in texture and stronger than wrought iron but less ductile.

Visually it is generally easy to distinguish between cast iron on the one hand and wrought-iron and steel on the other. Irregular shapes which could only be formed by moulding – as in a jelly mould – are of cast iron, while wrought-iron and steel sections are generally rolled and thus of constant section throughout their length. Lines of rivets point to wrought iron or steel. Welds point to steel, rather than wrought iron which is difficult to weld and was mostly used before welding was introduced.

It is almost impossible to distinguish at first sight between a rolled section of steel and one of wrought iron. The form of fracture gives the best indication, short of full metallurgical examination.

It is too easy to look on the evolution of structural iron and steel as an exclusively British affair. We tend to see it all as starting with the Ironbridge at Coalbrookdale of 1777-81 and the use of cast iron for columns and decorative gothic tracery in churches like St. Anne's, Liverpool, in 1770-72. Yet there is the giant cooker hood and chimney dated 1752, still in place, at the monastery of Alcobaca in Portugal (9.1 by 3.3 metres in plan). This rests wholly on wrought-iron beams and cast-iron columns, and antedates the British pioneers by nearly thirty years.

The Russian iron industry, dating from the time of Peter the Great, was certainly producing structural iron, cast and wrought, from early in the 18th century, although to a large extent under British and German technical control.[3]

In France, 18th- and early 19th-century structural iron was almost all wrought of the blacksmith's type and developed in parallel with, and it seems largely independently of, British cast iron, nevertheless achieving a roof span of over twenty metres in the 1780s.

These diverse achievements seem to have had no significant effect on developments in Britain. Here it was cast iron which first made the revolution in construction possible and it would be difficult to see this as anything other than an indigenous industry. The Coalbrookdale Bridge, not the first but certainly the most famous cast-iron arch, was followed by others of increasingly logical form. Some of the most elegant and, in terms of material, perhaps the most economical, were due to Thomas Telford, of which Tewkesbury Bridge (1823-26) spanning fifty-two metres is perhaps the finest survivor. It is interesting to note that, at that time, the whole construction of this bridge could be shown on two working drawings, one of the masonry and foundations and the other – fully detailed – of the ironwork.[4] Repetition of standard elements led to economy both in detailing and construction.

In buildings the main outlet for cast iron, where again standardisation held

3. J. G. James, 'The Application of Iron to Bridges and Other Structures in Russia to About 1850', (trans. Newcomen Society), advance copy, 1983.
4. These drawings are held in the library of the Institution of Civil Engineers.

the key to economic successs, lay in textile mills, where fire had been a major hazard with the timber floors. Following the pioneering work of Bage, Strutt and others in the 1790s and early 1800s, 'fireproof' mills with cast-iron columns, cast-iron beams and brick jack arch floors became virtually universal and continued almost unchanged for several decades.[5]

Concurrently with the mills, cast iron became popular for large span beams in important buildings, where previously installed long timber beams had been found to sag. Another use was for repetitive decorative units, even quite large ones. How many people realise that the Doric pilasters of Carlton House Terrace facing the Mall of 1827 are cast iron and not rendered brick?[6]

The magnificent conservatories which date from 1820 to 1850 and are frequently referred to as cast iron structures, were mainly dependent on wrought iron as were all the major spans of the Great Exhibition Building of 1851.

As well as in arch form, for which its compressive strength made it eminently suitable, cast iron was used extensively for bridge beams in the first half of the 19th century. For these, reliable tensile strength was vital and it was here that, in spite of widespread proof-loading, the brittleness of the material started to show up as a major hazard. This was particularly the case with railway bridges. In fact it was the needs of the railways in the 1830s and 1840s which ushered in the age of structural wrought iron and led to the eclipse of cast iron.

It is not surprising that in this frenzy of entrepreneurial activity some schemes went wrong, while others were brilliantly successful. In the mid to late 1840s, for instance, Robert Stephenson was responsible for, among many other projects, the huge and very successful Newcastle High Level Bridge with tied cast-iron arches, the disastrous Dee Bridge at Chester with trussed cast-iron beams and the innovative Britannia and Conway tubular bridges of riveted wrought iron.[7,8] Whatever the exact cause of the collapse of the Dee Bridge, it is clear to us today that the thinking behind its design was seriously at fault in more ways than one. Equally it is almost unarguable that the research and testing for the tubular bridges advanced structural knowledge more than any other single event in the last 250 years. At the same time these two bridges were an unqualified success.

Starting with a brief to build a bridge of enormous span for which there was no existing precedent, Stephenson's team developed the great tubular beams through which the trains would run – a totally new structural concept.

The three principal men behind the Britannia Bridge story were William Fairbairn, Eaton Hodgkinson and Stephenson himself. In little more than two years of research and testing, they moved from the erroneous belief that wrought iron is weak in compression like cast iron is weak in tension, to isolating and overcoming the problem of plate buckling. The testing was very extensive, ranging from simple tensile and compression tests on various materials to experiments on models of an increasing scale which established the best form of tube and how to analyse its strength in wrought iron, a

Commemorative jugs, homage to the Steel Makers

5. H. R. Johnson and A. W. Skempton, 'William Strut's Cotton Mills 1793-1812', (trans. Newcomen Society), Vol. XXX, 1956.
6. Gloag and Bridgwater, *A History of Cast Iron in Architecture* London, 1948.
7. J. C. Jeaffreson, *The Life of Robert Stephenson FRS*, Longman, 1864.
8. L. T. C. Rolt, *George and Robert Stephenson*, Longman, 1960.

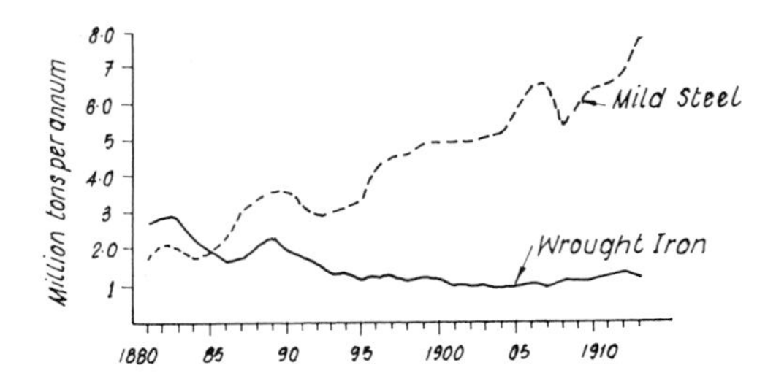

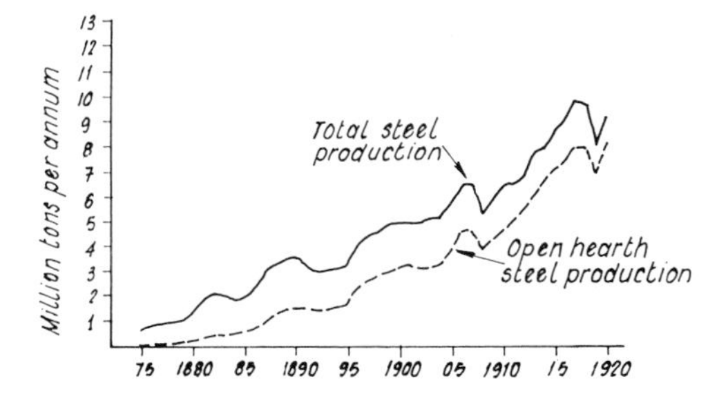

Top: figure showing transition from dominance of wrought iron to dominance of steel in Britain. *Bottom*: Steel production in Britain 1875-1920, showing increasing proportions of open hearth to Bessemer Steel (both J. C. Clarke)

material hardly ever used before for beams. The design also demonstrated for the first time how the bending strength of beams could be greatly increased by making them continuous over intermediate supports.[9,10]

It was very largely the research, testing and analytical studies for the tubular bridges which made riveted wrought iron into the pre-eminent structural material of the second half of the 19th century, while the collapse of the Dee Bridge did more than anything else to kill off cast iron as a material for anything other than direct compression. Cast iron, quite rightly, continued to be used extensively for columns until at least about 1900 but seldom for anything other than minor beams after 1850.

In practical terms, the structural steel period started partly with rails but mainly with ships. In the 1860s high-tensile steel was used for ship building on a limited scale but apparently the practice was dropped. In 1877 'HMS Iris' was launched with a hull of mild steel and in 1879 the 'Rotamahana' was built incorporating the same feature, to be followed in 1880 by the P and O 'Ravenna' also in mild steel. In 1881 the last wrought-iron ship 'The City of Rome' was built. Some of these dates may be a little foggy but it is clear that 1880 was the watershed as far as wrought iron and steel for ships was concerned. The gain with steel was clear in that Lloyd's Register allowed steel scantlings four-fifths the thickness of wrought-iron ones.[11]

In 1877 the Board of Trade approved mild steel for bridges and the Forth Bridge was built between 1883-90 using some 50,000 tons of open-hearth steel. In 1885 the output of mild steel in Britain exceeded that of wrought iron for the first time.

For buildings steel took over gradually, the first *wholly* steel framed building probably being a warehouse of 1896, with the most famous early one, the Ritz Hotel, not being built until 1904.[12]

With the possibility of large-scale production from the late 1850s one may well ask why steel did not supersede wrought iron earlier. The answer must be partly the natural gap between technical possibility and commercial exploitation but also a lack of confidence in early steel. The high-tensile strengths found possible in the early days were associated with brittleness and even mild steel produced by the Bessemer process often proved unreliable.[13] The growing dominance of open-hearth steel production after 1880 was probably the biggest factor in making engineers happy to replace wrought iron by steel. By 1900, sixty per cent of steel produced in Britain was open-hearth steel and, by 1909, Lloyd's Register would not approve any other process. Today Bessemer steel is effectively a thing of the past.

Reinforced concrete had several false starts. The principle was demonstrated in Britain by Sir Marc Brunel and General Pasley in their experiments on reinforced brickwork in the 1830s.[14] Then came Lambot's famous boat and Monier's flower pots in France in the 1840s and, back in Britain again, Wilkinson's patent of 1854.[15] In France interest seems to have been more or less continuous from the 1850s but in Britain, after Wilkinson's coachman's house of 1865, the technique was ignored until Hennebique arrived from

9. Edwin Clark, *The Britannia and Conway Tubular Bridges*, Day and Son, 1850.
10. R. J. M. Sutherland, 'The Introduction of Structural Wrought Iron', (trans. Newcomen Society), Vol. XXXVI, 1963-4.
11. Ewan Corlett, *The Iron Ship*, Moonraker Press, 1974.
12. S. B. Hamilton, 'Building Materials Techniques', in Singer, Holmyard Hall and Williams (eds.), *A History of Technology*, Vol. IV, Oxford 1958.
13. J. F. Clarke and F. Storr, 'The Introduction of Mild Steel into the Ship-building and Marine Industries', occasional papers in the history of science and technology, Newcastle-upon-Tyne Polytechnic, 1983.
14. C. W. Pasley, 'Observations on Limes, Calvaraeus Cements, Mortars, Stuccos and Concrete', Weale, 1847.
15. Joyce M. Brown, 'W. B. Wilkinson (1819-1902) and his place in the History of Reinforced Concrete', (trans. Newcomen Society), Vol. XXXV, 1962-3.
16. A. W. Skempton, 'Portland Cements 1843-1887' (trans. Newcomen Society), Vol. XXXV, 1962-3.
17. W. O. Alexander, 'The Utilisation of Metals' in *A History of Technology*, T. Williams (ed.), Vol. VI, Oxford, 1978.

France in the 1890s, after which it flourished.

The reason for the lack of enthusiasm in Britain was almost certainly the remarkable success of riveted wrought iron from 1850 onwards. Structurally you could make almost anything with it and there was no shortage of the raw material. British engineers did not need to innovate any further. From 1850 almost until the airships of the 1920s and 1930s we just exploited our earlier brilliant advances with riveted iron, while all the new structural thinking went on elsewhere, notably in USA, France and Germany.

This may be an over-simplified explanation. There was another factor which affected the commercial success of reinforced concrete and that was the development of cement technology, but here again, after the pioneering work in Britain up to around 1850, the main developments were in Germany. Thus Skempton[16] refers to the German breakthrough of the 1870s, and the German and Swiss Standard Specification of 1887 as the effective beginning of reinforced concrete.

The period from 1900 to 1940 was in many ways the heyday of the structural frame for buildings in Britain whether in steel or reinforced concrete, but in either case usually with heavy masonry cladding. Curtain walls, as in the Peter Jones store and the exposed reinforced concrete structures of Owen Williams, were the exception. Bridges were a mixed bag of steel and reinforced concrete with steel predominating for the larger bridges, the main exception being Mouchel's concrete road bridge over the Tweed at Berwick.

The exciting challenges for structural designers in this period were neither in buildings nor bridges but in airships and aircraft. For these, the material was aluminium.

Aluminium, reinforced concrete and structural steel were developed in parallel in the first half of this century but with little commercial connection between the three. For this reason they are best considered as three isolated 'ages'.

Aluminium became commercially available following the almost simultaneous, but independent, finding of a cheap reduction process in France and USA in the 1880s. However, its structural use dates from the accidental discovery of age-hardening of aluminium alloys by Alfred Wilm in Gemany in 1909. His attempts at hardening had little immediate effect, but when the same specimens were tested again a week later the improvement was found to be substantial. This led to Duralumin, produced by Durenner Metalworke and used extensively for Zeppelins in the First World War.[17]

Other alloys followed, but aluminium had very little impact on building structures or bridges in the period before the Second World War. Nevertheless, great advances were made in the analysis and understanding of metal structures at this time, mainly in connection with rigid airship frames. If you were a clever, imaginative engineer in the 1920s and 1930s, this was clearly the best field to work in.

The period of post-war reconstruction from 1945 (or effectively 1950) to 1970, had much in common with the years 1830-50. Both were times of

Top: the airship R.100 with structural framing of advanced design in aluminium.

Sir Benjamin Baker/Sir John Fowler, Engineers; the building of the Forth Bridge, c. 1882-90. (*top*) living model illustrating the structural principles. (*bottom*) contemporary construction photograph

structural transition, with intense entrepreneurial activity, a welter of new concepts and strong external pressures on the engineering profession. In both periods engineers believed in what they were doing and there was great enthusiasm.

Structurally there were exciting new techniques to grapple with. For steelwork welding, familiarised through war-time ships, was rapidly replacing the rivet. Glues had advanced so that weather-resistant lamination of timber was not just practicable but was being actively promoted and widely used. Concrete had become scientific and strengths previously unheard of had become possible even with quite low cement contents. Precasting of concrete had reached an almost religious level of acceptance. Then there were the new semi-structural materials like Stramit board, woodwool, metal deckings and glued honeycomb sandwich panels; all of which could be used, sometimes compositely, with more conventional structural materials.

The basic creed of the time was economy of materials, with a calculated close-tailoring of form to suit known – or at least anticipated – needs.

Perhaps the technique which fitted the period best was prestressing. Intellectually, this was, and still is, the most satisfying of all structural methods. The approach is positive. With all other techniques one provides strength to support loads, but with prestressing one puts in actual forces to counteract the forces due to loads. Not only was this approach in keeping with the mood of the period but, at a time of steel shortage, it used less steel and, what is more, steel which was not controlled by licence. Further, it provided the ideal method of joining precast concrete units, thus making possible long beams assembled on the string-of-beads principle out of small, largely identical castings.

No sooner had large panel precast concrete construction been established for high and medium rise housing than engineers turned to brickwork and concrete blocks to prove that such cellular structures could equally be built with economy in thin-walled masonry. They also showed that these materials could be reinforced or prestressed like concrete.

Towards the end of this period a wave of new composites started to emerge: glass reinforced plastic, glass reinforced cement, and, in rarefied fields, carbon fibre.

Perhaps 1950-70 was not quite as exciting a period as 1830-50 but the more we focus on it the more remarkable some of its aspects look.

In the last fifteen years designers have been tending to lick minor wounds as far as materials and engineering innovation are concerned. The public blames architects and engineers of the post-war years for the leaking of flat roofs, the ugly appearance of concrete and for a wide variety of structural dangers and inadequacies. Many of these 'faults' could better be put down to not understanding the need for maintenance, while others could be cured quite readily if it was not more convenient to use structural and material problems as an excuse for social change.

However tempting it is to do so, one cannot dismiss all criticism as over-reaction. In all periods, especially innovative ones, some things must go wrong. It is the lessons arising from these on which we must concentrate.

Twenty years ago we looked on reinforced concrete as a material – or more precisely as a 'composite' – which needed no maintenance if properly constructed. Now we realise that due to carbonation its life must be limited unless it is protected in the way we protect timber. What is more if painted regularly its image might be greatly enhanced; few if any complain of the appearance of the painted stucco of Carlton House Terrace.

We now know that there is no such thing as a maintenance free material. Further, it is clear that those materials which we have looked on as not needing maintenance, like most plastics, are the most difficult – if not impossible – to treat once they lose their youth.

Another lesson is that the minimum material philosophy of the 1950s is not necessarily the most economical. It has led to a lack of adaptability not shared by less sophisticated structures of earlier periods.

One of the biggest problems of precast, and in particular prestressed, concrete structures is that of inspection, both during construction and subsequently. No one has found a satisfactory way of checking that a prestressing cable is intact and we have now had enough problems due to corrosion to know that there is a need for such checking.

Ease of inspection and ease of repair (or ease of partial replacement) should be the main aims in our use of materials in the future. It is no good planning a limited life-span for buildings or bridges. Either they will become redundant for extraneous reasons or society will want to keep them.

Today, there are some extremely interesting new materials available. Not only are there readily available fibres with tensile strengths in excess of that of high-tensile steel wire – Kevlar for instance – but there are rarer materials of even greater strengths.

The important point here is not so much high material strength as our new ability to design composites, often by lamination, with varying properties in different directions and in different positions within a structural element. Strength, elasticity and plastic yielding can all be made susceptible to variation by design. We are beginning to have composite materials with synthetic fibres in polymer matrices which automatically change shape in relation to load. Dr. Anthony Kelly gave an indication of these possibilities in a recent talk to the Royal Society of Arts.[18]

It may be too early to talk of 'intelligent' structural materials but in the future there should be many actively responsive ones. Structural designers may face a far greater revolution in the future than that brought about by the introduction of iron. Not only will they have to maintain and enlarge their mastery of present materials but they may also need to design new and highly sophisticated composite materials with which to build their structures. For this they will need to acquire a far greater understanding of chemistry than most engineers possess today.

Top: brick and tile machine, c. 1860. *Bottom*: brickmaking, mid 19th-century

18. Anthony Kelly, 'Composites: Industrial Innovation via New Materials', Proc. Royal Society of Arts, No. 5264, Vol. CXXXIV, Nov. 1986.

TLE.
ENT.
ILE.
COCKERELL & Co

DR. DENIS SMITH

Sir Joseph Bazalgette and Public Health Engineering

Above:
Sir J. W. Bazalgette, c. 1819-1891
Opposite:
Sir Joseph Bazalgette, Engineer: District Railway and low-level sewage works, Blackfriars (*top*), projected view of Embankment at Victoria (*bottom*)

> The majority of the inhabitants of cities and towns are frequently unconscious of the magnitude, intricacy, and extent of the underground works, which have been designed and constructed at great cost, and are necessary for the maintenance of their health and comfort. *J. W. Bazalgette, March 1865.*[1]

The growth in the size and population of towns in early Victorian Britain led to enormous problems and the concept of public health emerged as a result. Life in overcrowded, ill-ventilated dwellings, without pure drinking water or main drainage – so well captured in the writings of Charles Dickens – was increasingly seen to be a public scandal. Matters reached crisis proportions when Indian cholera reached Britain in the 1830s and 40s. Cholera struck in the autumn and usually killed about half those who contracted the disease. The tackling of these problems was to engage the professional skills of the doctor, scientist, lawyer, architect, civil engineer and politician. Not that these professionals were always to work harmoniously together. One of the leaders of the sanitary reform movement was Edwin Chadwick, a lawyer with a poor impression of engineers, who wrote to a councillor in 1844:

> For all purposes it would be of the greatest importance that you should get the advice of trustworthy engineers, of whom I am sorry to say there are marvellous few – a more ignorant or more jobbing set of men . . . I have rarely met with.[2]

Municipal engineering, as a profession, undoubtedly arose out of tackling the public health problems in the middle of the 19th century, and one man in particular made a unique contribution to raising the status of sanitary engineering. He was Joseph William Bazalgette (1819-91), whose family were of French origin, his grandfather having arrived in England in 1784, after leaving France for the West Indies.

Joseph was born in Enfield, Middlesex, on March 28th 1819. We know little of his childhood except that he was privately educated. His engineering career began in 1836 with a pupilage in civil engineering under Sir John McNeill in Northern Ireland. In 1842 he set up in private practice as a consulting engineer in Great George Street, Westminster. In 1845 he married an Irish girl, Maria Keogh, and the following year was elected a Member of the Institution of Civil Engineers. These were the years of the railway mania, when he worked himself into a total breakdown of health. During 1847-48 he moved out of town to recuperate.

1. Joseph William Bazalgette, 'On the Main Drainage of London and the Interception of the Sewage from the River Thames', *Proc. ICE*, Vol. 24, Mar. 14th 1865, pp. 280-314.
2. Letter dated Oct. 9th 1844, Chadwick to Cllr. John Shuttleworth in Manchester. See G. M. Binnie, *Early Victorian Water Engineers*, Thomas Telford Ltd., 1981, p.172.

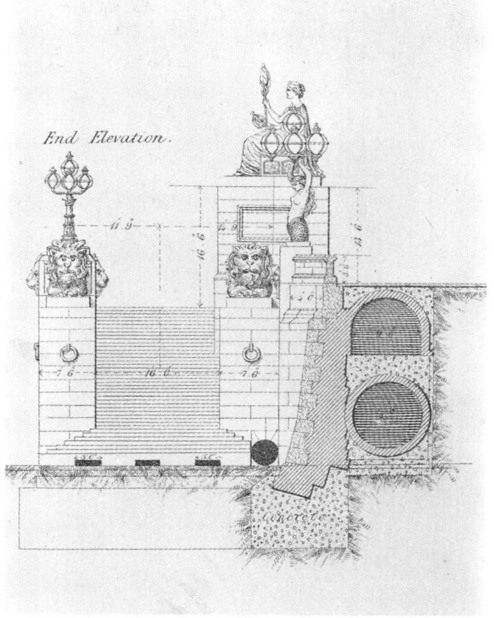

Top: contemporary cartoon (*Punch*). *Bottom*: Victoria Embankment, sectional drawing

Meanwhile, in London, the first steps had been taken towards establishing an appropriate administrative apparatus – an essential prerequisite to tackling the problems. In 1847 Central Government intervened and set up a Royal Commission to investigate 'what special means might be requisite for the improvement of the health of the Metropolis'. They recognised that the exisiting seven drainage districts, comprising many parishes and vestries, could not take a sufficiently broad view of the problem. This led to an Act of 1848 and the inauguration of the first Metropolitan Commission of Sewers. The body of commissioners, including Edwin Chadwick, had powers to levy a rate to pay for works and, perhaps more importantly, to employ salaried civil engineers to design and supervise the construction of such works. However, the term of office of each commission was limited to two years and there were six such commissions between 1848 and 1855.

When Bazalgette had recovered from his illness, he returned to London and his career in Local Government engineering began. On August 16th 1849 he was appointed assistant surveyor to the second Commission at a salary of £250 a year. He was to remain in public service for forty years until his retirement in 1889. Bazalgette was appointed engineer in 1852 and remained so up until the last Commission of 1855. By this time the essential nature of the public health problem in London was clearly understood and a strategic plan had been devised. It had been established that cholera could only be contracted by ingesting the micro-organism and this was done by drinking polluted water. The natural drainage channels collecting domestic and industrial effluent flowed down the valley sides to the Thames. Bazalgette graphically described the problem:

> the whole of the sewage passed down sewers from the high ground at right angles to the Thames, where at high water it was pent up in the sewers, forming great elongated cesspools of stagnant sewage, and then when the tide went down and opened the outlets that sewage was poured into the river at low water at a time when there was very little water in the river.[3]

The plan was, therefore, to construct extensive east-west sewers to intercept the sewage before it reached the Thames and carry it, by means of outfall sewers, to remote reservoirs in East London.

It is important to note here that the supply of drinking water was in the hands of private waterworks companies and an Act of 1852 compelled them to remove their intake from the Thames to points above Teddington Lock by 1855. The metropolitan commissioners had agreed to Bazalgette's plan for independent intercepting systems north and south of the Thames, but they did not have time to implement the massive scheme of works. In 1855 the Government passed the Metropolis Management Act and on January 1st 1856 a totally new body, the Metropolitan Board of Works, superseded the old commissions of sewers and took charge of public works over an area of 117 square miles of the capital.

3. *Parl. Pap.*, 1889, XXIX, p.335.

The Metropolitan Board of Works

Although public health issues were the principal motivation for setting up this body, it also undertook a range of other responsibilities for the environment of London, including roads, bridges, tunnels, ferries and other issues relating to safety such as the storage of dangerous materials and the provision of a capital-wide fire brigade. Bazalgette, as engineer to the last Metropolitan Commission, was asked in January 1856 to continue his oversight of sewer matters, operating from his existing office at 1 Greek Street, Soho. On January 25th Bazalgette was chosen from among nine candidates to be engineer to the Board at a salary of £1,000 a year. He brought with him three of his assistants from the 1855 Commission, thereby promoting the maximum technical continuity in the change-over to the new administration. Bazalgette methodically enumerated the questions he had to answer which were:

1. At what stage of the tide should sewage be discharged into the river?
2. What is the minimum fall of interception sewers?
3. What is the quantity of sewage and its flow pattern during twenty-four hours?
4. Should rainfall be mixed with the sewage?
5. How are the sizes of the sewers to be determined?
6. What type of pumping engine and pumps are best suited to sewage lifting?[4]

Bazalgette immediately picked up the reins of data-collection and design work on the main drainage system, but there were to be many frustrating delays and it wasn't until the spring of 1858 that Bazalgette presented his final *Report*[5] to the Board. In August 1858 the Board obtained an Act of Parliament enabling site work to commence on January 31st 1859. As engineer to the Board, appearing before select committees, Bazalgette was to spend a significant proportion of his time considering Parliamentary Bills, both those of the Board and private Bills, monitoring their effect on the public.

The plan illustrated shows the main drainage system of London. Construction work was undertaken by competitive tender from a number of contractors, and the preparation of specifications and drawings for massive subterranean works, pumping stations and machinery, reservoirs and embankments, kept Bazalgette and his team extremely busy during the 1860s. It was also the responsibility of the engineer's department to supervise construction on site. By 1865 Bazalgette's department comprised three assistant engineers, twenty draughtsmen, five engineer's clerks, an engineer's accountant, two surveyors and fifty-nine clerks of works.

The great brick-built intercepting sewers were laid to a fall of two feet per mile and, though this enabled them to be self-cleansing, it also necessitated building sewage lift pumping stations near the outfalls. The outfall on the southern drainage system was sited, after lengthy discussion, at Crossness on

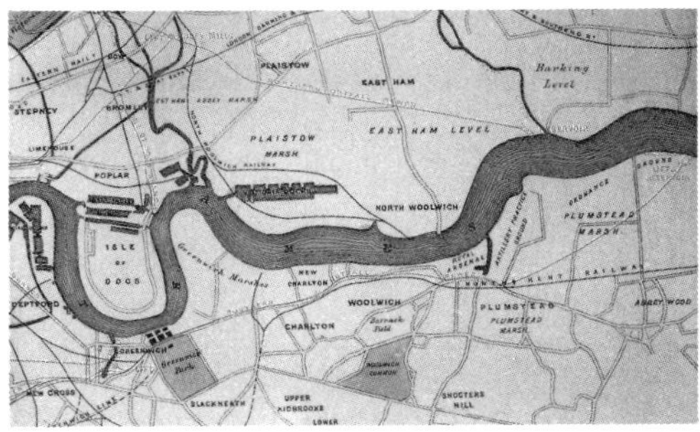

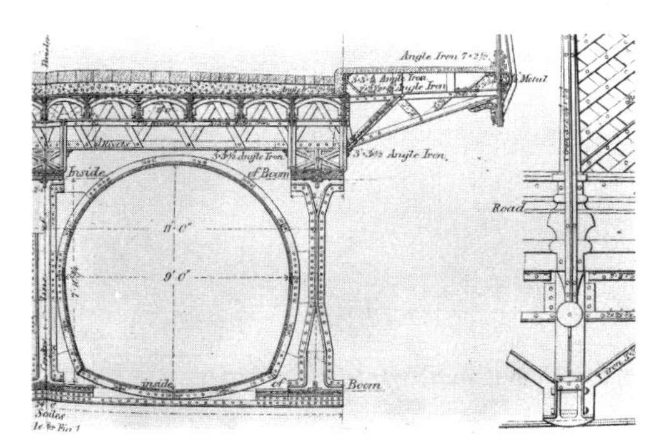

Crossness Pumping Station: opening ceremony c. 1865 (*top*), main eastern drainage system, (*centre*), bridge section of northern outfall sewer (*bottom*)

4. Ref. 1, p.288.
5. Report to MBW: 'The Main Drainage of the Metropolis', Apr. 6th 1858. Bazalgette's Report was presented in conjunction with the consulting engineers G. P. Bidder and T. Hawksley.

Top left: Victoria Embankment, section showing service duct low-level sewer and railway. *Top right*: northern outfall sewer under construction at Old Ford

the Erith Marshes, while the one on the north was at Beckton near Barking. At the end of both the southern and northern outfall sewers were large covered reservoirs and the design philosophy was that the sewage would be stored here and then released, untreated, into the Thames at the beginning of the ebb tide. Construction work south of the river progressed more rapidly than that on the north. The southern high level sewer and its branches drained an area of about twenty square miles which included Tooting, Streatham, Clapham, Brixton, Dulwich, Camberwell, Peckham, Norwood, Sydenham and part of Greenwich. Bazalgette said that the lower part of this system had made the area 'as dry and as healthy as any portion of the Metropolis'. The sewers converged on Deptford Creek where a pumping station was built to lift the sewage eighteen feet into the southern outfall sewer. Messrs. Aird and Son were the building contractors and the four rotative beam engines, each of 125 H.P., were built and installed by Slaughter, Gruning and Company of Bristol. The works was operational by May 1864.

From Deptford, the southern outfall sewer runs the seven-and-an-half miles to Crossness. The sewer, eleven-feet-and-six-inches in diameter and eighteen inches thick, was built of brickwork by William Webster, the principal contractor to the Metropolitan Board of Works. It cost £310,000 and was completed by June 1862. The Deptford and Crossness works, at either end of the southern outfall sewer, were kept in communication by means of a telegraph wire run through the invert of the sewer. Work progressed and Bazalgette's eighth *Annual Report* to the Board in July 1864 conveys an impression of the magnitude of the task:

fair progress has been made in the construction of the Main Drainage

> Lines of Sewers . . . and several Pumping Stations, Reservoirs, Outlets, etc. during the past year, the weather having been, as in the preceding year, favourable to their progression . . . it is also satisfactory to me to be able to state that, whilst a large amount of tunneling has been completed on the South Side of the River, under canals, railways, houses, and through treacherous soils, filled with water . . . no one section of these works has failed, whilst the damage to property . . . has been unimportant, and the casualties to the workmen not numerous, and with but few of a fatal character.[6]

The Crossness works occupied a site of twenty acres and work began there in the autumn of 1862. Again, the contractor was Webster who undertook to build the engine house and reservoir for £300,000. The impounding reservoir is six-and-an-half acres in extent and is covered by a roof supported by brick columns and Jack-arches spanning fifteen feet. The magnificent engine house includes excellent examples of Victorian cast ironwork where the structural and decorative functions are superbly integrated. The four large beam engines, and the ironwork, were supplied by James Watt and Company for £44,900. They survive today, although in an altered and somewhat derelict state. The Crossness works was a massive piece of Victorian civil engineering comprising 160,000 cubic yards of excavation, 82,000 cubic yards of concrete, 43,000 cubic yards of masonry and 7,000 rods of brickwork. The works, and indeed London's main drainage, was opened by HRH the Prince of Wales on April 4th 1865, when he named the four beam engines: 'Victoria', 'Prince Consort', 'Albert Edward' and 'Alexandra'.

An important factor in choosing the Crossness site was its remoteness from central London, with the result that the staff had to be small, self-contained and community-housed on site. The Metropolitan Board provided 'pretty villas for the enginemen and chief officers, and about twenty neat cottages for the workmen with a large handsome school, which also serves for a chapel and lecture-room'.[7] The staff in 1866 comprised a superintendent at £300 a year, forty-nine workmen and a schoolmaster at £80 a year. The Board maintained medical coverage at their pumping stations by paying a retaining fee to local general practitioners. Unlike many consulting civil engineers, Bazalgette had a continuing responsibility for the works he designed and constructed and was concerned with staff changes, maintenance, development of plant operational techniques and the renewable contracts for the supply of coal and other materials.

The northern main drainage system comprises high, mid, and low level interception sewers. Work on this northern section was delayed by the plan to combine the low level sewer with the works of the Victoria Embankment. The Embankment itself was further complicated by the works of the Metropolitan District Railway which brought Bazalgette and John Fowler into close collaboration. The northern mid level sewer is twelve miles long and runs from Kensal Green in the west, along Oxford Street, through Bethnal Green to Old Ford in the east where it connects with the northern

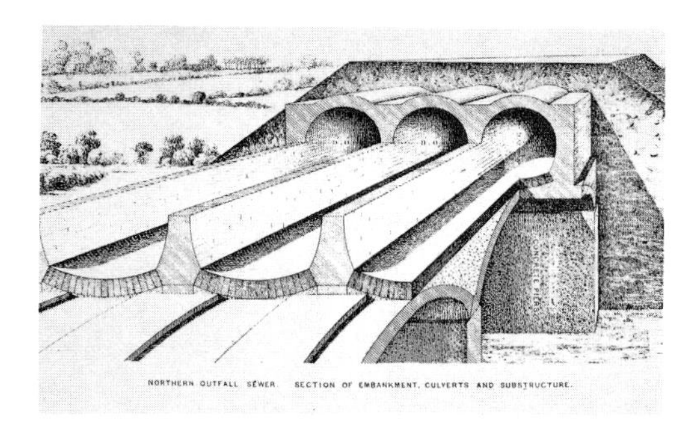

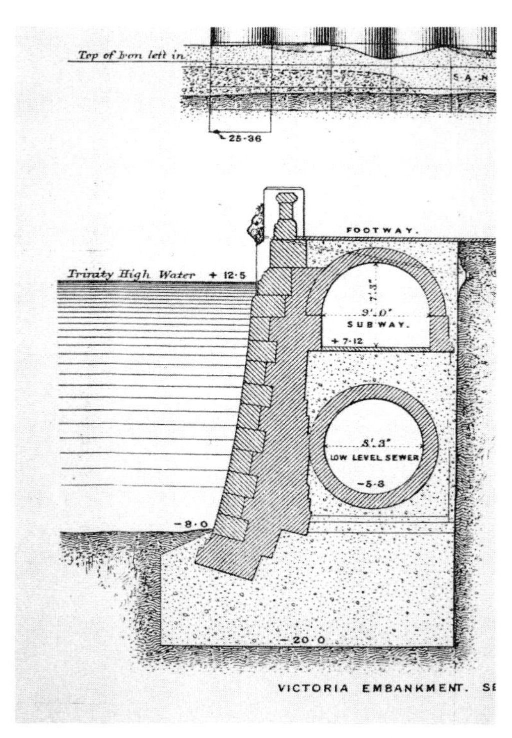

Top: outfall, brick barrels in the Embankment. *Bottom*: Victoria Embankment, sectional drawing

6. JWB's eighth *Annual Report* dated July 4th 1864.
7. James Thorne, *Handbook of the Environs of London*, Murray, London, 1876, p. 202.

Top: Abbey Mills Pumping Station, completed 1868. *Bottom*: Crossness, southern outfall work.

outfall sewer. It was a major work built by Thomas Brassey. William Webster was the contractor for the low level sewer and, in the spring of 1865, was working on Stratford Marsh where the pumping station at Abbey Mills was to be built. As he had men and plant on site he was given a contract to build the deep foundations for the building, and Bazalgette managed to convince the Board that it would save a whole summer season if Webster were to complete the building on a fixed time-related budget. This he did, roofing-in the building within just over a year from starting – an impressive feat.

Abbey Mills is an excellent example of the lavish care that the Victorians bestowed on their functional architecture. The building is cruciform in plan with polychrome brickwork, stone dressings, turrets and a mansard slate roof surmounted by a central octagonal lantern. The original pumping plant comprised eight beam engines, two in each arm of the building, and these engines, boilers, and structural ironwork were supplied by Rothwell and Company of Bolton in Lancashire for the sum of £54,570. There was also an house on site for the station superintendent and cottages for the staff. The function of this station is to lift the sewage into the adjacent northern outfall sewer whence it gravitates to the reservoir at Beckton. The topography of this part of East London is such that the outfall sewer is above ground and of brick construction contained in an earth embankment. There are two brick barrels from Old Ford to Abbey Mills and three thereafter to Beckton. Where the sewer had to cross roads, rivers, canals or railways it was constructed of wrought-iron plate supported by rivetted plate girders. The northern section of the main drainage system became operational with the opening of Abbey Mills Pumping in the summer of 1868.

Bazalgette summarised the scale of his main drainage scheme in 1865 when he said:

> There are about 1,300 miles of sewers in London, and eighty-two miles of main intercepting sewers. Three-hundred-and-eighteen-millions of bricks, and 880,000 cubic yards of concrete have been consumed, and three-and-a-half million cubic yards of earth have been excavated in the execution of the Main Drainage works. The total pumping power employed is about 2,380 nominal H.P.[8]

To complete the first phase of London's main drainage a fourth pumping station had to be built to cope with the outlying districts to the west. This was to be the Western Pumping Station at Pimlico. Webster began work on site in July 1873 and the station was opened in August 1875. The original pumping plant consisted of four beam engines by James Watt and Company, each of ninety H.P., supplied with steam from eight boilers. These engines lifted sewage eighteen feet into the low level sewer. Bazalgette built three further pumping stations during his career with the Metropolitan Board. The first was the Effra Pumping Station at Vauxhall opened in 1879 and a sister plant, known as the Falcon Pumping Station, opened the same year. His last was the Isle of Dogs Pumping Station, opened in 1888, designed to lift sewage and storm water from the low-lying island area and deliver it to Abbey Mills.

8. Ref. 1, p.314.

Once the main drainage system was working, Bazalgette and the engineer's department only maintained the eighty-two miles of intercepting sewers and the two outfalls. The remaining sewers were handed over to local district boards and vestries but, as Bazalgette explained:

> in order to make it one complete system, they are bound to send in plans for any sewers they propose to construct to the Metropolitan Board of Works and get its approval of those plans. By this means one complete system is always maintained although the works are constructed by a number of different district boards.[9]

During the period from 1875 through to the end of the Metropolitan Board, Bazalgette could devote more time to municipal engineering matters other than drainage. But throughout his period as engineer to the Board he was allowed to undertake a range of free-lance consultancy work. He was frequently requested to report on the drainage schemes of towns, as the following list indicates:

EPSOM (1858)
LUTON (1858)
NETLEY (1859)
FELTHAM (1862)
SHREWSBURY (1862)
BRISTOL (1863)
CHELTENHAM (1863)
WESTON SUPER MARE (1865)
OXFORD (1865)
HASTINGS (1866)
FOLKESTONE (1866)
OXFORD (1866)
CAMBRIDGE (1866)
DON VALLEY (1866)
TOTTENHAM (1867)
GLASGOW (1868)
HERNE BAY (1869)
NORTHAMPTON (1871)
SKIPTON (1871)
BIRMINGHAM (1872)
SCARBOROUGH (1872)
DORKING (1872)
CROYDON (1873)
BECKENHAM (1873)
NORWICH (1873)
FARNHAM (1874)
NORWICH (1874)
HAMPSTEAD (1874)
MARGATE (1874)
RUABON (1875)

This list does not claim to be complete but it at least serves to indicate the extent of Bazalgette's reputation as an expert with unrivalled experience. He was also consulted about the drainage of the stables at Windsor Castle (1868) and of the city of Pesht in Hungary (1869). In the 1880s, during the last years of the Board and of Bazalgette's career, attention was given to the treatment of the sewage at the outfall works at Crossness and Beckton prior to its discharge into the Thames. In July 1888 Bazalgette said that works were in progress 'for the purpose of using chemicals at the outfalls, and precipitating the solid matter and turning the effluent into the river in a somewhat purified condition'.[10] He experimented with pressing the residual sludge to reduce its bulk and had hoped to find an agricultural fertilizer market for the material. Having failed, he said: 'We are having hopper vessels constructed for taking the sludge right out to sea.' The first of the fleet arrived in the Thames in June 1887 and was named the 'Bazalgette'.

We have seen that the root of the public health issue was the pollution of the water-cycle. Throughout his career with the Metropolitan Board, Bazalgette had argued for powers to bring the private water companies under the control of the Board and in 1884 he remarked:

> In 1880 it was proposed to purchase the interests and property of the water companies, and place the water-supply under the municipal authority . . . A purer and more copious supply of water on constant

Top: main drainage works, delivery of building materials by Thames barges. *Bottom*: main drainage works, concrete batching

9. Ref. 3, p.335.
10. Ref. 3, p.329.

Top: Crossness reservoir and engine house under construction (*TWA Archive*). *Centre*: Abbey Mills Pumping Station, iron work under construction. *Bottom*: Abbey Mills Pumping Station under construction, April 1867

> supply and at high pressure is demanded, and whether this is to be attained by purchase, or by some regulation of the present Water Companies' powers, it is obvious that each year's delay will only increase the cost and the difficulties involved.[11]

But this was not to be achieved in Bazalgette's time. It had to await the formation of the Metropolitan Water Board in 1903 and, even then, was not fully integrated with main drainage until the formation of the Water Authorities in the 1970s.

But public health engineering evolved to embrace more than just water-supply and main drainage during Bazalgette's career. In 1884 he outlined his concept of the issues involved, saying:

> In contemplating any comprehensive improvement or extension of large cities, the following are some of the questions which present themselves for consideration:
> 1. What should be the width of the streets?
> 2. To what height should the houses be restricted?
> 3. What should be the minimum air-space allotted to each individual in the houses?
> 4. What proportion of the area of a city should be set apart for its lungs and recreation grounds?
> 5. What public buildings and markets, and what water-supply, sewerage and means of lighting should be provided?
> 6. What should be the regulations to be enforced in order to secure the effectual combustion of fuel, and to prevent the contamination of the atmosphere by smoke?[12]

Bazalgette, with all these issues, had been concerned with improving the built environment of London. In the middle of the century much had been done, by private associations, to provide rented housing for the poor and, after the Artisans' and Labourers' Dwellings Act of 1875, a responsibility was placed on the Metropolitan Board. By 1884 Bazalgette could say:

> Twelve areas, situated in different parts of London, embracing an aggregate area of forty acres, in which the houses were over-crowded and declared to be unfit for human habitation, have been already dealt with by the Metropolitan Board of Works, at a cost of £1½ million.[13]

And on air-pollution in the same period he said:

> In London 5,800,000 tons of coal are consumed per annum, being at the rate of nearly one-and-an-half ton per head of population, in addition to the two million tons used in the manufacture of gas; and bearing in mind that each ton of coal consumed generates 56,000 cubic feet of carbonic acid gas . . . the mode of dealing with this product becomes a subject of grave importance.[14]

It will be seen then that public health engineering was becoming a discipline with an overall view of the factors influencing the health of urban dwellers. In Bazalgette's own words:

11. J. W. Bazalgette, 'Presidential Address', *PRC. ICE*, Vol. LXXVI, pt. 2, Jan. 8th 1884, pp. 2-26.
12. *Ibid.*, p.20.
13. *Ibid.*, p.21
14. *Ibid.*

Above left: Sir Joseph and Lady Bazalgette with children and grandchildren. *Above right*: Crossness, site visit, Bazalgette in left foreground

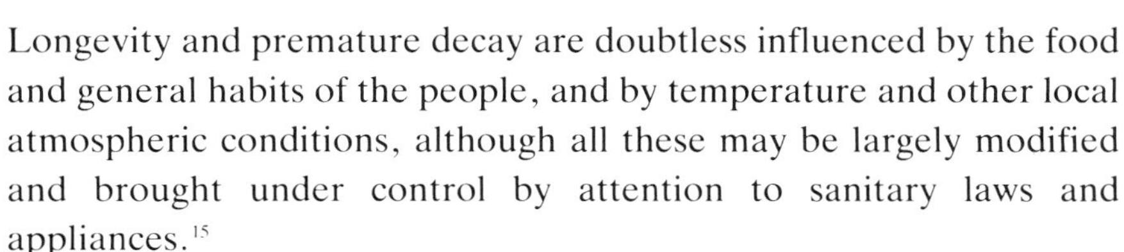

> Longevity and premature decay are doubtless influenced by the food and general habits of the people, and by temperature and other local atmospheric conditions, although all these may be largely modified and brought under control by attention to sanitary laws and appliances.[15]

In March 1889 the Metropolitan Board of Works was replaced by the London County Council and Bazalgette took the opportunity of retiring, 'after forty years of arduous and responsible work in the public service'.[16] At the time of his retirement Bazalgette was the highest-paid officer of the Board at £2,000 per annum and he retired on two-thirds of this salary. His career shows him to have been a most capable manager of men and of millions of pounds of public money, as well as being a competent designer of public works on a grand scale. Moreover, his work was always subject to the minute scrutiny of the committees – so necessary in the field of public works.

His career was marked by various honours. In 1871 he was made, somewhat appropriately, a Companion of the Bath, and in 1874 he was knighted by Queen Victoria at Windsor Castle. His fellow professionals accorded him the high honour of electing him President of the Institution of Civil Engineers in 1884 and, perhaps not surprisingly, his presidential address discussed, 'those engineering works which promote the health and comfort of the inhabitants of large cities, and by which human life may be preserved and prolonged'.[17]

He retired and died on Sunday March 15th 1891. His many obituaries spoke of his unique contribution to public health. In his office at the Metropolitan Board he trained several influential public health engineers of the next generation and we owe him a great debt for the standards of municipal engineering now taken for granted in many parts of the world.

15. *Ibid.*, p.24.
16. MBW Minutes, Feb. 8th, 1889, p. 326.
17. Ref. 11, p.4

PROFESSOR CHRISTOPHER FRAYLING

The Strange Case of the Duke of Wellington's Funeral Car

'No successful results can ever be attained', wrote Henry Cole in February 1851, 'until the designer and the engineer know each other's business'. He had come to this conclusion, as part of his wide-ranging critique of the governance and performance of the School of Design at Somerset House which had been founded fourteen years earlier. This view was expanded in the *Journal of Design and Manufacturers*, which Cole had co-edited with Richard Redgrave since March 1849 and which was mainly concerned with the 'complete reformation' of the School of Design. The *Journal* was pledged to the view that *if only* art educators and industrialists could get together, and *if only* design students could be encouraged to work 'within constraints', the School would at last begin to effectively promote 'improvement in the ornamental and decorative manufactures of this country'. It was a view shared by the equally optimistic Samuel Smiles, who, in *The Lives of the Engineers*, written as a kind of coda to the Great Exhibition, asked what England was:

> without its tools, its machinery, its steam-engine: its steam-ships, and its locomotives? Are not the men who have made the motive power of the country, and immensely increased its productive strength, the men above all who have . . . made the country what it is?

Prince Albert was perhaps making a connection between the two men when (according to Queen Victoria) he used to jest: 'We must have steam, get Cole!'

Up until 1851, the School of Design had been mainly concerned with drawing 'from the flat' a selection of antique objects, which were, incidentally, to form the basis of the Victoria and Albert Museum many years later. The students were not considered grown-up enough to draw things that moved, so the 'life class' was looked upon, by successive regimes, with great suspicion. In fact, the whole question of the 'life class' provoked the first recorded student 'occupation' in the history of higher education, which took place at the beginning of April 1845. For the School, which had been founded to teach members of 'the manufacturing class' to communicate through the medium of drawing, kept turning into an *art school*, and the governing bodies (which included several members of the Royal Academy) were determined to stamp *that* tendency out. Apart from anything else, the RAs didn't want the competition. And so, from 1837-51, the pendulum had swung from 'design' as the language of communication for those involved in manufactures, to 'design' as the art of drawing, with no one able to stop it swinging. As Cole concluded, it just went to show that you could half educate people if you were trying to educate them wholly (most educational institutions did that all the time), but what you couldn't do was deliberately half educate people. It simply wouldn't work. The solution was to talk to the manufacturers themselves about what sort of 'design' *they* wanted, and to take it from there.

In suggesting this, Cole was in fact questioning one of the basic assumptions on which the whole of 'design education' in this country had been built. For, when the Select Committee of 1836 recommended the establishment of a central School of Design, it did so on the grounds that 'the contemplation of noble works in fresco and sculpture is worthy of the intelligence of a great and civilised nation'. The results of this 'contemplation' – through the teaching of drawing – could then be *applied* in the form of 'ornament' to the objects of manufacture. In other words, the nurturing of 'taste' came first, the application of that 'taste' to industry (an altogether lower form of 'intelligence') came second. Hence all the models, casts, prints and sculptures which were, for its entire life, virtually the only 'equipment' to be purchased by the School of Design. 'Design' was considered by both the Select Committee and the teaching staff of the School, to be something distinct from production, something that could be learned as a language in itself. Hence the need to separate the new School of Design from the concept of the Mechanics' Institute – a concept which stressed the value of 'learning by doing' and which placed a great emphasis on teaching the rudiments of science and mechanics to industrial workers. A network of Mechanics' Institutes already existed in 1837, and was particularly strong in 'the northern industrial cities', but, following the Select Committee of 1836, the decision was that the School of Design should go it alone. When the *Art Journal*, that great defender of the Somerset House system (if that isn't too strong a word), heard of the appointment of Henry Cole as General Superintendent of Practical Art in 1852, it noted ruefully 'this is the beginning of the end . . . ' In future, the *Art Journal* thought, 'design' might even have something to do with *trade*: a prospect almost too horrible to contemplate.

It was in his *First Report of the Department of Practical Art* that Cole for the first time publicly acknowledged that it was no simple matter introducing industrial machinery (such as Jacquard looms) into the studios of the improved School of Design.

> . . . much caution has been, and must continue to be, exercised in conducting these classes, and the difficulties of establishing them require to be fully and fairly acknowledged. Among them may be stated the indifference and ignorance of the public, the discouragement given to the admission to the Schools of Design of any other but artisans, the yearning for 'fine' in contradistinction to 'decorative' art among the students, the pressure for employment which prevents continuous study, the scepticism of the existence of the principles of art, and, lastly, the want of confidence on the part of manufacturers that the training can be imparted, or that if imparted, the results of it

The Wellington tree sketched on the field of Waterloo (*Illustrated London News*, c. 1852)

Opposite:
Top: The Duke of Wellington's Funeral Carriage as drawn by the School of Design, Marlborough House. *Bottom*: the Funeral Tent at the Horse-guards (*Illustrated London News*, c. 1852)

are wanted by the public.

This 'want of confidence' he had noted during an elaborate fact-finding mission among some of the leading 'art manufacturers' of the day. Garrard's had thought that, apart from public ignorance, the greatest drawback was the demand for low prices; Wheeler's, the silversmiths of Clerkenwell, had considered that though their workmen were not able to draw, they would probably not be willing to learn, so Cole should in future concentrate on apprentices if employers could be persuaded to give them time off; Hunt and Rockell had felt that their silversmiths and chasers would certainly benefit from instruction in design, but feared that not many of them would attend the School – 'a workman', they observed, no doubt correctly, after the fatigue of a twelve or sixteen hour day, 'is not in a fit state to acquire art', but they too had apprentices who might profit from the experience. Even among the students, Henry Cole had found 'very little willingness' for practical training, and gave as an example one from Spitalfields who had been given a design scholarship for silk weaving, which he soon abandoned in order to start a career as a portrait painter. Therefore, concluded Cole, it would take quite some time before a 'more correct' feeling towards industrial design could be expected. This 'more correct' feeling would have to originate with contributors to the manufacturing process – and yet it was precisely these employees who the manufacturers were not prepared to release. The notion of a 'designer' as a separate species was, of course, a thing of the future.

Part of Cole's *First Report* was made up of papers written by the various professors working at the School. One of these was the great architect and design theorist Gottfried Semper, who was in charge of a class dealing with 'the principles and practice of Ornamental Art applied to Metal Manufactures'. Semper had sent in a characteristically thorough 'System of Instruction', which offered advice on how to reorganise *all* the design classes. They should, he wrote, be based on *practice in the workshop*; students should be used as 'assistants' to the professors in their commissioned work; and, above all, the amount of time spent on 'copying' should be drastically reduced: 'many a talent has been spoiled by having been too long a time engaged with copying and studying from models and even from nature'. When Cole published some of Semper's recommendations in his *Report*, he personally amended this sentence to read 'those copies would have much more interest for the student if done in connection with some idea which the student had in his mind' – which doesn't have quite the same force as the original, to put it charitably. Even Cole, it seems, was finding it difficult to imagine how on earth the practice of 'copying' – 'from the flat', 'from the antique' or whatever – could be dislodged from its position at the centre of the curriculum. Had he changed his mind since the days of the *Journal of Design and Manufacturers*? Was he, following his enquiries among the 'art manufacturers', becoming disillusioned? Or was it simply that he was too busy to give his undivided attention to the School, at a time when he was (with Redgrave) reorganising the complex national system of *elementary* art education all over Britain?

BLESSED ARE
THE
LORD
VITTORIA
S^T SEBASTIAN
FORTUNA

The Audience Chamber at the Horseguards the night before the funeral

Whatever the reason, it would be half-a-century before Semper's 'System of Instruction' was introduced to South Kensington, and even then it was to be filtered through the ideology of the Arts and Crafts movement.

Nevertheless, in the autumn of 1852, Cole was presented with a terrific opportunity to publicise the new style of management and the new intentions of the School, when in his official capacity he was offered the commission to design and make the Duke of Wellington's funeral car. As he was later to recall, he seized the opportunity because it would make more widely known 'the facilities afforded to all classes of the community for obtaining education in art'. The day of the Iron Duke's funeral had been proclaimed a bank holiday and it was anticipated that at least one-and-an-half million people would be watching the spectacle.

Originally, the design of the funeral car was entrusted to the Royal undertaker, William Banting of St. James' Street, who was to achieve a certain notoriety a decade later when he went on a high-protein diet, lost three stone, and made his name synonymous with slimming. But the selection of designs he displayed before Prince Albert – consisting of ornate hangings over a simple wooden frame – were all rejected, and so the Lord Chamberlain instead approached the School, which by then had moved to Marlborough House. Banting, who was furious, later claimed that he had been given 'imperfect instructions as to what was really required'. According to the official *Report* of the decision to involve Marlborough House, 'the Board of Trade considered that the Department was not yet adequately organised to undertake such work officially, but permitted the assistance to be given . . . as a private transaction': an ominous start.

Henry Cole went straight out to measure the arch of Temple Bar and decided, with Redgrave, that the carriage should be made of solid bronze – as befitted the solidity of the Duke's reputation: 'a real substantial work- . . . thoroughly simple, as was the character of the Duke, the only decorations being the pall, the armorial bearings and the names of his great battles'. Redgrave made a series of drawings of 'the general design of the car', Gottfried Semper drew 'the structure, with its ornamental details', and Cole noted in his *Diary* 'sketching car'. These drawings were all displayed before Prince Albert two days later, and Cole, after his audience, told Semper that 'the Prince liked many parts of your design', and wrote in his *Diary* 'settled and approved'. Apparently, Prince Albert had exclaimed 'this is the thing'. Under guidance from Redgrave, Semper and Octavius Hill (who was at that time in charge of the classes in 'woven fabrics, lace, embroidery and paper staining'), the senior students of the metal-working, porcelain painting and textile classes got to work on executing the details. Some of the ornamental bronze 'figures of Victory' were modelled by Mr. Whittaker, a scholar, and Mr. Willes, a day student, while the embroidery of the heraldic devices was done by 'some fifty female students' working round the clock.

It was to be an extraordinary construction, twenty-seven-feet long, ten-feet wide and seventeen-feet high, and the Department had only *three weeks* in which to have it ready and on the road. Since the arch of Temple Bar (as Cole had discovered for himself) was barely seventeen feet, the car was 'so arranged that, by the application of some ingenious machinery, the whole could be lowered a couple of feet while passing under the Bar, and raised again after entering the City'. The six bronze wheels were cast by firms in London, Sheffield and Birmingham – 'an instance', said the newspapers, 'of the remarkable rapidity with which the most elaborate works can be manufactured'. Actually, relationships with the manufacturers did not always go quite so smoothly: Cole was to recall on November 24th 1852:

> We resolved whatever there was should be real, and not a sham; but were defeated in this by the *disobedience* of two of the manufacturers entrusted with the castings. Helmets had to be fitted in particular spaces; one had to be modelled and was sent to Birmingham to be repeated six times in bronze. *Would you believe it!* The manufacturer had the irreverent audacity to put aside the model altogether, and to substitute a helmet different in shape, and so big that it could not be used. We sent to another manufacturer a lion's head of a particular model: he returned it a sort of *pug dog's* head, too large. Such is the wilful ignorance of Art and moral disrespect for authority among some manufacturers displayed at a moment so critical; and thus we where driven to affix painted papier mâché helmets: but these were the only shams . . .

The Duke of Wellington's Funeral Carriage, c. 1852

Above the frame, there was a pediment seven-feet wide, modelled by Whittaker and Willes, 'presenting a mass of gilt carving, enriched with circular panels, within which the names of the Duke's principal victories were emblazoned'. And above *that* was a bier, six-feet high and four-feet wide, designed by Hill, manufactured in Spitalfields and covered 'in a gorgeous canopy of rich Indian kinkhol'. This bier was arranged on a turntable, so that 'upon arrival at the west front of St. Paul's Cathedral, it can be readily moved round'. As the *Illustrated London News* pointed out:

> the mere manufacture of this car has been wonderful proof of the English capacity, such as the deceased himself was always one of the first to honour. Those who saw it in its full magnificence in the funeral pageant, would scarcely believe it . . .

The final assembly took place in a huge ordnance tent, erected on Horseguards Parade. And after a last-minute decision to cover the top storey with a canopy ('up to the very dawn of the day', wrote Cole, the students 'were at work on the embroideries'), the Duke of Wellington's last journey commenced – on time. As twelve black dray horses (supplied by Booth's Gin Distillery) heaved the funeral car from under the tent, Cole and Redgrave both heaved a long sigh of relief.

'London never yet was in such a state of ferment and excitement', wrote the *Illustrated London News* two days after the State Funeral, on November 20th 1852. The funeral car itself was 'the object of universal admiration, even

Top and Bottom: Artillery sketch books, c. 1810

as it was drawn along with sufficient rapidity to prevent a scrutiny of its design'. It showed just 'how fine was the contrast between the style of decorative art, and that adopted at the tawdry fittings-up of the Invalides for the reception of Napoleon's ashes in 1840'. The School of Design had been originally set up in order to show those 'tawdry' French a thing or two, and the revenge was very sweet . . .

But a week later, on November 27th, after all the ferment had died down, the *News* quietly added that 'a great deal remains to to told'. The twelve dray horses from the private sector had done magnificent service, but 'the ponderous weight of the car can be seen in the traces left by the wheels'. And on the Mall, nearly opposite the Duke of York's column, there had been a disaster. For, on reaching the Mall, the enormous weight of the car caused its wheels to sink into a small gutter. Pandemonium broke loose. Militamen galloped to and fro, orders were given, advice tendered. It took quarter-of-an-hour and the combined effort of the twelve horses, numerous soldiers, police and bystanders to free the carriage from the mud. The car, undamaged, then continued sedately on its way.

Temple Bar was successfully negotiated, the arrival at St. Paul's went without a hitch, but the sheer weight of the funeral car had let Cole and Redgrave down – that and the wheelbase which was apparently far too narrow. It must have been a bitter blow to Cole, who himself was beginning to reflect that: 'Triumphal cars belong to a past age – *the artilleryman's gun carriage* or the *soldiers themselves* carry their comrade to the grave in these days.' Perhaps, after all, the form of the car should have followed its function (as Owen Jones, a lecturer at the School, might well have put it).

To make matters worse, the quarter hour delay had enabled at least one eye-witness to have a good long 'scrutiny of its design'. 'It was a moving sight', he later wrote in his *Recollections* of the funeral, 'which even the horrible South Kensington catafalque with all its tawdry vulgarities, could not altogether deprive of its solemnity'. When Cole lectured on this particular design project, at Marlborough House, four days after the great day, he was very much on the defensive. He began his lecture with the words, 'Whatever may be its merits and defects, I may say that the car would not have been produced, if our *special classes* here had not existed', and used the occasion to promote the sterling work of Semper and Hill under the new management. And he finished by stating that the School had really done remarkably well, considering that they had 'just three weeks to produce a work which would reasonably occupy a whole year . . . '

It is very tempting indeed to use the image of the funeral car stuck fast in the gutter of the Mall as a symbol of all the confusions which surrounded official thinking about design education at the time of the Great Exhibition – forty-five years before the School of Design became the Royal College of Art. But the confusions were, of course, far more complex than that. 'Design' was still thought to be an added aesthetic value *applied* to everyday things – even by those, the vast majority, who did not think it meant drawing plain and simple (its original usage). The political economists of the day, such as Andrew Ure and Charles Babbage, had described manufactures in terms of universal laws and norms, which, in the hands of members of the Department of Practical Art (later Science and Art), were to be turned into laws and rules for the teaching of drawing to 'all classes of the community'. The social class issue itself was complex: the Royal Academy wanted to differentiate itself from the School of Design – hence the banning of life drawing which might encourage hapless artisans to have ideas above their station – which itself wanted to be differentiated from the Mechanics' Institutes, where they weren't concerned enough about 'taste'. Add to these the general cultural reaction against the implications of industrialisation – neatly symbolised by Pugin's Mediaeval Court housed inside Paxton's iron and glass construction of the Crystal Palace – and there are more than enough historical contexts within which to interpret the strange case of the Duke of Wellington's funeral car. Pugin's Mediaeval Court begat an Arts and Crafts philosophy which reacted against the normative tradition of the first half of the century, replacing it with a new tradition which was critical of the machine for what it could do to human beings. It was this Arts and Crafts philosophy (complete with a caricature of something evil called 'the factory' or the 'manufactory') which dominated design thinking within art education for the half-century following the Cole revolution. Cole himself was to give up even trying to build bridges between the School of Design and the world of the manufacturers, concentrating instead on the provision of teachers for his new national system: under Redgrave and Cole, the School became the National Art Training School, housed at South Kensington in the heart of 'Albertopolis'. Albert, one imagines, would not have been amused by this turn of events, but Cole, for his part, was happy that at last the system could be controlled down to the minutest detail.

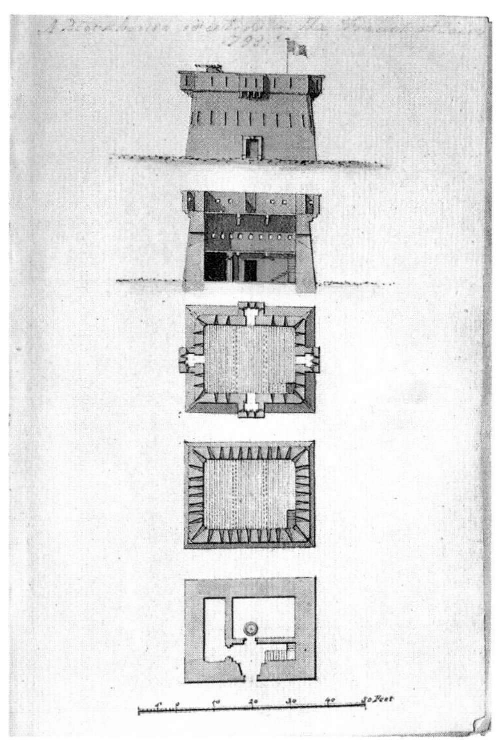

Anonymous watercolour by a military engineering student

But still, the image of the funeral car stuck fast in the gutter of the Mall must have dramatically reminded Cole (and his readers with longish memories) of the statement he had made when a restless 'outsider', fiercely critical of the School of Design for producing students who seemed to have both feet planted firmly in the air: 'no successful results can ever be attained until the designer and the engineer know each other's business'.

DON HOLLAND

Construction as a Prime Export

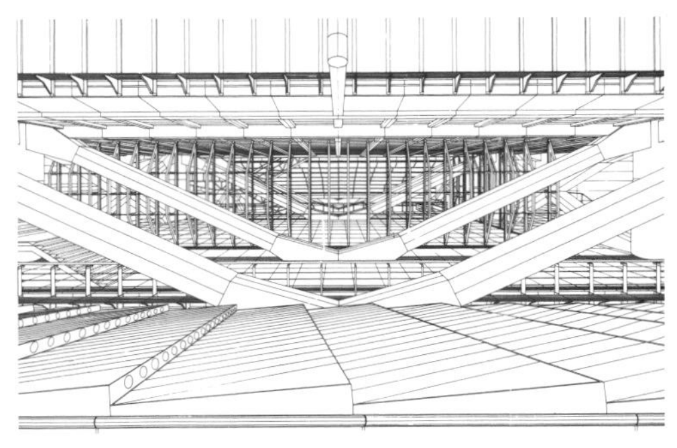

Above:
Ove Arup and Partners, Engineers; Foster Associates, Architects; Hong Kong and Shanghai Bank, c. 1986, sun scoop
Opposite:
Top: R.M. Ordish, Engineer; Handyside and Company, Contractors; bridge over the River Pruth at Czernowitz, c. 1870. *Bottom:* SS Great Eastern loading the Atlantic Cable, c. 1865

The opening of the Royal College of Art in 1837 took place at the time that the industrial revolution was getting into its stride. George Stephenson's 'Locomotion', the first locomotive to work on a public railway, was built in 1825 and was the fore-runner of many other developments in locomotive design over the succeeding decades. The evolution of the railways which occurred with such fervour in the middle of the last century presented the great engineers of that period with enormous opportunities to provide the permanent way which was to bring railway transportation to every corner of the British Isles.

It is interesting to note that this tremendous undertaking was carried out almost entirely by private capital put up by a large number of companies formed for the especial purpose of railway operation, which, over the succeeding century, by purchase and amalgamation, formed the handful of railway systems extant at the time of the creation of British Rail.

Not only did the building of the railways give opportunities to, and greatly enhance, the reputation of many distinguished Victorian engineers, it was also the spring board for other important aspects of the construction industry. The role of the contractor – working by definition under contract to the client but following the instructions of the engineer – became more clearly defined and the relationships which are the foundation of today's *modus operandi* evolved. But more than that – the traditions of 'following the work' by the essential work force evolved.

The building of the railways was undertaken, at the 'pick and shovel' level, by a remarkable breed of men. They worked in large gangs, usually drawn from the same area, and travelled the country digging the cuttings, forming the embankments and laying the track at a speed which, by any standards, was remarkable. They were well paid, but then it was their just reward for the hard work they did. Working conditions were tough and many died from accident and illness, but by the early 1840s they looked for fresh challenges and were to find them abroad.

The first foreign railway on which British navvies were to work in any number was the Paris to Rouen route in 1841. Having regard to the prestige the French railways enjoy today, it is surprising that in the middle of the last century the French lagged behind in railway building, lacking both capital and engineering techniques.

Indeed, British participation went further than the construction aspects:

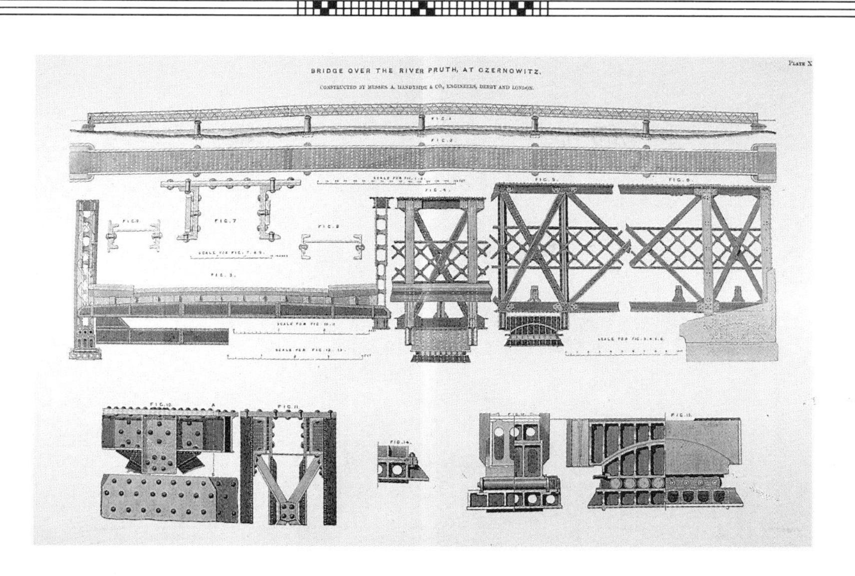
BRIDGE OVER THE RIVER PRUTH, AT CZERNOWITZ.
CONSTRUCTED BY MESSRS. A. HANDYSIDE & CO., ENGINEERS, DERBY AND LONDON.
PLATE X

Above and opposite: Robert Stephenson, Engineer; constructing the Victoria Bridge, Canada, c. 1859

two-thirds of the money to finance the eighty-two-mile long Paris to Rouen line came from the British money-market; the engineer was British (Locke); and the contractors (Brassey and William Mackenzie) were Scottish. Half the 10,000 strong labour force was British, since Locke was anxious to have experienced men to ensure completion on time. The workers who assembled for this job spoke thirteen languages and evolved a patois – part Franch, part English and part something else – which seemed to fit the bill in as much as the line was constructed in spite of many major technical problems having to be solved.

After the railway boom died down in the United Kingdom, railway building became a service much in demand abroad, and in the 1850s Brassey worked in Italy, Denmark, Norway and Austria, as well as in France, taking his British management, engineers and gangers with him, and relying on labour recruited locally through labour contractors except when particularly extenuating circumstances arose.

In addition to Europe, British techniques and work forces were deployed in 1852 to construct the 539-mile long Grand Trunk of Canada from Quebec to Lake Huron. No local labour was available for this task and 3,000 British navvies were shipped over the Atlantic. The work was undertaken under very cold conditions and many of the work force suffered badly from frost-bite and cholera. Because of these vicissitudes, Brassey was forced to use an American steam excavator, probably for the first time in railway construction, but in spite of his efforts, by 1859 he had lost nearly a million pounds on the construction of this railway, a very great deal of money at that time.

During this period, the Crimean War had broken out and, by 1854, 30,000 British troops, together with a large French army, were laying siege to Sevastopol. The conditions these soldiers faced in the Russian winter – which, as many commentators at home and in the Crimea did not hesitate to point out, seemed to have come as a surprise to the generals – were arduous in the extreme and proper transport facilities were essential, since there was only one bad road from the port seven miles away which disintegrated completely in the winter. Supplies could only be moved by pack horse or 'pack men', but three British railway contractors were to solve the problem. Brassey together with his brother-in-law Petts and an MP named Peto, offered to build a railway at cost, shipping from England everything needed to build and operate the line. The Civil Engineer Corps – a civilian body – was formed under a chief engineer named Beattie and recruited experienced men for a six-month contract. They were to be well paid and equipped and set out just before Christmas 1854. Twenty-three vessels were mobilised to carry the men, the vast quantity of materials and the machinery to the Black Sea.

The small armada arrived in February 1855 and the progress achieved in construction astonished everybody including *The Times* correspondent William Russell whose dispatches had done so much to alert the British public to the conditions prevailing in the Crimea and the shortcomings of the military

commanders. The railway was a bit rough, even the contractors agreed, but it was serviceable and laid at a fast pace, up to a quarter of a mile a day. Twenty-nine miles were laid in all, vast quantitiues of supplies were carried, the army was relieved and Sevastopol fell in September 1855. The contractors came out of the war well, and the *Illustrated London News* said that these works of construction demonstrated that the men 'who made England great by their skill, enterprise and powers of organisation were of a far different calibre from the officials the Government employ'.

These examples of railway construction in many parts of the world a hundred years ago demonstrate the innovative skills of managers, engineers and contractors, their entrepreneurial attitudes and their willingness to 'follow their trade' wherever and whenever needed.

Victorian Britain was the heyday of the British Empire. Development of the vast lands of Africa came about for many reasons: the land was there to be developed, British pride in good government and, not least, an altruistic belief on the part of the British that they had something to offer the indigenous populations. The Christian missionary societies were also actively proselytizing and, since they came mostly from these islands, the many opportunities for trade in what was a captive market were actively exploited and British trade prospered.

Trade meant communications and it was natural that the successful British engineers would satisfy the needs that arose. In addition to work in Africa, ports and harbours, irrigation schemes, railways and bridges were designed by British engineers throughout the world, and most of the construction works were undertaken by local labour under expatriate supervision. During the middle of the last century, construction works were undertaken in Europe and India by James Rendel, whose family name is kept alive to this day in Rendel, Palmer and Tritton, consulting engineers, in Cape Town, Rio de Janeiro, Genoa and La Spezia in Italy. Rendel was not alone in providing engineering design skills from this country.

India was of course a very special part of the Empire. The British had established themselves there particularly through the East India Company which was formed as a trading corporation in 1599 and given a charter by Queen Elizabeth I the following year. The Company had its own army within a few years, and British rule followed British trade over most of India. Trade needed ports and extensive communications, as did the military, and there followed over the centuries the creation of a vast infrastructure – one of the legacies the British left behind – including a 2,500 mile long railway system covering the whole of India. The Lansdowne Bridge crossing the Jumma River was, in its day, the longest single-span bridge in the world. The Great Trunk Road ran for 1,000 miles from Calcutta on the Bay of Bengal to Peshawar in what is now Pakistan. Although the irrigation systems of the Punjab were of long standing, they were considerably extended when barrages were constructed across the River Indus. The billion-dollar Indus Basin Scheme, which became necessary following the partition of India and

Above:
Ove Arup and Partners, Engineers; Sydney Opera House.
Opposite:
Top: Lt. Colonel J. Kennedy, Engineer; Taptee Viaduct, Bombay, Baroda, Central India Railway, c. 1860. *Bottom:* Robert Stephenson, Engineer; Messrs. Handysides constructing yard at St. Petersburg, Russia, c. 1870

Production platform for the Shell/Esso Brent Oilfield

Pakistan, was designed in the 1960s by British consultants Coode and Partners to share the waters of the five rivers in the Punjab and maintain the extensive irrigation schemes on which an area half the size of Britain depended.

British preoccupation with the Indian subcontinent spilled over into the Middle East. The waters of the great rivers of the Green Crescent, the Tigris and Euphrates, were controlled by works designed by British consulting engineers and constructed by British contractors in the years before the Second World War. Extensive flood protection schemes were similarly provided in Iraq and control of the waters of the River Nile, flowing through Egypt and the Sudan, was also very much the concern of British engineers.

Many of the construction works associated with these massive schemes provided employment for thousands of local labourers throughout Africa, the Middle East, India and the many other lands where British influence was to be found. British industry greatly benefited by the near captive market for capital goods that was created.

The near-unique procedures for construction in the United Kingdom as enshrined in the Institution of Civil Engineers' Conditions of Contract – which provides the tripartite arrangement of client, consulting engineer and contractor – have had wide acceptance in those parts of the world where British influence has been strong. The skill, engineering judgement, and independence which British consultants have deployed around the world on behalf of their clients has been recongnised by their continuing employment in over 150 countries far beyond the ex-colonial territories. The value of the work which has been entrusted to members of the Association of Consulting Engineers now exceeds £43 billion and their invisible earnings exceed £500 million per annum. Their expertise is manifest in designing and supervising engineering works covering the whole spectrum of civil construction. While applauding their substantial contributions to the construction industry, it must be stated that very little of the work which British consultants engaged in overseas is actually carried out by British contractors. This is primarily because competitive tendering is almost axiomatic in most parts of the world, which means that local contractors are able to price work more cheaply than offshore contractors who have to mobilise into the territory to undertake the work by expensive process which in all probability makes the UK contractor's tender uncompetitive, though there have been exceptions to this state of affairs.

The massive increase in the price of oil in the 1970s had a profound affect on the whole world, the reverberations of which are still with us and are likely to affect events for years to come. The price of oil in 1970 was US \$1.80 a barrel. It rose to US \$14 in 1978 and peaked at US \$40 in 1980. Most of the Western world's oil came from Saudi Arabia and states on the shores of the Arabian Gulf. The vastly increased oil revenues provided the rulers of the oil-producing nations with wealth beyond their wildest dreams. Developments on a massive scale were commissioned and construction companies from many parts of the world found that their services were in great demand. Much of the work undertaken was concerned with the creation of infrastructures for countries where the standard of living had been low by Western standards. Once the demand for roads, airports, harbours, housing and the trappings of a Western life-style had been satisfied, considerable developments were undertaken to provide 'downstream' activities for the oil-producing nations against the day when the finite oil reserves would finally be exhausted.

Much of this work was completed by the mid 1980s and coincided with the near halving of the price of oil which was being obtained only three or four years previously.

The fluctuations in the price of this most basic commodity has had most profound effects world-wide and particularly on countries without indigenous fuel supplies. The South American states, Africa and the Far East, which in the past have provided many opportunities for the construction industries, have had to undergo considerable retrenchment, finding themselves in serious financial straits with their external debt increasing considerably. The spin-off arising from these circumstances has seriously affected the amount of major construction work available to contractors under direct grant or multilateral aid. This has brought about the need for a new expertise on the part of the contractor who has had to assemble finance for major schemes in addition to men, materials and machinery. This new approach to construction is now widespread and most major construction undertakings worldwide are so financed.

Another important development, which is now of growing importance, particularly in the Far East, is a move towards the privatisation of many undertakings. But there is no doubt that the construction industry will take this development in its stride, as it has done with the changes in the market-place throughout the last hundred years. This particular circumstance will bring into the forefront clients and employers whose motivation will be profit orientated to a higher degree than has sometimes been the case in the past, and the construction groupings who will be building for these new authorities will be led by contracting groups and joint ventures. Within these groups will be found, not only the construction skills needed, but also the design capabilities and financial skills necessary to complete the package deals through the arrangement of private and governmental funding.

However things many change, the construction industry will adapt to new circumstances – and do so overnight if necessary – and continue to demonstrate that construction *is* a prime export.

Ove Arup and Partners, Engineers: Hong Kong and Shanghai Bank: seawater intake tunnel

Above:
Developing the geometry for the shell roof forms
Opposite:
The Opera House on completion, c. 1973
Page 146:
Development of the roof form, c. 1957-63
Page 147:
Top left: cladding the shells. *Top right*: shell construction. *Bottom left*: positioning ridge segment. *Bottom right*: shell A2 pediments and lower segments
Page 148:
Isometric view of roof erection (*above*), the scooped out auditorium seating (*below*)
Page 149:
Glazing the shells
Page 150:
Night view (*above*), roofs under construction (*below*)
Page 151:
Interior of Opera House

PROFESSOR DEREK WALKER

The Sydney Opera House

A building that took fifteen years to emerge from its chrysalis is worth more than passing reference, especially when its physical beauty still dominates the skyline of one of the most beautiful harbours in the world and where the poetry of its conceptual form has figured on every poster extolling the virtues of Australia since the Opera House opened in 1973.

I remember attending a talk by Ove Arup when he was placed in an unacceptable, dilemma, in the dark days just prior to Utzon's resignation. Whilst maintaining an unswerving love for Utzon's concept, which had seduced politicians and public alike, he was plagued by a growing awareness that Utzon was not prepared to face the problems of combining buildability and user requirements in the extraordinary silhouette which had become the Opera House.

The politics and confusion surrounding the sad saga left Arups with many problems: the departure of a wayward genius, an escalating budget and a research and development requirement quite unprecedented for a building at that time. It was not the perfect scenario to complete the programme, especially when this was coupled with politics both fiscal and artistic which started to dominate progress. Yet at least it was reassuring to know that Arups was probably the only practice in the world who could continue on the project and bring it to a successful conclusion.

The memory I retain of Ove in crisis is the clipped, no nonsense, slightly weary delivery as he turned around and around in his hand the wooden sphere illustrating the geometric order of the designs for the shells. Ove knew instinctively that the intuitive grasp of architectural and engineering fusion sat easily in his hands – the quality was going to be fine because the old man said so! It was quite anachronistic, a re-enactment of the old virtues, the same logical, precise, professional, dispassionate concern for a good product.

Knowledge is still the engineers solace in crisis, and fortunately Ove was able to tweak the invention and research needed to achieve solutions. The legacy of Stephenson, Paxton and Brunel surfaced instantly and Arups did what they normally do when confronted with complexity and a seemingly endless series of problems – they solved them.

Sydney Opera House remains, affectionately, the built symbol of the 1960s.

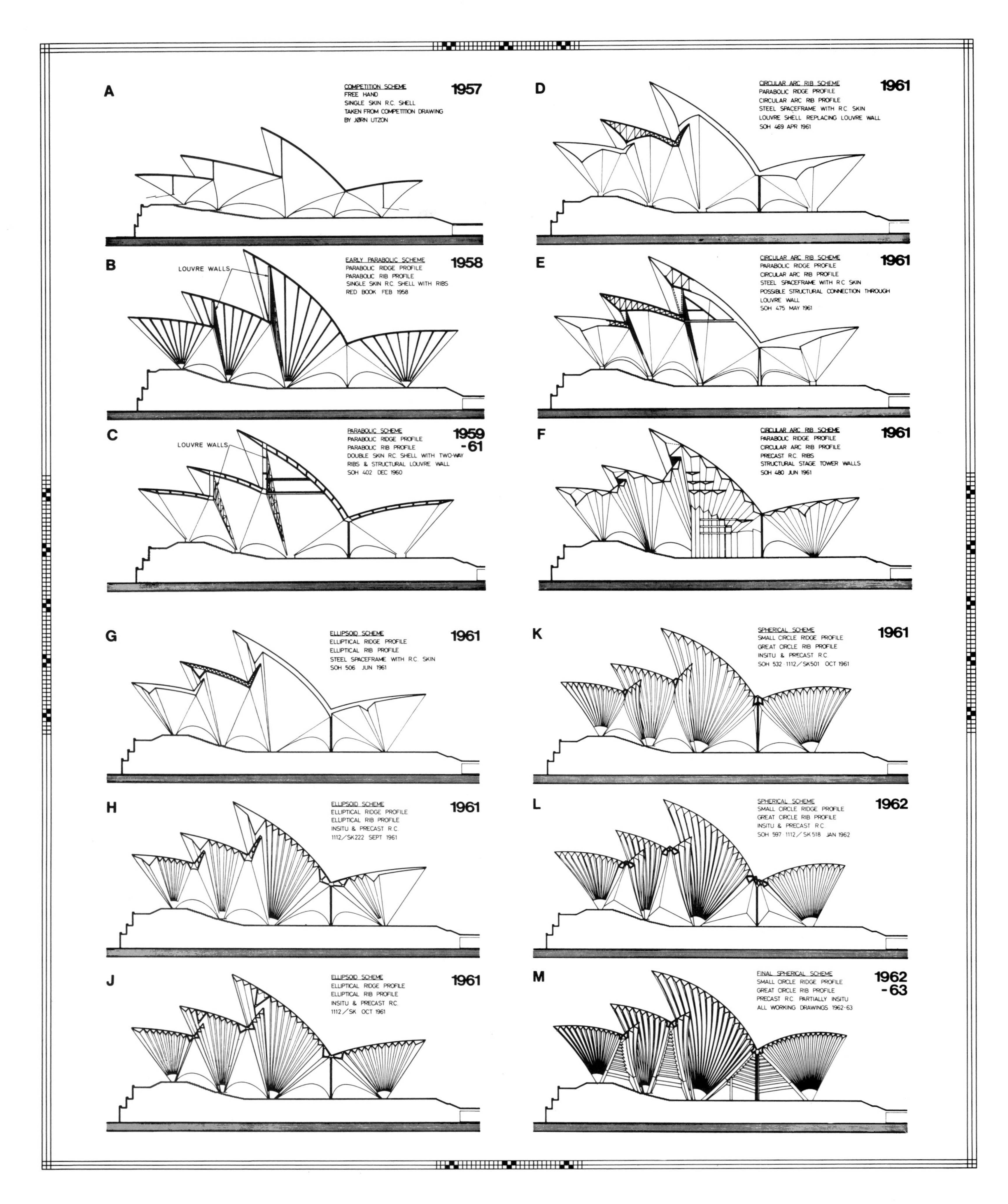
A
COMPETITION SCHEME
FREE HAND
SINGLE SKIN R.C. SHELL
TAKEN FROM COMPETITION DRAWING
BY JØRN UTZON
1957
B
LOUVRE WALLS
EARLY PARABOLIC SCHEME
PARABOLIC RIDGE PROFILE
PARABOLIC RIB PROFILE
SINGLE SKIN R.C. SHELL WITH RIBS
RED BOOK FEB 1958
1958
C
LOUVRE WALLS
PARABOLIC SCHEME
PARABOLIC RIDGE PROFILE
PARABOLIC RIB PROFILE
DOUBLE SKIN R.C. SHELL WITH TWO-WAY
RIBS & STRUCTURAL LOUVRE WALL
SOH 402 DEC 1960
1959
-61
D
CIRCULAR ARC RIB SCHEME
PARABOLIC RIDGE PROFILE
CIRCULAR ARC RIB PROFILE
STEEL SPACEFRAME WITH R.C. SKIN
LOUVRE SHELL REPLACING LOUVRE WALL
SOH 469 APR 1961
1961
E
CIRCULAR ARC RIB SCHEME
PARABOLIC RIDGE PROFILE
CIRCULAR ARC RIB PROFILE
STEEL SPACEFRAME WITH R.C. SKIN
POSSIBLE STRUCTURAL CONNECTION THROUGH
LOUVRE WALL
SOH 475 MAY 1961
1961
F
CIRCULAR ARC RIB SCHEME
PARABOLIC RIDGE PROFILE
CIRCULAR ARC RIB PROFILE
PRECAST R.C. RIBS
STRUCTURAL STAGE TOWER WALLS
SOH 480 JUN 1961
1961
G
ELLIPSOID SCHEME
ELLIPTICAL RIDGE PROFILE
ELLIPTICAL RIB PROFILE
STEEL SPACEFRAME WITH R.C. SKIN
SOH 506 JUN 1961
1961
H
ELLIPSOID SCHEME
ELLIPTICAL RIDGE PROFILE
ELLIPTICAL RIB PROFILE
INSITU & PRECAST R.C.
1112/SK222 SEPT 1961
1961
J
ELLIPSOID SCHEME
ELLIPTICAL RIDGE PROFILE
ELLIPTICAL RIB PROFILE
INSITU & PRECAST R.C.
1112/SK OCT 1961
1961
K
SPHERICAL SCHEME
SMALL CIRCLE RIDGE PROFILE
GREAT CIRCLE RIB PROFILE
INSITU & PRECAST R.C.
SOH 532 1112/SK501 OCT 1961
1961
L
SPHERICAL SCHEME
SMALL CIRCLE RIDGE PROFILE
GREAT CIRCLE RIB PROFILE
INSITU & PRECAST R.C.
SOH 597 1112/SK 518 JAN 1962
1962
M
FINAL SPHERICAL SCHEME
SMALL CIRCLE RIDGE PROFILE
GREAT CIRCLE RIB PROFILE
PRECAST R.C. PARTIALLY INSITU
ALL WORKING DRAWINGS 1962-63
1962
-63

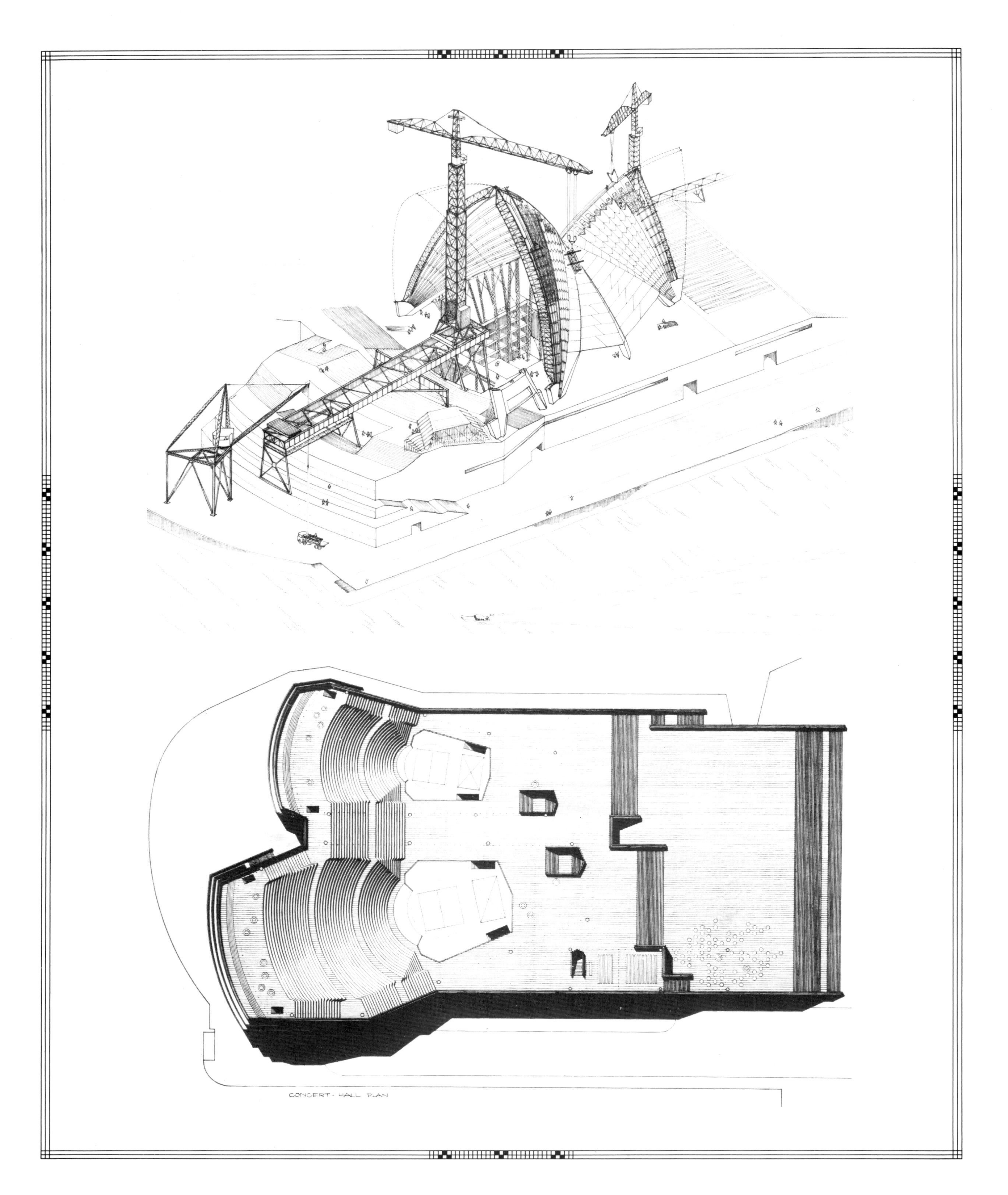
CONCERT-HALL PLAN

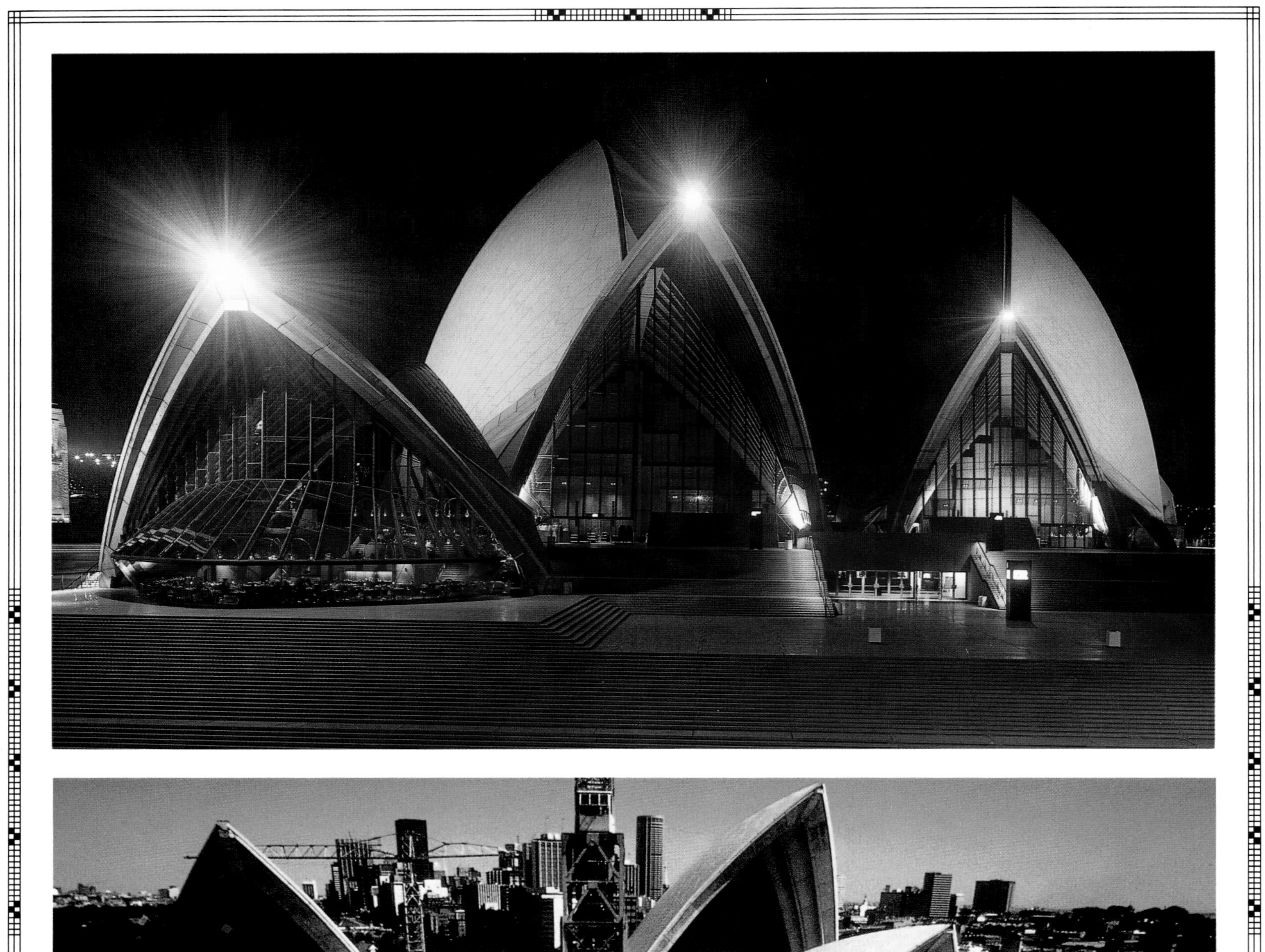

HORNIBROOK

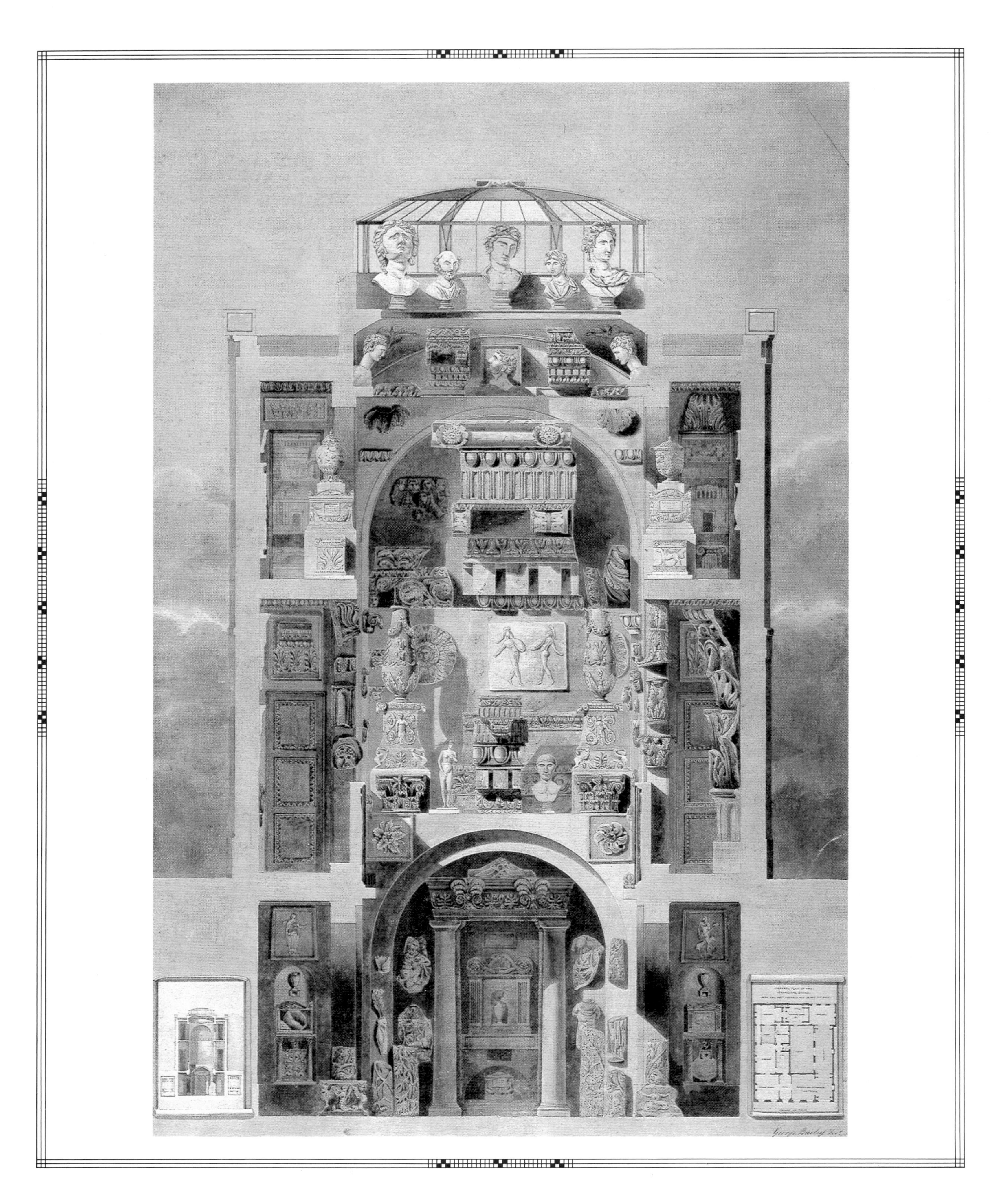

JAMES GOWAN

The Engineering of Architecture

Above:
Top: Michael Graves, Architect; Sunar Showroom, London, c. 1986. *Bottom*: Freeman Fox, Engineer; Ralph Tubbs, Architect; Dome of discovery, festival of Britain, c. 1951
Opposite:
Sir John Soane, Architect; 13 Lincoln's Inn Fields, London, 'The dome in 1810' (drawn by George Bailey)

> But the magnitude of that achievement would soon have been eclipsed and forgotten did not every detail remind us of it, by reflecting Brunel's infallible eye for proportion and his sense of grandeur. His exquisite sketches of the architectural detail of tunnel mouth, bridge, or viaduct, of pediment or balustrade remain to reveal, not Brunel the engineer but Brunel the artist at work. *Isambard Kingdom Brunel.*[1]

If a scrupulous distinction were made between architecture and engineering, it would be that one is concerned primarily with art and the other, utility. When one activity invades the territory of the other, it does so at considerable risk. In the catalogue of the current Soane exhibition at the Dulwich Gallery, Michael Graves tells us that:

> My long vault of the Sunar showroom from side to side is nothing like Soane's. My vaults become more Roman than his, by virtue of going from side wall to side wall. He didn't do that. He made an attic building by these wonderful tiled vaults. They are so strange – strange in a wonderful sense. You hear all kinds of stories about whether they're structural or not, or phoney. In a way it doesn't matter but one of the features of the Dulwich vaults is that one hasn't seen them before, they are inventive and they don't seem quite right. Soane makes us look again at that Roman vaulting system which he employs at other places as well.[2]

This commentary is illustrated with an interior photograph which features flat-arched beams and clerestoreys topped with a low-slung lantern and much glare. The effect could be said to be the contrary of what Soane stage-managed. Soane directs one's attention downwards by an orchestration of shadows and bright planes. In Graves' interior the lantern holds the spectator's gaze and the notion that this construction is more Roman than Soane's is only fleetingly entertained, for Graves' arrangements display the semantics of an attachment to a rigid frame.

Much of the architecture of Sir John Soane has about it a terrible simplicity and, in contradistinction, overlays of considerable refinement. The great recurring notion derives from Rome, the pantheon and the hypogeum: both prototypes of the most splendid and awesome kind. The paradox is that Soane made them work for most day-to-day activities and invevitably the chief component of his architecture was the arch. This obsession resulted in a

1. L.T.C Rolt, *Isambard Kingdom Brunel*, Longman, p. 141
2. Giles Waterfield, *Soane and After*, Lavenham Press, p. 92-3.

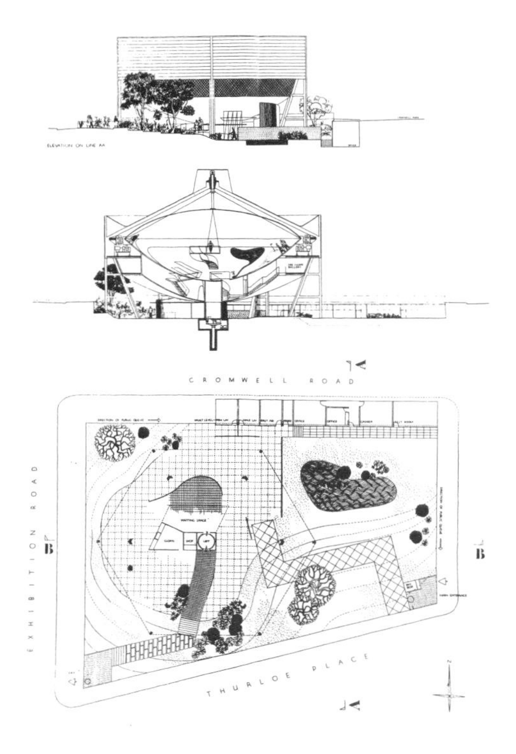

Colonel Hazelhurst, Consulting Engineer; Felix J. Samuely, Structural Engineer; Powell and Moya, Architects; Newton-Einstein House Project, 1950, elevation, section and plan (drawings by James Gowan). An experimental building project to demonstrate to the public the various effects of the Coriolis force

display of great serial invention, not unlike Mies and his excursions with the Chicago frame. Syntactically, the two architects could be said to be both simple and complex. In the event, both proved difficult to follow.

Three years after the end of the 1793-1815 war, Soane was asked by the Surveyor-General to advise on the design of the new Parliament-funded churches. His response, a model of clarity and common sense, ran thus:

> That the interior of the churches to be within the compass of an ordinary voice should not exceed in length ninety feet and in breadth seventy, that the square and parallelogram are the most economical forms. That the structure as respects the walls should be of brick, and no greater quantity of stone used than is required to assist their construction, or to render the exterior characteristic, and for the requisite pavements.

A hint of incompatibility between architecture and engineering is latent in the comment:

> That the gallery in small churches be sustained by iron pillars, but in those of large size their supports should be partly of stone and continued to the roof, and should it be objected that the use of iron alone has not sufficient character and appearance of stability, it may be enclosed in the manner best adapted to prevent obstruction.[3]

But sometimes the architecture of Soane belies the straightforwardness of the text. A model of the interior of an unbuilt church[4] places an arcosolian vault overhead, instead of underground, and the clerestorey carries a corbelled arcade of a type that might be expected around a courtyard, shock tactics that one had come to associate with Le Corbusier; the house perched on *piloti*, the garden on the roof. In the Council Chamber for the Freemasons[5] each wide flat arch is carried, not on a solid wall or abutment, but on two decorative pilasters flanking a sash window with a fireplace underneath. This conjunction is ambiguous; amusing and disquieting, adroit and perverse.

When he described the Lloyd's Building at a recent and lively RCA talk, Richard Rogers was at pains to correct misinterpretation. 'The outcome was not a matter of novelty. Its assembly was sustained by a good deal of sense and more than that, a link with the past and a precedent .... the mediaeval cathedral.' He was referring to the structural exposure of this style of architecture, and the proposition he was making appeared to be that the new and the old were kindred and true in their nakedness. It is a point of view which the eye finds difficult to accept and is a convenient historical analogy, if one is prepared to jump over twenty volumes of Ruskin. To the sage, architecture was structure enhanced by decoration and hand craft, hardly an engagement that Lloyd's can claim to be concerned with, even peripherally. Indeed, Rogers is on record as saying that his early experience with on-site improvisation, a newspaper damp-course, turned his attention and allegiance to component construction.

Lloyd's and Beaubourg have their roots in the 1950s at the Architectural Association – Archigram notions of fun and fairgrounds and their stylish

3. Arthur T. Bolton, *The Works of Sir John Soane*, St. Lukes Printing Works, p.91.
4. *ibid.* p.90.
5. *ibid.* p.116.

structures ... good looks, economy, and speedy prefabrication. One did not make the link then that these pleasure-buildings were not peasant art but the work of versatile engineers who kept a low profile. Not so long ago, British TV took Cedric Price and Archigram incorporated on a 'Tutti-Frutti' trip to Paris where they had a light-hearted time pinpointing the bits of Beaubourg that they had put on paper in times past. I remember having an extended argument with Charles Jencks about morality in art and he held the opinion that it did not exist. On authorship, it was simply a matter of who got there first and he was probably right.

Mark Swenarton wrote recently of a symposium[6] staged by the fundamentalists Krier, Terry, Adam *et al* and observed that the key words were 'authority', 'discipline' and 'tradition'. Apparently, structure was referred to by Quinlan Terry when he advised against cavity walls in favour of the stout and solid. It is good advice, up to a point, as solid walls have a nice simplicity about their make-up. Asking as little as possible from the British building industry is a sensible strategy. Walter Segal, part architect, part engineer, had the measure of that with his light, ephemeral structures perched on the surface, not within: the good earth. What the fundamentalists do not face up to is that, operationally, classicism was creaking very badly at the start of this century. Stone walls laced with steel and cramps ran counter to the ground rules of masonry construction and massive walls encroached upon the floor space esteemed by developers.

But there have been periods when architecture and engineering have been disposed in sweeter accord. The stadium at Wembley, the Regent's Park pool for penguins and Boots Nottingham Factory are stylistically as adroit as any of the buildings of the new white architecture. The mention of style blurs any discussion on engineering, but it exists not only to confuse. Brunel could not have bettered the graceful lean mechanics of the Clifton Bridge, yet had trouble with the pylons whose bulk had to be given a form and, surprisingly, he chose to make them Egyptian in appearance; thus making them appear lighter and more graceful. Art is often concerned with deceit and one presumes that mathematics is not. Anyway, that was how artists of the 1930s and those at the Bauhaus, in particular, saw engineering – as rational, enviable and objective. Speaking of this period Gropius says that:

> the object of the Bauhaus was not to propagate a style, system, dogma, formula or vogue, but simply to exert a revitalising influence upon design. We did not base our teaching on any preconceived ideas of form, but sought the vital spark of life behind life's ever changing forms.[7]

Even so, when Leslie Martin arranged his book[8] on the furnishings of the English flat, with Sadie Speight, it was principally about style and a monolithic, brave-new one at that.

The engineer of this period, or perhaps I should say the architects' engineer, was Felix Samuely. Small, not impressive in manner, a little irascible and none too dexterous with the English language, he had taught a generation of young architects at the Architectural Association very well

Above:
Top: Ove Arup and Partners, Engineers; Piano and Rogers, Architects; Centre Georges Pompidou, Paris, view of the piazza, c. 1976. *Centre*: Ove Arup and Partners, Engineers; Richard Rogers Partnership, Architects; Lloyd's Building, general view, c. 1987. *Bottom*: Felix Samuely and Partners, Engineers; Cedric Price, Architect; Joan Littlewood's 'Fun Palace', c. 1961

6. *Building Design*, Morgan Grampian Press, Mar. 1987.
7. Walter Gropius, *The New Architecture and the Bauhaus,* Faber and Faber, p.24.
8. J.L. Martin and S. Speight, *The Flat Book*, William Heinemann.

indeed. In the year of 1950, the talents of Philip Powell and Hidalgo Moya were joined to win the competition for the Skylon, a cigar-like structure held up by wires and steel pylons, not Egyptian in style, indeed severely purposeful. The design started off as a horizontal slim balloon, filled with helium and held to the earth by two wires running parallel, one from each tip. Samuely was impressive in action, particular and pernickety; if a beam needed fifty millimetre bearing, then that was it. Konrad Wachsmann distinguishes between refinement in engineering and pragmatism with a comment he made in his book on the Crystal Palace:

> However, Paxton had achieved a better adaptation of the statical loads, for the thickening and tapering of the members he used correspond more closely with the forces acting than do Eiffel's parallel lattice girders, with their two-dimensional components.[9]

Technical experimentation is more of a necessity than a fad. A miracle was needed to rearrange the shambles of World War II, and science and the new men described by C.P. Snow seemed to offer a dynamic salvation. Apparently, World War I had similar consequences. Maurice Casteels comments upon these in his splendidly illustrated survey of *The New Style* printed in 1931.

> The War it was that brought into being these problems that made a change essential. The War altered the whole situation, psychologically, as we all know and from the practical point of view as well ... a more sober architecture arose, an architecture of straight lines whose keynote was utility, an architecture that excluded imitations of past styles.[10]

In 1950, by book and broadcast, J.B. Priestley had given warmth and optimism to the cold calculations of socialism and, on the island site in front of the Victoria and Albert, the Science Museum was sponsoring an experimental building to demonstrate the effects of the Coriolis force. The engineer was Colonel Hazelhurst who specialised in fairground structures and the trials took place in a large screened roundabout, screened off for the purpose.

The Festival itself was dominated and enlivened by engineering gestures; the Skylon, Hungerford Bridge and the Dome of Discovery – Ralph Tubbs' masterwork with Freeman Fox. Leonard Manasseh won the open competiton for the restaurant with the novel idea of using an off-the-peg agricultural Dutch barn; prefabrication and romanticism – the best of both worlds, it seemed. This mechanical preoccupation and its general application to a variety of uses continued into the school building programme which started, with great promise and elegance, in rural Hertfordshire at Cheshunt. These initiatives extended into Powell and Moya's housing at Pimlico, then rising from the ground in bright yellow brickwork. Heated by surplus energy from Battersea Power Station, the enterprise and its engineering lay unseen in a tunnel below the river and, if less dramatic than that of the Brunels' at Blackwall, it was very much a part of the spirit of the time.

Felix Samuely and Partners, Engineers; Powell and Moya, Architects; 'The Skylon', Festival of Britain, c. 1951
Opposite:
Top: I. K. Brunel, Clifton Suspension Bridge, c. 1864. *Bottom*: Ove Arup and Partners, Engineers; Tecton, Architects; Penguin Pool, London Zoo, c. 1934

9. Konrad Wachsmann, *The Turning Point of Building*, Reinhold p. 24.
10. Maurice Casteels, *The New Style*, B.T. Batsford, pp. 20-21.

FRANK NEWBY AND DAVID COTTAM

The Engineer as Architect – Sir Owen Williams

Above:
Owen Williams, Engineer
Opposite:
Top left: Boots Wet Goods Factory, c. 1930-32. *Bottom left*: Boots factory interior. *Top right*: Daily Express, Manchester, c. 1935-39. *Bottom right*: Daily Express, London, c. 1929-31

There is little doubt that the schism between structural engineering and architecture became most visibly apparent during the 19th century when the roles of architect and engineer became clearly defined; the former concentrating on the stylistic issues of building with the latter developing new structural technologies and applying them to the design of ambitious building and engineering structures. By the turn of the century, many of the structural technologies developed by engineers for 'engineering structures' began to make significant inroads into architectural design itself, with the introduction of steel and concrete frame systems to support floors, roofs and external walls traditionally designed by the architect. Certainly this inevitable development began to place important question marks over the architect's precise role as a building designer. Many observers began to argue that the engineer would ultimately take over the architect's role, or that the increasing use of modern techniques would be the catalyst for the long-awaited reunification of the two disciplines.

If ever there was a time when either of these two possibilities could transpire in Britain it was during the inter-war period when functionalism as a design theory came of age. Proponents of functionalism argued that traditional forms of architecture were completely alien to the modern world. Modern architecture, it was claimed, could only be achieved when designers abandoned their stylistic approach to design and started to design their buildings with a scientific bias, creating new forms by frankly accepting the new realities of modern structural techniques. The age of functionalism in British architecture would therefore seem to have presented engineers with a great opportunity in architecture. Few took up the challenge. Instead the new architecture was developed by enthusiastic architects who became increasingly dependent on the services of structural engineers. One engineer who met the challenge was Sir Owen Williams. An unbiased assessment of his work will therefore prove useful in examining the concept of 'The Engineer as Architect'.

Williams' engineering career began following conventional lines with an apprenticeship to the Metropolitan Tramway Company and his part-time study for a degree in civil engineering at London University. Owen Williams first started work with the British office of the Indented Bar and Concrete Engineering Company in 1911. Architects and clients went directly to these reinforcement supply firms for their designs and estimated costs. Structural

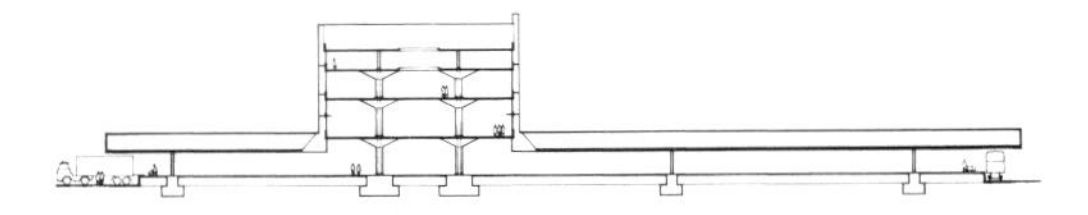

Above:
Top: Boots Dry Goods Factory, c. 1935-38, section showing concrete hangers. *Bottom*: Boots Wet Goods Factory, Packing Hall

theory was well established and British Standard Specification had first been published in 1904 and updated in 1907 and 1910.

Williams moved to the British office of the Trussed Concrete Steel Company in 1912 and by 1913 was chief estimating engineer (aged only twenty-three). If his creative abilities were not stretched by this experience, his work during the First World War effectively filled the gap. Between 1917 and 1919 he headed a Government project at Poole to research, design and supervise the construction of concrete ships; a project which extended his design abilities by forcing him to design intricate concrete forms using minimum quantities of materials.

His first major project came in 1921 when he was appointed consulting engineer to the architect Maxwell Ayrton of John Simpson and Partners, for the design of the principle buildings comprising the British Empire Exhibition at Wembley which included a permanent national sports stadium. It was at this exhibiton that concrete made its officially recognised *début* in Britain. The architectural forms produced at Wembley, however, were highly traditional in character, each building adopting variations on a neo-classical theme. Williams' principal role in detailing the architect-designed facades was to coax *in situ* and precast concrete into forms which resembled traditional masonry techniques. Despite the clear failure of Wembley to capitalise on the architectural potential of concrete, the buildings were well received in the architectural press and Williams' contribution earned him a knightood at the early age of thirty-four. He became convinced that if concrete was to develop its own distinctive architecture it was necessary for him to continue collaboration with architects and build upon the lead established at Wembley. In an article published in 1924 under the title 'Concrete as a Partnership of Engineering and Architecture', Williams explained the reasons why mutual collaboration was essential:

> The engineer and architect have a long road to travel before their separate roles can be played by one man. Till that end is achieved the fullest expression in concrete cannot be attained. But the goal may be reached more quickly by sympathetic co-operation on both sides. The engineer must realise that sound architecture is only sound engineering and the architect must believe that sound engineering is the only sound architecture. Beauty of design must not be considered the sole property of the architect, nor must the engineer assume exclusive possession of the theories of stability. The eye of the architect may often be a more truthful guide than the slide rule of the engineer. On the other hand, the theories of the latter may achieve something more perfect than the architect can, because the engineer is in closer touch with the demands of the material.[2]

By the end of the 1920s Williams completely abandoned this approach to architectural design, believing that only the engineer was capable of producing modern forms of concrete architecture; clearly his co-operation with Ayrton did not prove as successful as he had hoped. One has only to look at the designs of the bridges on which they worked to see how far they diverged from the functional solution. The Findhorn Bridge in Scotland in 1924, their first, is made up from Vierendeel trusses – not the most economic solution. The faceted concrete, however, became a familiar part of Williams' later designs. Their second bridge over the River Spey has a simple arched form while their third bridge at Wansford, again in 1925, is a combination of mass concrete arch and faceted parapet. Generally speaking, in those bridges which possessed arched forms the functional engineering solution was left to express itself, while flat deck bridges tended to be overstructured with applied decoration. The acid test of the effectiveness of their collaboration, however, was the work on those building design projects on which they were jointly engaged and were given the opportunity to develop a form of concrete architecture. In the two buildings they designed between 1928 and 1930, a farm and a warehouse, the architecture was surprisingly more reactionary than their earlier work at Wembley. In both these buildings a clear distinction was made between the architect-designed traditional brickwork facades and the engineer's simple frameworks they concealed. These projects clearly indicated Ayrton's unwillingness to address the problem of using the structural characteristics of reinforced concrete as determinants of architectural form. There is little doubt that the complete failure of these buildings to approach Williams' objective of a modern form of concrete architecture provoked him to abandon collaboration with architects. What is interesting is that Williams' talents were not used by others and that his decision to register as an architect in 1929 appears to have directly resulted from his experience of working with Maxwell Ayrton.

Above:
Top: Empire Pool, Wembley, London, c. 1933-34. *Bottom*: Dollis Hill Synagogue, London, c. 1936-38

However, by the end of 1929 he had already acquired two important commissions – the Dorchester Hotel and the London Daily Express Building. These were accompanied by a series of articles in which he outlined the philosophy he was to apply to his own work as an architect. On the specific subject of architect/engineer collaboration perhaps his feelings are best summed up in a reply he gave to a lecture Ayrton delivered to the Royal Institute of British Architects on the subject of 'Modern Bridges'. Replying to Ayrton's recommendation that architects and engineers collaborate Williams said:

> I do not believe an architect as an architect can collaborate with an engineer as an engineer . . . You have the opposition of two philosophic ideas . . . you can either maintain practicality, carry it to the extremist point. With a philosophical basis you will in this way produce the finest form of art, that is to say art is the capacity to do a job, having regard to every condition. Practicality is a method of achieving the effect without making the effect a method of achieving itself. On the other hand you have the doctrine that by effect, conscious effect, you can deliberately achieve beauty. To my mind this is very similar to a man who sets up in life and says 'I shall be a very beautiful character' and you say to him 'Be honest first and if you are

Above:
Maxwell Ayrton, Consultant Architect; (*top*) Spey Bridge, Newtonmore, Scotland, c. 1924-26, (*bottom*) Findhorn Bridge, Tomatin, Scotland, c. 1924-26

honest you will be beautiful, but do not attempt to be beautiful and dishonest'. And if you think of architecture and engineering one trying to be practical and the other trying to say 'We have a God-given mission to be effective', these two things are actually opposing doctrines which cannot collaborate.

As mentioned earlier, Williams' first project as engineer/architect was the Dorchester Hotel. He envisaged the cellular bedrooms sitting on deep concrete flat slabs and supported on large column heads at wide centres to cope with planning requirements at ground level. The scale of structure was impressive.

For an engineer to be given perhaps the most prestigious building design project in Britain at that time was a spectacular achievement, particularly as Williams' expressed intention was to use reinforced concrete in a modern functionalist manner. The national press seized on the story. Headlines such as 'ENGINEER INSTEAD OF ARCHITECT' and 'UTILITY IN NEW BUILDINGS – THE ENGINEER AND ARCHITECT – WHO WILL BE MASTER?' abounded, with correspondents claiming that the appointment of Williams for the design of the building represented a direct challenge to the architectural profession and to traditional building forms.

However, after construction work began, Williams resigned because his client, Malcolm McAlpine, had asked him to restrain his Modernist tendencies in the design and allow an architect to help him with the decoration. Few changes occurred and the scheme as built is substantially in accordance with Williams' design both in its planning and structure. To Williams the entire episode demonstrated the superficial nature of contemporary architectural design.

At this time he was commissioned for the first of what was to become a series of newspaper offices, where the major problem was designing long span floors subjected to large loads. The first in 1921 was for the Daily Express Building in London in which he used concrete frames at twenty-four foot centres to span over fifty feet and to cantilever a further thirteen feet on either side. Although an architect had originally been involved, it was perhaps Williams who suggested the black vitrolite glass cladding which was highly acclaimed.

His first major architectural success which seemed to endorse the validity of his philosophy was the Boots Wets Factory at Beeston, Nottingham. It was here in 1930 that Williams created what many consider to be one of Britain's earliest examples of Modern architecture. At the time of its completion in 1932, it was cited by many people within the architectural press as a prophecy of the type of architecture that would become increasingly dominant as designers returned to the sanity of 'science, reason and order'.

The most important aspect of this building is the complete subservience of almost every feature of its design to the efficiency of the structural layout. The entire building was conceived as a simple four-storey reinforced concrete flat slab structure arranged on a rigid rectangular grid layout with vast light wells carved out of it. Around the atrium spaces and extremeties, the floors cantilever out beyond the flared head capitals of the flat slab columns. Externally the floors were expressed simply on the elevations by their short projection, beyond the fully glazed walls. The roofs over the atria were spanned with pitched steel latices on top of which was cast a glass prism and concrete roof. In essence it was a concrete and glass building. Although it had clear precedents in Ford's American car factories of the early post-war years, it was perceived as being a highly advanced feature of British industrial building design. Two additional features contributed to the widespread acclaim Williams received. First was his extensive use of flat slab construction. Although this technique was quite well developed by this time, the scale was quite new. More importantly, however, was the way in which Williams took this technique of concrete construction and maximised its architectural potential. In this respect he was certainly breaking new ground in British architecture. Second, was the favourable climate of opinion, for many architects were looking for a more functionalist architecture. In this respect the Boots Building could not have been completed at a more appropriate time. The fact that Williams was an engineer made his achievement doubly significant. One commentator wrote:

> it is difficult for a trained architect to be of the true functionalist faith, his aesthetic training and temperament make it almost impossible. And thus it is hardly surprising that Britain's most outstanding functionalist building has not been designed by an architect at all, but by an engineer.

Montrose Bridge, Scotland, c. 1927-30

Williams' next major building after the Boots was the Empire Pool at Wembley. This was designed to house one of the world's largest swimming pools on part of the site of the 1924 Wembley Exhibition. The concrete frame structure that Williams designed comprised a series of three pinned frames, spanning the unprecedented distance of 236 feet, which supported a stepped glazed roof and the terraced seating for spectators. The most prominent feature of these frames are the massive concrete counterbalances which Williams exposed on the elevations as a series of projecting fins. The structural intention was that these fins, originally semicircular on elevation, would act as counterbalances to reduce the horizontal reaction at the ridge. Recent structural analysis has indicated that this couterbalance effect is minimal, but suggests that the semicircular form is better suited than the rectangular form which was the choice eventually adopted. Could it be, therefore, that Williams changed his design to produce a more economical form for the builder to construct? Or was it his aesthetic judgement?

It is not clear whether Williams admitted this difference between minimum content and cost. What is incontrovertible, however, is that in Williams' later buildings a radical shift in his design approach can be seen to have taken place. This is most noteworthy in his 1935 design for the Boots Drys Building; a structure that was originally intended to be a mirror image extension of the Wets Building. Instead of an extension Williams designed a completely

BOAC Maintenance Headquarters, London Airport, 1950-55

independent structure, the design concept of which differed markedly from the earlier scheme. The most superficial comparison between the two highlights these differences. Whereas the Wets Building appears as a large homogeneous glass box, the Drys Factory adopts an asymmetrical massed form with a multi-storey spine structure flanked to each side by different sized single-storey elements. Despite the excellence of the structural design at the Boots Dry Building, however, elevationally it appears rather conventional. Indeed, it is the type of building that could easily have been designed on conventional lines – the architect designing the architectural form and the engineer producing a subservient structural system to accommodate it. In short, Williams appears to have adopted for this building a conventional approach to design by accepting a convenient distinction between the architectural and engineering components, an approach wholly in contrast to his earlier pronouncements.

Similar observations can be made about many of Williams' other buildings of the same period, although most of these failed to live up to his own high standards for different reasons. Many of these unsuccessful schemes were either small in scale or were buildings for which he had been obliged to hide his engineering genius behind brickwork facades. A good illustration of the former is the Dollis Hill Synagogue in London. For this building Williams produced a folded plate reinforced concrete structure for both walls and roof. Although technically advanced for its time, architecturally the building appears highly contrived both structurally and stylistically. As one commentator wrote:

> Sir Owen in his design for a Synagogue has let me down. Architects are bad at engineering but engineers are very good at architecture – provided always that they are not aware that it *is* architecture. Sir Owen has been consciously putting art on his Synagogue and he seems to be aware that it is art.

There seems little doubt that for small buildings of this type Williams' functionalist design principles had limited application because the scale of the problem presented no overriding structural issues. His successes were therefore mostly limited to large-scale projects where structural considerations dominated the design process, thus allowing Williams to display to maximum effect his architectural skills in reinforced concrete design. Other than the Boots Wets Factory, the schemes which stand out as his most accomplished are the series of buildings he designed for the Daily Express newspaper (1931-1939) and the BOAC hangars at London Airport (1952).

Following his Daily Express Building in London, he designed two others in Manchester and Glasgow, each sharing a similar organization and stylistic pattern. The plans are arranged with the printing presses at basement and ground-floor level, upper floors given over to office space and the topmost storeys tiered back from the building line. The reinforced concrete structure of each varies but elevationally they are all completely encased in glass with the concrete frame elements clad in black glass. Architecturally the most exciting of these buildings was the Manchester scheme. On an island site in the centre of the city, Williams produced a spectacular structure complete with a 'shop front' view into the press hall from pavement level. The flat slab structure included a series of columns and ring beams at every floor level but at each curved corner to the building vertical supports were omitted. This has in recent years caused some minor distortion to the glazing mullions at these positions but architecturally the decision was right and the building, viewed as a whole, has a refined quality which has allowed it to retain its image as a 'modern' building right up to the present day.

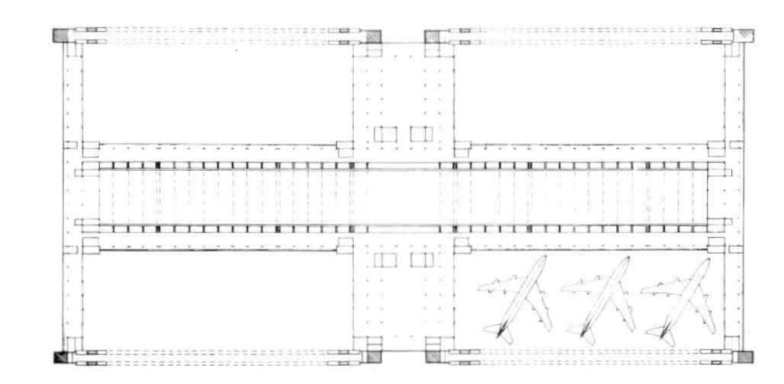

BOAC Maintenance Headquarters, plan, c. 1950-55

More typical of Williams' heavy reinforced concrete architecture is the much later BOAC Maintenance Headquarters at London Airport, built in 1955. Although little-known, this building stands alongside his early Boots Wets Factory as one of the most accomplished schemes of his career. The plan, covering one million square feet, was in essence very simple, comprising four hangar pens separated by a cross plan form which contained office accommodation and engineering workshops. The most spectacular feature of the design is the vast arched entrance frames with huge concrete counterbalances to each of the four hangar pens. These were executed *in situ* in reinforced concrete to a span of 336 feet and are reminiscent of his work on the Empire Pool.

However, his major work during 1951-59 was as a civil engineer designing Britain's post-war motorways. His bridges on the M1 are simple economic flat slabs of up to three-foot-six-inches thick, sitting on central circular columns with column heads – similar in a way to his proposed bridge over the Thames of 1932. The non-structural *in situ* concrete parapets mask the structure, investing it with a heavy ungainly appearance. Bridges for the second stage of the motorway were rendered visually more exciting by the use of inclined circular supporting columns. His viaduct over the River Ouse echoes his 1925 arched road bridge over the River Spey.

Williams no doubt gained satisfaction in returning to civil engineering and to functional structures. His pre-war practice as architect/engineer provides a fascinating study, as he was then responsible for totally designing a few significant industrial buildings. Having a flair for planning and an appreciation of the potential for long spans of reinforced concrete, he played a major role in bringing about a new dimension to concrete framed structures in the 1930s. Whether the application of a glass skin to a functional structure creates architecture is debatable.

However his early buildings are extant and there to be appreciated.

Above:
Production platform on tow to North Sea oil-fields
Opposite:
Shell Esso's Brent C platform in rough seas

DR. EDMUND C. HAMBLY

The North Sea Challenge

Imagine the effect of an hundred-foot high wave breaking over the Houses of Parliament. Now imagine how you would design a structure to resist wave after wave, when standing in 150 metres of water in the middle of the North Sea. The development of North Sea oil fields has presented engineers with one of their greatest challenges, comparable to the construction of the Pyramids in ancient Egypt and the railways in the last century.

During the last twenty years there has been an investment in the British sector of the North Sea of about £70,000 million (at 1987 values). The development of the oil and gas fields has been remarkably successful, both in financial terms and as an engineering achievement. The technical achievements have been as impressive as the developments in space technology and microelectronics.

One of the attractions of the off-shore industry is that it is international in outlook and involves teams of engineers with a wide variety of backgrounds. Each development has involved an integrated team of petroleum engineers, reservior engineers, chemical engineers, civil and structural engineers, naval architects, marine engineers, master mariners, geophysicists, oceanographers, meteorologists and so on. It is difficult to identify individual engineers who deserve special mention because of the huge scale of the projects and the numbers of people making important contributions. Many engineers have been given exceptional responsibilities in terms of financial control and technical developments, and they have thereby gained an individual sense of achievement and fulfilment.

This article concentrates pretty much excusively on the actual structures, because that is where my interests lie. However, it is important to remember that an oil field investment of £1,000 million might be distributed thus:

Topside facilities	£400 million
Oil wells	£200 million
Platform structures	£200 million
Pipelines and terminals	£200 million

The terminology can sound strange. The 'drilling rig' is the machine which turns and lowers the 'drill pipe' with the drill that bores the oil well on its end. The drilling rig is part of the 'topsides' which include all the oil and gas production equipment as well as the accommodation facilities supported above the sea. The structural floors of the topsides are the 'decks', while the pipes coming up from the oil wells and pipelines from the sea bed are called

SHELL/ESSO
BRENT-C

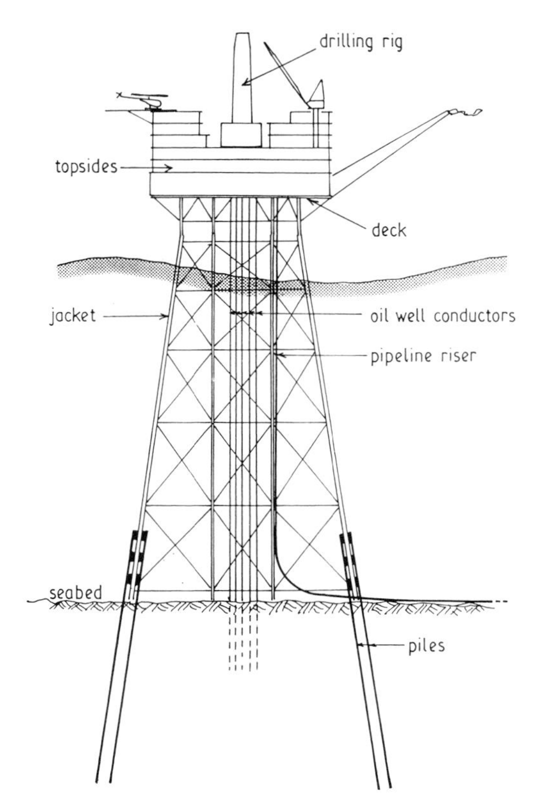

Above:
Parts of an offshore platform
Opposite:
Top:looking through the structure of British Petroleum's Magnus platform
Bottom: top sides of Mobil's Beryl B platform

'conductors' and 'risers'. The lines of safety valves at the well heads are often called 'christmas trees'. The whole installation is called the 'platform' because it provides a platform for the topsides to work on. The lower structure, on a steel platform, is usually referred to as a 'jacket', because on the early small steel platforms the decks were supported directly on piles driven into the sea bed, and the jackets enveloped the piles and held them in position. On the giant platforms in the northern North Sea the deck is supported by the jacket, which is supported at its bottom by the piles. Some platforms are referred to as 'gravity' platforms because they are held in place against the wave loads by their weight, instead of by piles. Most of the gravity platforms are made of prestressed concrete, though steel gravity platforms also exist. A 'semi-submersible' is a floating mobile structure with the deck supported on columns which go down through the waves to large buoyant hulls below the surface. A 'jack-up' is a mobile structure which consists of a hull that is raised above the waves by jacking itself up its legs which stand on the sea bed.

It is difficult to imagine the enormous scale of a North Sea platform when you are in an ordinary sized building. A platform could be 300-metres high from the sea bed and the deck could cover an area the size of a football field. The topsides may accommodate 200 men and the equipment requires a power plant big enough to supply 10,000 homes. The legs of the largest steel jackets are large enough to drop a double decker bus down inside, while the legs of a concrete platform are large enough to contain a house. The piles that pin a steel jacket to the sea bed may penetrate one hundred metres below it. The British Steel Corporation prides itself on having supplied more than 2.2 million tons of steel plate and tubing to the off-shore oil industry.

The forces that the structures have to support, or resist, are also enormous. A concrete gravity structure would weigh 6,000 MegaNewtons (MN), equivalent to 600,000 tons. It could be subjected to loads of 600MN (60,000 tons) by a thirty-metre (hundred-foot) high wave, accompanied by a load of 20MN (2,000 tons) from an hundred miles per hour wind.

The off-shore industry has provided engineers with a flood of new problems requiring innovative solutions and, as a result, there is considerable variety in the projects and their structures. Each platform can be thought of as a component of a giant machine involving oil wells, drilling rigs, separation equipment, supply boats, helicopters and so on. An innovation in any one of the components will thus affect the performance of all the others.

One of the attractions to engineers of off-shore design has been the challenge of working out the solutions for themselves without cumbersome traditions and rule books. The engineers have had to undertake special research projects and analyse problems using first principles, in order to determine the complicated loading and structural behaviour. Many of the structures in the North Sea were designed with a code of practice fractional in comparison with that for bridges, even though the off-shore structures have greater variety of form and much greater uncertainty of loading. The good

Above:
British Petroleum's Magnus platform, self floating, being submerged and up-ended on location
Opposite:
Top: pile driving on Shell/Esso's Brent A jacket. *Bottom left*: British Petroleum's Magnus Oilfield, emplacement of jacket section of production platform. *Bottom right*: two men inside the Menck steam hammer, capable of dropping a weight of 125 tons a distance of 1.8 metres onto the pile 30 times per minute

performance of the platforms is an indication of the quality of the engineering involved.

The hostility of the North Sea environment presents engineers with a major challenge in building and installing the platforms. Designing and building a stationary structure on land presents problems, but nothing like those of moving fabrications weighing tens of thousands of tons and installing them, possibly a different way up, at a very exposed location off shore. The engineers' choice of method of construction and installation strongly influences the form of the final structure and topsides. The variety of different methods has increased as larger and more versatile construction plant have become available, and there has been a leap-frogging evolution of methods.

The small steel platforms in the southern gas fields were light, simple structures which were transported to site on a barge and placed on the sea bed by crane. Alternatively they were slid off the barge into the water and then, by controlling the flow of water into flotation tanks, were turned upright and sunk to the sea bed at the right location.

As the platforms have got larger, the same techniques have been used. However, the relative merits of the two techniques have vied with each other as the cranes and launch barge methods have individually developed and improved. These rapid developments have required a close interaction between engineers building platforms and those involved in building the construction equipment.

The first giant steel jackets in the northern North Sea, on BP's Forties Field, were designed in 1972 to be installed by a purpose built submersible raft. The raft, weighing 10,000 tons, was used as a barge to support the 16,000-ton jacket during the tow to the field. The raft was then flooded with water under carefully controlled conditions, and the jacket was tilted bottom-down and set in place. The raft was then detached and used again. Other jackets, such as Shell/Esso's Brent A and BP's Magnus, were designed as self-floaters which had legs of large volume to act as the buoyant raft during installation. These jackets were towed and installed in the same manner as BP's Forties jackets, except that there was no raft to detach afterwards. Self-floaters have the advantage that they do not involve the expense of a separate raft or launch barge, but the disadvantage that the structures have to be strengthened to resist the large wave loads that act against the legs, enlarged for buoyancy on the tow.

Launching jackets from barges has also made great strides. Each stride has been accompanied by extensive research with impressive model and computer simulations. Conoco's Murchison jacket weighed over 25,000 tons when it was launched from its barge in 1979. Prior to the launch, an aluminium model was used to perfect the procedures for controlling the buoyancy tanks, setting the structure the right way up and placing it on location. When a jacket is launched it is slowly tilted over one end of the barge and then, at a critical stage, it starts to accelerate and dives into the water.

HEEREMA

2
12500

In the 1960s a crane lift of several hundred tons was considered a major achievement. By 1969 the record had reached 680 tons, and by 1974 a crane ship could lift modules of 2,000 tons. By 1981 the record had increased to over 3,000 tons and by 1987 the lifting capacity of the largest crane barges has risen to around 9,000 tons. These enormous increases in lifting capability create new possibilities for the design of jackets and topsides. A jacket can be made lighter if it does not require all the strengthening members needed to survive a launch, or the large-floating tanks needed for self-floating installation. The reduction of these structural members reduces the wave loading which again enables the main bracing and legs to be made lighter. As a result a platform can be much lighter (and less expensive) if there is a crane available that can lift it off a barge and set it on location.

The benefits of large capacity cranes to topside design are even more significant than for jacket design, because it is now possible to fabricate a deck on land as one or two huge integrated modules which can be 'hooked up' (pipework and electrical systems connected together) and precommissioned on land. In the earlier developments, the topsides had to be designed as ten or more modules of about 1,000 tons each, and a great deal of work had to be done off shore during hook-up and commissioning. This was not only extremely expensive, since hundreds of men had to be provided with temporary accommodation off shore, but caused serious delays. Furthermore, each module had to have its own structure, so that it could be lifted into place, which meant a great deal of additional steel.

The methods of pinning the jacket to the sea bed with piles have also made enormous strides, enabling still further improvements to jacket design. The early jackets in the Southern Basin were fixed with piles about 0.9 metres in diameter, which were driven down to a depth of twenty to thirty metres by a steam-driven hammer, with an energy input of about eight metre-tons per blow. By the mid 1970s, the piles on Shell/Esso's Brent A platform were 1.8 metres in diameter and were driven by a hammer delivering 220 metre-tons per blow. Each pile can support 4,300 tons. Steam hammers have the disadvantage that they must drive the piles from above the water level, so that on a platform with deep piles the shock waves from the piling hammer may have to travel downwards for more than 200 metres before they move the bottom of the pile downwards. Hydraulic hammers were developed which could work underwater and follow the pile heads down to the sea bed where the piles are attached to the jacket. On BP's Magnus platform the 2.1 metre diameter piles were driven ninety metres into the sea bed by an hydraulic hammer. The large energy inputs of modern hammers not only enable jacket designers to support larger loads on fewer piles, but they are also less likely to be halted in their work by a hard stratum at depth. As a result time is saved off shore and there is much less risk of a jacket being damaged by a severe storm before it is properly pinned to the sea bed.

When the oil companies developed the northern oil fields in the early 1970s, the engineers showed great confidence when designing jackets for the

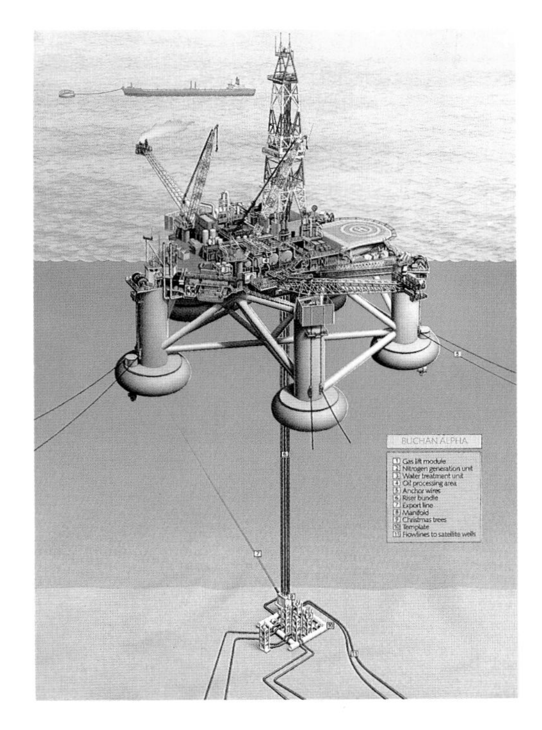

Above:
Top: British Petroleum's Buchan floating production platform. *Bottom*: Conoco's tension-legged platform, Hutton Field
Opposite:
Top: Philips Petroleum Group's Maureen Steel gravity platform is towed to the field. *Bottom*: Shell Esso's Cormorant A concrete gravity platform towed from Ardyne Point for mating with its deck

Production platform leaving on tow to the Shell Esso Brent Field in the North Sea

enormous forces from thirty metre high waves in water depths of one hundred metres and more. The maximum wave load on a jacket in one hundred metres of water is in the order of one hundred MegaNewtons (10,000 tons). About half of this load results from the effects of the wave on the structure, and the other half from its effects on the oil well conductors, pipeline risers and other equipment attached to the structure. Over the years the arrangement of the steel tubes has been progressively improved in order to reduce their interference with the waves, thus enabling the structures to be immersed in deeper and deeper water. The development has been somewhat similar to that followed by Eiffel in the last century when he perfected the shape of his railway viaducts (and the Eiffel Tower) to reduce their resistance to wind forces.

Fatigue of welded steel joints has presented off-shore engineers with a challenge uncommon in land based structures; during a service life of thirty years, a platform could be subjected to fluctuating loads from 100,000,000 waves. Such a large number of cycles of loading can cause welded steel to fatigue and so reduce the strength of the joints by tenfold or more. Fortunately most of the waves are small and cause stresses less than a tenth of those due to the maximum wave. The vulnerable points for fatigue on the off-shore platforms have been the welded joints between the steel components. All the joints are slightly different and it has taken many years to test a selection of representative joints by subjecting them to 100,000,000 or so cycles of loading. The problem has been somewhat different to that experienced in aircraft design where vulnerable points have been located within the machined components, rather than at the joints, and it is possible to test prototype components to destruction before they are used. Some off-shore platforms are now being designed with special steel castings at critical joints to reduce the problem.

Not all steel platforms are pinned to the sea bed with piles. Phillips Petrolem Group's Maureen platform, which was installed in 1983, resists the enormous sideways forces from waves with its dead weight of more than 110,000 tons and its three enormous splayed feet. The simplicity of its concept, like a giant three-legged stool, is satisfying and the advantage of three legs over four, five or more, is that the structure is always stable and does not rock on the uneven sea bed. The three enormous feet also formed flotation tanks, which made it possible to tow the platform to the North Sea location with all the topsides of the integrated deck completed and operational. During operations the three tanks provide a storage facility for the oil. The fundamental difference between the piled steel jackets and the steel gravity platform demonstrates how completely different solutions exist for design problems and how engineering benefits from lateral thinking.

Concrete gravity platforms have developed at the same time as steel jackets. The qualities of prestressed concrete, as opposed to steel, have made it ideal for heavy structures which can store large quantities of oil. The tanks are used as storage reservoirs during periods when the oil cannot be pumped to pipeline or tanker. Concrete gravity platforms had an impressive start in 1973 with the Ekofisk Tank, in the Norwegian sector, which weighs more than 230,000 tons. Today there are sixteen giant concrete gravity platforms in the North Sea. Shell/Esso's Brent Field has three concrete gravity platforms and one steel. The oil from Brent is pumped to the Cormorant A concrete gravity platform which acts as a junction for pumping oil from nine different oil fields to Sullom Voe.

Gravity platforms have the advantage over steel jackets that they can be towed out to the field with the topside facilities complete, or nearly complete. After the concrete base structure has been built, it is towed to deep in-shore water and submerged below the completed topsides, which might weigh 20,000 tons or more and are supported on barges. The concrete base is then deballasted and rises up with the topsides in place. At the critical stage, when the deck is being floated over the top, the concrete structure may be submerged 150 metres into the water with less than five metres above the surface. (The situation can be simulated on a small scale with a submerged beer bottle with only five millimetres of the neck above the water.) The concrete platform with topsides is then towed to the field and submerged again until it rests firmly on the sea bed. The advantage of gravity structures over piled steel jackets – being able to install completed topsides – has, in the last few years, been overshadowed by the increasing capacity of crane barges, since it is now possible to install the nearly-completed topsides on top of piled jackets. However, the increased lifting capacity of cranes has likewise led to greater flexibility in the installation of concrete gravity platforms.

The enormous weights of steel and concrete platforms, while being spectacular, create costly problems when working in deep water. Many engineers have invested great effort into finding different ways of producing oil; ones which do not incur such enormous supporting structure costs.

The first oil from the British sector of the North Sea was produced in 1975 from the Hamilton Brothers' Argyll Field using a floating production platform over well heads on the sea bed. The floating production platform is a converted semi-submersible drilling rig. The system has the great advantage that it does not involve investment in fixed platforms and can be moved when the oil runs out. As it happens, Argyll has been able to produce much more oil than was originally anticipated. The same concept has been adopted by others, including BP when they developed the Buchan Field which came on stream in 1981. The technical achievements of these systems include the well heads on the sea bed and the flexible riser pipes that connect the well heads to the floating platform. The risers have to resist the wave loads as well as accommodate the motions of the platform. On Buchan, the riser system contains nineteen pipes and is kept in tension by a complicated system of pulleys and shock absorbers which compensate for the motion of the platform. On the rare occasions when very bad weather is forecast, the riser pipes are pulled up into the platform, to be reconnected when the weather improves.

A tension leg being assembled within a corner column of Conoco's Hutton field tension-legged platform

THOR

A markedly different floating production platform, called a Tension Leg Platform (TLP), was installed by Conoco on the Hutton Field in 1984. The TLP has been developed by naval architects and engineers as a method for producing oil from very deep water. The platform is held in place by sixteen vertical legs which are kept in tension by the buoyancy of the platform. The tension legs prevent heave (vertical motions) of the platform and restrict surge and sway (horizontal motions). The riser pipes from the oil wells at the sea bed neither have to accommodate vertical motions of the platform nor be disconnected in severe weather, while the well heads are located on the platform instead of the sea bed. The development of the tension legs has been a major task, since they are precision-made machines as much as structural components. Each tension leg of 155 metres height is made up of fifteen elements, screwed together and connected to the corner columns of the platform by articulated joints. The corner columns also contain equipment for connecting together the elements of the tension legs and lowering them to the foundation templates, piled to the sea bed.

Sea bed production, without floating platforms, has also been developed by oil companies as an alternative method of producing oil from very deep waters. On the Central Cormorant Field, Shell/Esso installed an Underwater Manifold Centre (UMC) in which all the well heads and controls are situated in a large steel structure weighting 2,200 tons lying on the sea bed. The UMC is remote controlled from the Cormorant A platform four miles away and has been designed so that it can be maintained by using a special remote controlled, unmanned submarine.

The excitement and pride of the engineers involved in these giant projects has been infectious. On each project the frontiers of knowledge and expertise have had to be pushed forward to meet tight dead-lines. Each engineer has had to work closely and in harmony with many others in large international teams. Every aspect of the industry has moved forward.

The momentum of development has had to slow down in recent years as a result of the decrease in world energy demands in the early 1980s, followed by the fall in oil prices. The versatility of the engineers, which enabled them to respond so quickly to the challenge of the North Sea, has now been used to tackle other markets. But, before long, the urgency for new oil supplies will return again and engineers will then have to tackle the challenge of the Atlantic.

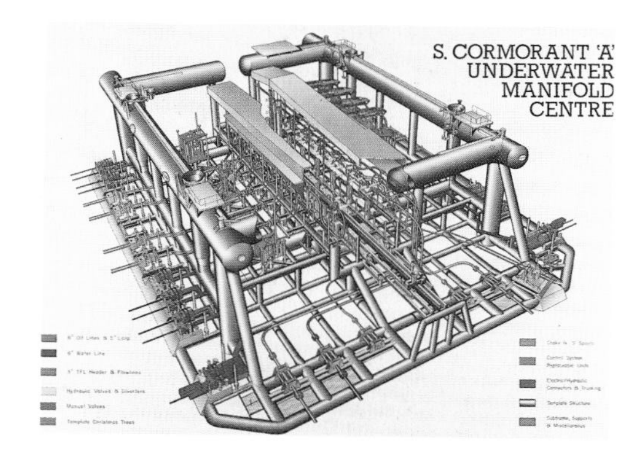

Above:
Shell Esso's underwater Manifold centre in the Cormorant Field. The UMC is 52 metres long and stands on the seabed

Opposite:
Topside module weighing 1,750 tons, being lifted into position on British Petroleum's Forties Field

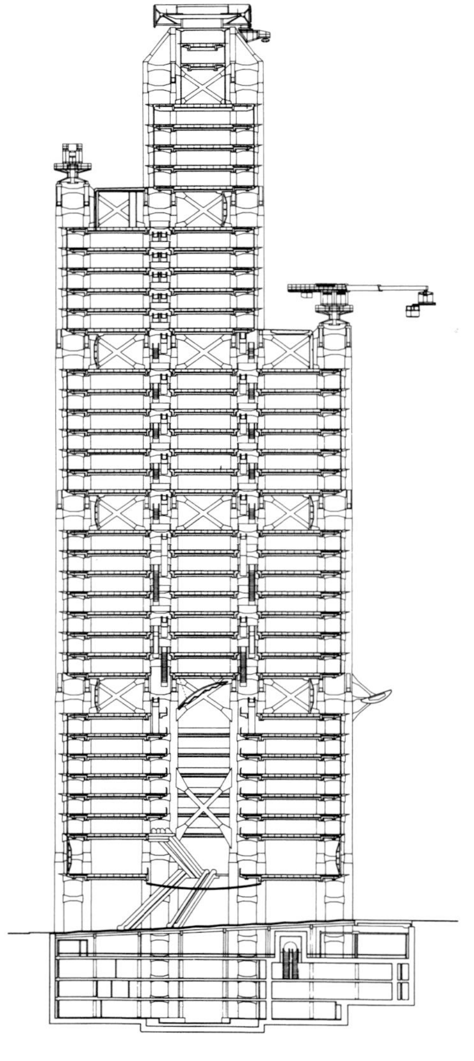

Above:
Ove Arup and Partners, Engineers; Foster Associates, Architects; the Hong Kong and Shanghai Bank, 1986, north-south section
Opposite:
The Bank under construction

PROFESSOR DEREK WALKER

Services and Structure – The Hong Kong and Shanghai Bank

The Hong Kong and Shanghai Banking Corporation opened their new headquarters building in April 1986 – one of the most talked about and technologically advanced structures in the world.

The design is tailored to meet the requirements of the bank for a headquarters building capable of fulfilling the exacting functional and technical demands of one of the world's leading financial institutions well into the next century. At the same time, it is of an architectural and engineering quality that reflects the status of the bank, and its confidence in the future of Hong Kong as the third largest financial centre in the world after London and New York.

The Hong Kong and Shanghai Banking Corporation's new headquarters building is one of the world's tallest suspension structure office towers. Since so few precedents exist for a structure of this type, it has been designed from first principles using the most advanced computer-aided techniques and with the benefit of the most exhaustive testing programme ever attempted for a building.

Standing 180 metres above ground with another four basement levels, it contains forty-seven floors in all, totalling nearly 100,000 square metres in floor area. Unlike conventional structures of this height, which depend on regularly spaced central cores, the bank's unique structural system devised by Foster Associates, architects, and Ove Arup and Partners, engineers, gives it a completely open floor plan. At ground floor level the public plaza which passes beneath the tower is interrupted by just eight steel masts that carry the entire weight of the superstructure down to bedrock and transfer windloading to the foundations.

Each mast, fabricated by British Steel in the UK and shipped piece by piece to Hong Kong for assembly, is made up of clusters of steel columns in fours, linked together by rectangular haunched beams at storey height intervals of 3.9 metres.

These links have the effect of turning the masts into a vertical Vietendeel structure of considerable stiffness. The thickness of the steel plate used for these columns, and the size of their diameter, varies in proportion to the diminishing loads placed on the structure at higher levels. At the base of the building each column has a diameter of 1.4 metres with a thickness up to one hundred millimetres, but at the top of the tower the columns are down to a

thickness of twenty-five millimetres and have a diameter of 800 millimetres.

The bank forms a rectangle in plan, approximately fifty-four by seventy metres, which is divided by two rows of four masts that have the effect of creating three bays running east to west. Vertically the structure is divided into five discrete zones by double-height suspension trusses, supported by the masts, and spanning 33.6 metres east to west with a further 10.8 metre overhang beyond the masts at either end. Every floor, made from *in situ* concrete and supported by steel secondary structures, is suspended from one of these trusses by tubular steel hangers which are connected to the central nodes of each truss and by two more rows of hangers attached to each outer node.

North-south stability for the structure is provided by two-storey deep X-shaped braces spanning between each suspension truss at either end of the building. Because of the exceptional conditions of the banking hall atrium, an extra brace three-storeys deep is provided there.

Vertical loads are carried down to bedrock, but horizontal windloads are transferred from the masts to a one metre thick concrete basement by means of the ground floor concrete slab. This is the only point at which lateral restraint is provided; the rest of the conventionally supported, concrete basement structure is independent of the steelwork.

Hong Kong's waterfront suffers from particularly difficult ground conditions. The water table is high and most of the sub-strata consists of completely decomposed granite, the behaviour of which is difficult to predict. For these reasons the basement structure is built within a one metre thick perimeter wall which extends from twenty-five to thirty-five metres down to bedrock and which has been grouted at its base to make it as watertight as possible. Within this perimeter the basement consists of reinforced concrete slabs spanning onto columns on a 7.2 by 8.2 metre grid.

Each of the four columns that make up a mast is based on a single reinforced concrete shaft which extends down and into solid granite with a maximum load-bearing stress of 5,000 KN/m^2. These foundations have been designed to take an increase in floor area of thirty percent in the building should future changes in the Hong Kong building regulations permit the set backs on the east face of the bank to be filled in.

The configuration of the superstructure masts prevents even distribution of the dead weight needed to resist ground movements caused by water table pressure. For this reason a series of permanent rock anchors have been installed in the granite to counteract any lifting tendency in the basement structure.

To meet the requirements of the bank for a structure with a minimum life-span of fifty years, the steelwork has been given a specially formulated anti-corrosion treatment. By using a polymer modified cement sand mixture, which reduces the permeability of the mix, and applying it to the steelwork by spraying it with stainless steel fibres, it was possible to achieve a layer of just twelve millimetres. Fire protection for the masts and trusses is provided by a

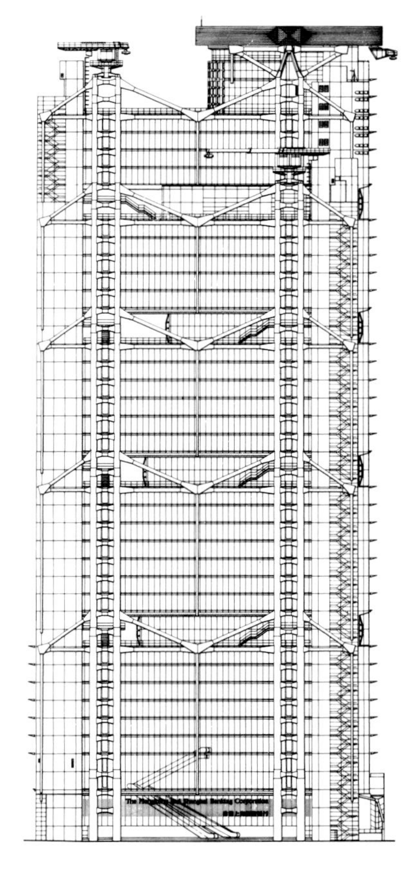

Above:
North elevation
Opposite:
View by night
Page 182:
Top of riser east elevation
Page 183:
Internal atrium view

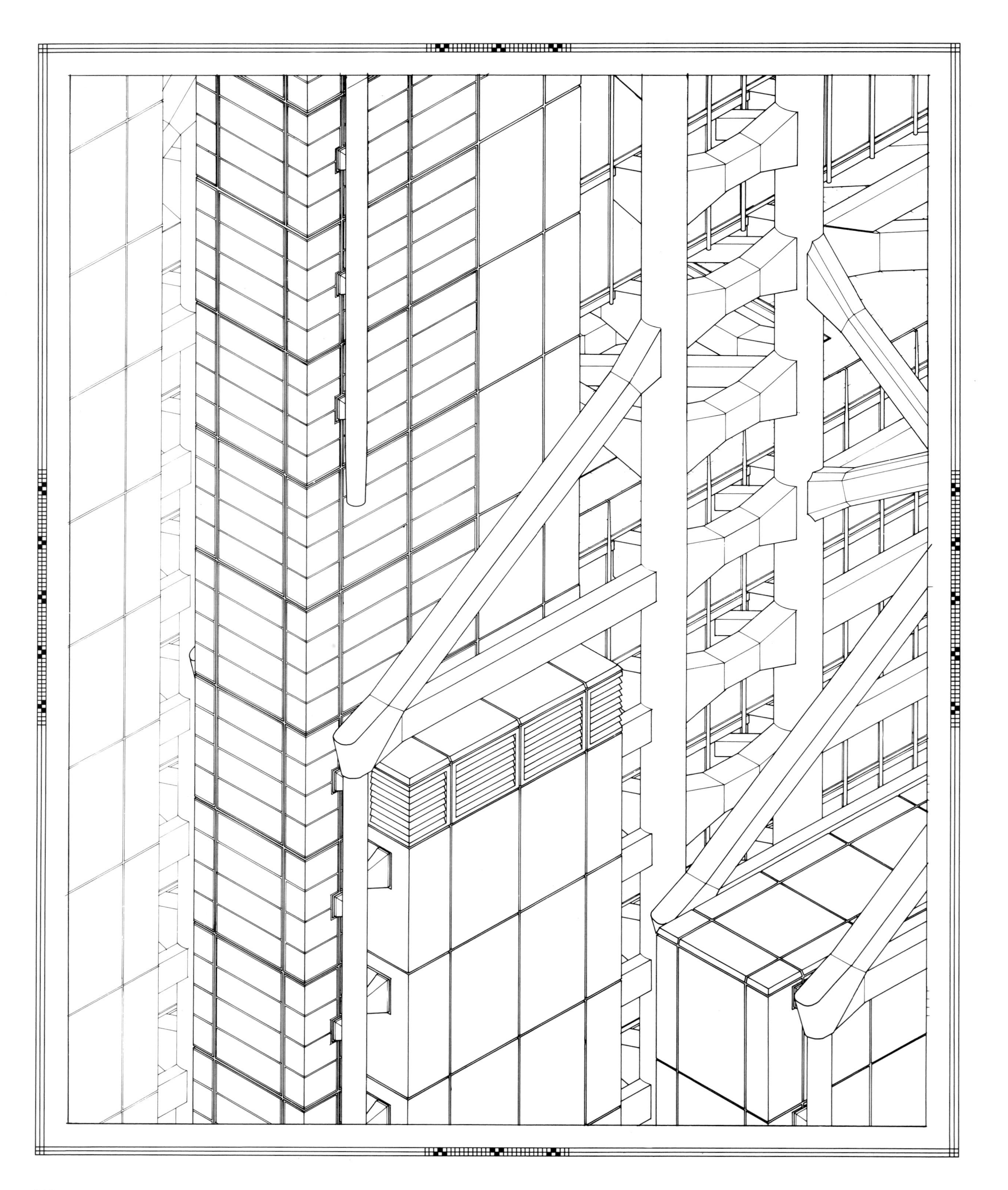

The Bank in context
Opposite:
Top: section through Statue Square. *Bottom*: isometric at mid level
Pages 186-7:
Top: looking towards Kowloon. *Bottom left*: looking towards Statue Square. *Bottom right*: the Bank vault

ceramic blanket and stainless steel mesh wrapped around each member, giving all primary and secondary steelwork a two hour fire rating.

Calculations for the stability of the main structure were reinforced by the testing to destruction of key steelwork components on a test rig in England.

Windloading in an area such as Hong Kong which regularly experiences typhoons was of particular importance to the structural design. It was necessary to obtain exact information on wind behaviour to predict the size and nature of every conceivable combination of windloads. For this purpose Professor Allan Davenport of the University of Western Ontario carried out a series of wind tunnel tests using a 1:500 proximity model including all structures present and projected around the new building, and another at 1:2500 to analyse wind conditions throughout the territory. It was the most exhaustive set of wind tunnel tests ever attempted for building. Based on this data the structure has been calculated to deflect to a maximum of 300 millimetres under the statutory equivalent static windload.

The building is fully geared up to face the challenge of rapid technological innovation within the banking industry thanks to the incorporation of two unique features – a floor-based distribution system for all electrical, telecommunication and air-conditioning services, and a prefabricated system of modular risers which house all vertical services distribution networks. Though other buildings do exist with raised floor services systems, none has one on such a complete scale as the Hong Kong Bank.

A floor stystem provides easy maintenance access, and allows for rapid changes in the configuration of services layouts as required. The flooring system developed for the bank uses a light-weight, aluminium honeycomb panel of a construction similar to that used in aircraft. Each panel is 1.2 by 1.2 metres and can contain either air-conditioning or electrical services outlets. Both types of outlet use a die-cast aluminium circular unit, positioned in one corner of the panel. There are three standard panels which can be rotated to position the outlet wherever it is required. Each floor panel is fitted with an expansion joint and an edge-wiper blade which facilitates movement and also provides an acoustic and air seal. The floor tiles can be adapted to take either carpet, rubber or marble finish, depending on which part of the bank they are used in.

Beneath the die-cast outlet is a segmented junction box, containing connections to power, telecommunications and telephone links which are distributed in under-floor trunking. There is spare capacity within the junction boxes to take further cable service connections should the need arise.

Vertical services runs are concentrated to the east and west sides of the building, feeding into prefabricated services modules. These contain all air handling plant and toilets for each floor, as well as certain modules which house diesel generators to provide a standby power source. A total of 139 modules down to soap dishes and taps were prefabricated and fitted out in Japan, before being shipped to Hong Kong for installation. The modules

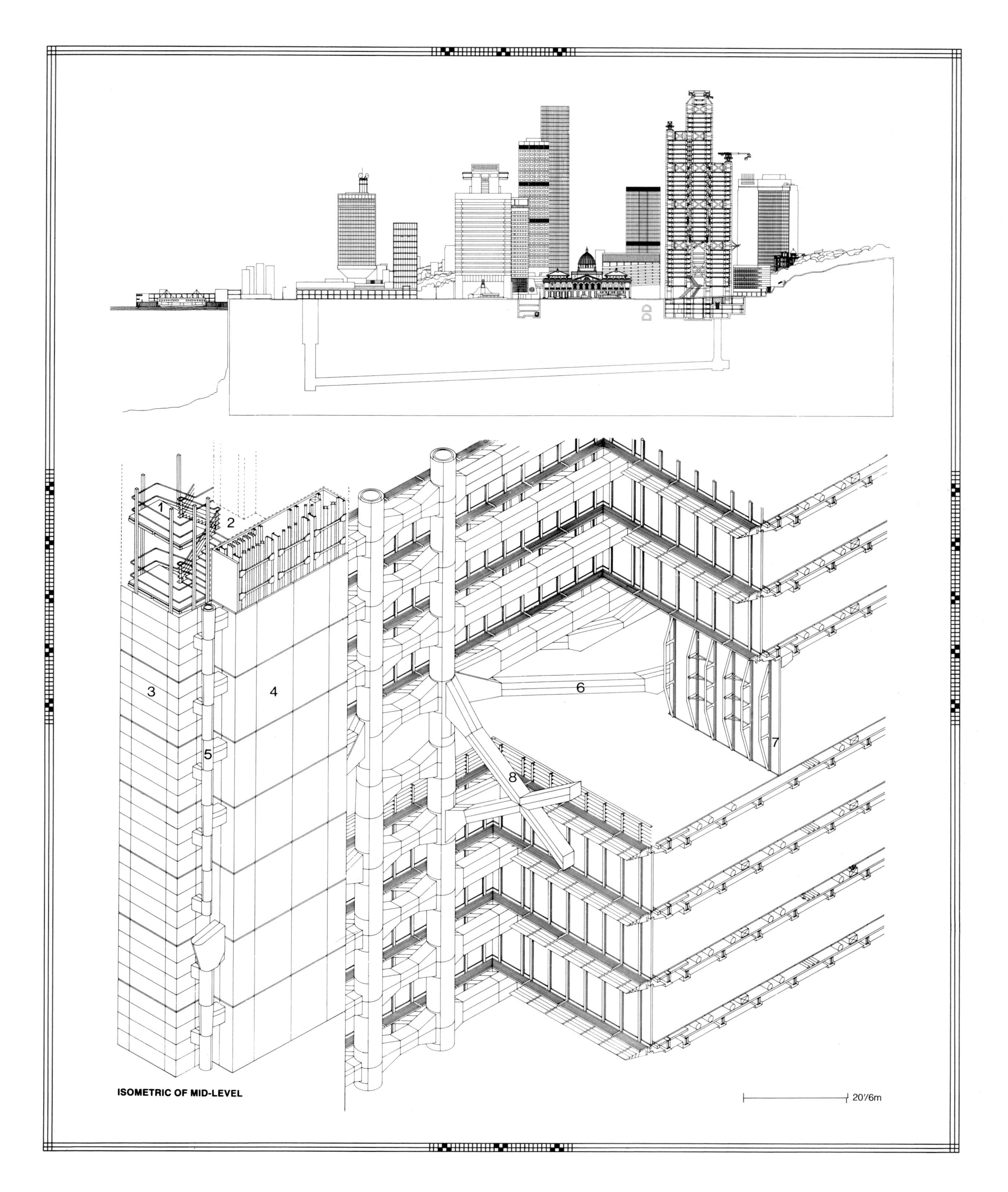
1
2
3
4
5
6
7
8
ISOMETRIC OF MID-LEVEL
20'/6m

weigh up to forty tons and take the form of steel trussed boxes, 3.6 metres wide by 3.9 metres high and either nine or twelve metres long.

Vertical services risers were prefabricated in steel frames two and three stories high and located adjacent to the modules. A total of eight kilometres of riser frames were installed in this way, thereby minimising the number of on-site connections required.

Hong Kong's extremes of humidity and heat make full air-conditioning essential for any modern office building constructed there. In common with most major buildings close to the waterfront, and because of shortages of

fresh water, sea-water is used as the primary coolant in the Hong Kong and Shanghai Bank's air-conditioning system, as well as for toilet flushing. This produces considerable energy savings compared with other cooling methods, but necessitated the construction of a 350-metre long sea-water intake tunnel seventy-five metres below ground level to bring sea-water into the bank's basement from the harbour and to allow for its subsequent discharge.

The 5.5 metre diameter tunnel was excavated through solid granite and has vertical metre shafts at either end. At the seaward end there is sufficient pumping and filtration equipment to deliver sea-water at the rate of 1,000

Cupples-Sunbury

litres per second. The water is pumped up into the bank's basement through a series of titanium plate heat exchanges and then returnend via the tunnel to the harbour. Two heat pumps utilise sea-water as a heat source when required to provide background heating. More usually sea-water is used as a condenser coolant allowing chilled water to be pumped up the building to localised air handling equipment contained in prefabricated services modules on each floor.

These modules are located on the west side of the building. Fresh air is drawn in through louvres in the outer walls, blown through chilled coils and then conditioned air is distributed by means of under-floor ducts.

One of the fundamental objectives in the design of the Hong Kong and Shanghai Bank's new headquarters was to create an attractive a climate as possible on the plaza running underneath the building. To this end extensive wind tunnel tests were carried out to investigate means of preventing wind turbulence. Also, just as important, the building has been planned to bring sunlight right into the heart of the tower, through the banking hall atrium and down to the plaza.

This has led to the installation of one of the bank's most unusual features, an array of mirrors known as the 'sunscoop system'. One sunscoop is positioned at the top of the atrium and the other, parallel to it, is attached to the exterior of the building on the south face. The exterior scoop tracks the sun's movements throughout the year and reflects the sunlight back to the internal array of mirrors safely positioned over the heads of people working on the double-height eleventh floor. The internal mirrors which form a curved screen running the whole length of the top of the atrium beam the light down the atrium through the glass soffit and onto the floor of the plaza below.

Each rack of mirrors is powered by a computer-controlled electric motor which allows them to track the sun's movements. The aim is not to cancel out the daily east to west fluctuations of the sun, which means that, as the day progresses, sunlight climbs down one side of the atrium, across the plaza floor and up the other side. Rather, the pre-programmed adjustments to the alignments of the mirror racks track the sun through its seasonal shifts, which in Hong Kong go through a forty-five degree switch between midsummer and midwinter.

A programme was developed which when fed into an IBM personal computer was able to direct the sun's rays along a similar path each day of the year. The racks make fractional adjustments each hour, the linear actuator controlling the mirrors move at the rate of 0.5 millimetres per minute and require a gearing ratio of 18,000:1 in the motor. The programme has been developed to direct rays on a path that will not only avoid dazzling people within the building, but which will also avoid as far as possible the internal obstacles of the structure, to deliver the maximum possible amount of sunlight where it is needed.

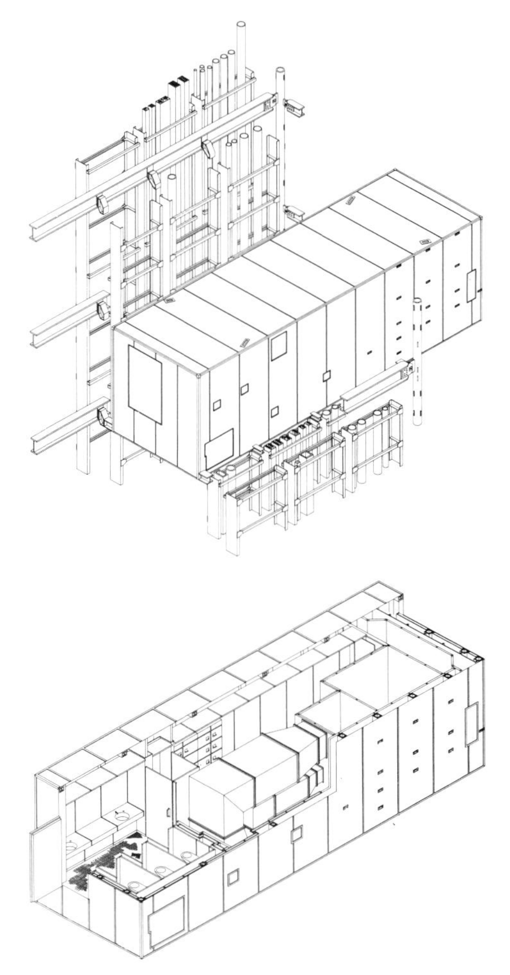

Above:
Top: axonometric view of modules and risers. *Bottom*: modular toilet and plant room
Opposite:
Construction details
Page 190:
Top: north elevation fragment. *Centre*: north-south section, detail through atrium. *Bottom*: section through a one-bay wide floor
Page 191:
Top left: view from the west. *Top right*: Directors' dining room. *Centre left*: principles of floor construction. *Centre right*: abstractions. *Bottom left*: the underbelly of the Bank. *Bottom right*: corner detail

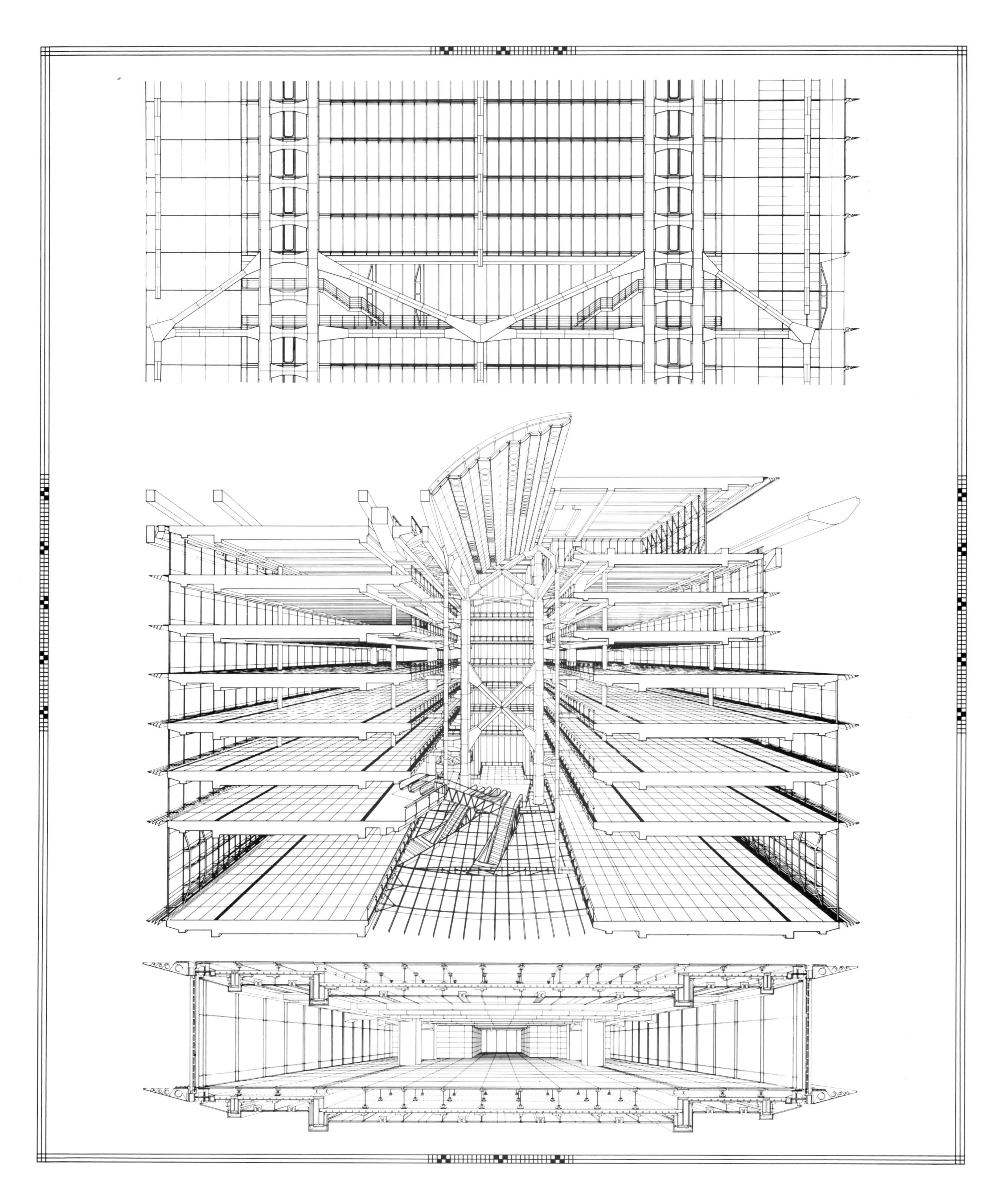

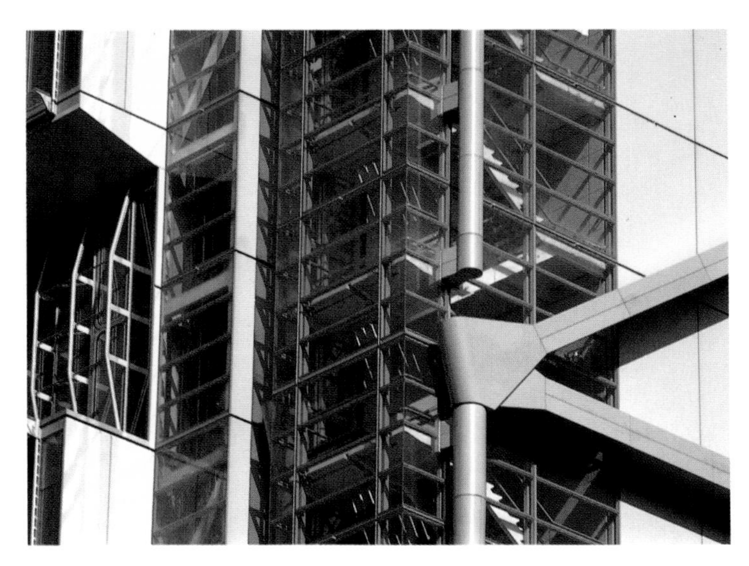

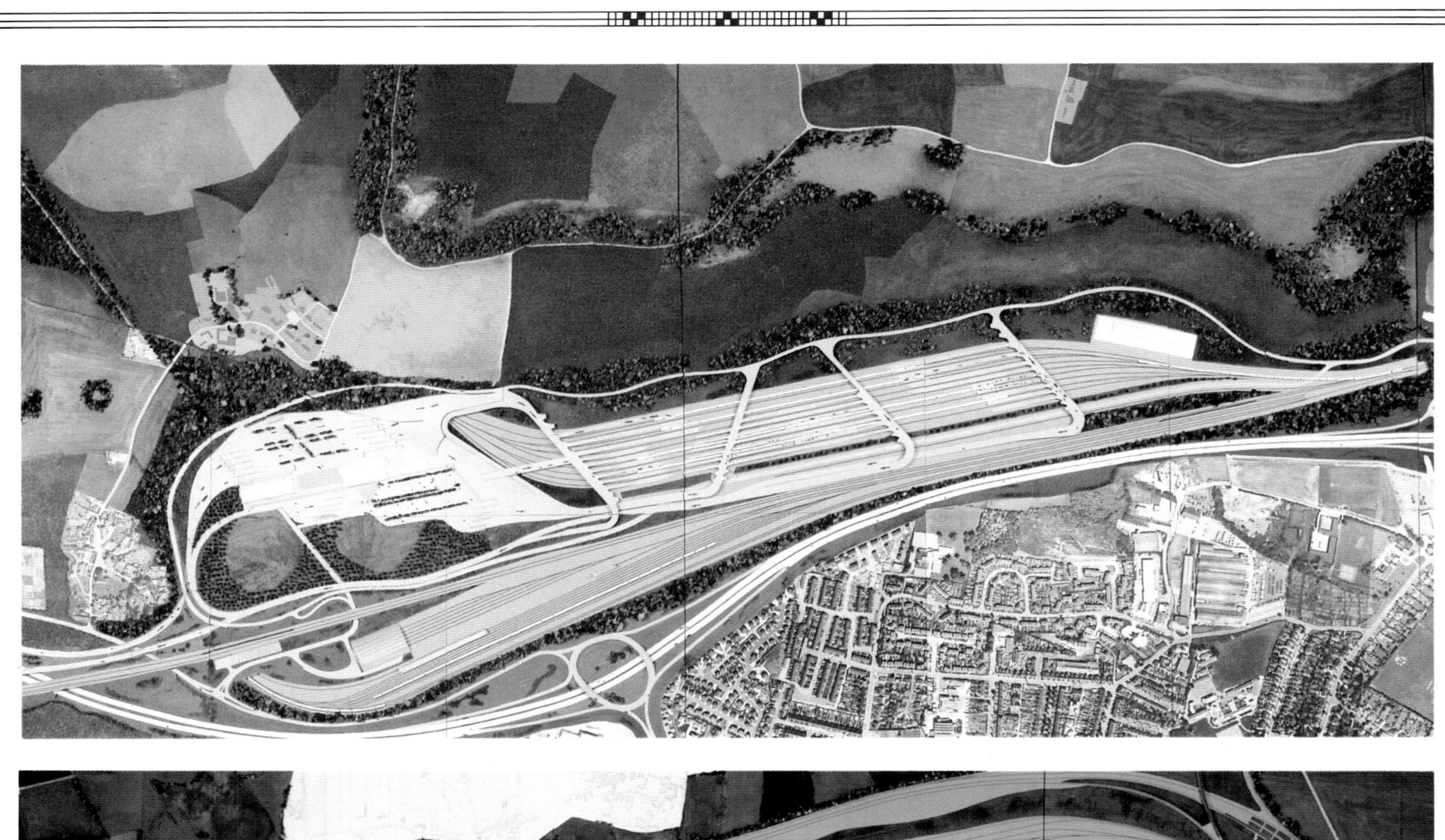

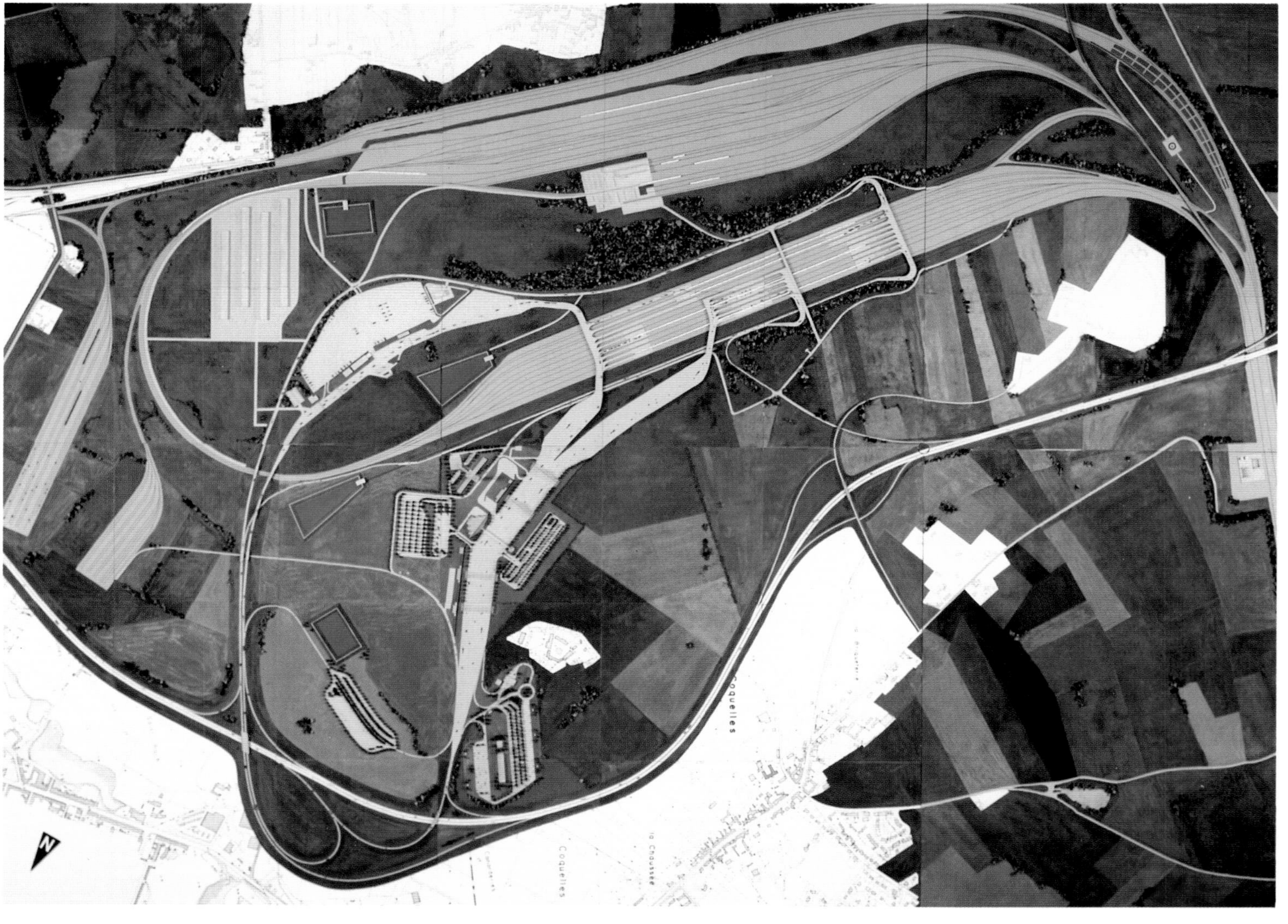
N
Coquelles

DONALD HUNT

The Grim Tale of the Channel Tunnel

Once upon a long time ago, a little known French mining engineer, Albert Mathieu, gave birth to the idea of 'le Tunnel sous la Manche' as a means of once more rejoining Britain with the Continent.

It was an imaginative concept and a daunting challenge to the state of tunnelling art in 1802, but Albert Mathieu was to pass into obscurity, denied the knowledge that his brain child would become a source of both inspiration and frustration to governments, statesmen, engineers, militarists and entrepreneurs over a span of nearly two centuries, and yet be no closer to realisation.

The pock-marked ground around Calais and Dover bears stark testimony to the hopes and endeavours of those who followed Mathieu, attempts which foundered through lack of political will rather than technical skill.

It beggars belief that fears of invasion from the Continent so influenced British defence strategy and political thinking that ideas for the construction of a permanent transport link between Britain and France, by tunnel or any other means, were successfully frustrated for a period of 153 years! It was not until 1955 that Britain's Ministry of Defence officially and finally exorcised these fears.

Despite military and political jingoism those 153 years witnessed a proliferation of ideas and schemes for fixed cross-Channel links: a variety of bored, immersed tube, and floating tunnels; bridges – both floating and fixed; causeways and combinations of tunnels and islands – all were promoted with great enthusiasm by eminent 19th-century engineers, such as Thome de Gamond, Tessier de Mottray, Schneider and Hersent, Hector Horeau, Dr. Prosper Payerne, James Wylson, John Hawkshaw, William Low and Francis Brady, to name but few. Several of these 19th-century schemes bore a marked similiarity to latter-day proposals.

These years of military blight also saw the first practical attempt to build the tunnel from both sides of the Channel. In 1872 the British Channel Tunnel Company was formed, followed by the French Channel Tunnel Company in 1875. Both companies carried out considerable geological and technical research, including test bores and pilot tunnels. However, in 1880 Sir Edward Watkin, Chairman of the South Eastern Railway Company, came on the scene with all the entrepreneurial driving force associated with our Victorian forebears. He joined forces with Colonel F.E.B. Beaumont,

Opposite:
Top: model proposed, Folkstone UK Channel Tunnel terminal site, 350 acres. *Bottom*: model proposed, Frethun France Channel Tunnel terminal

An overall view of the Brunton machine intended for driving the tunnel between England and France, c. 1874

the inventor of the first rotary tunnel boring machine, which in the same year carried out a successful test-boring of over half-a-mile at Abbots Cliff near Folkestone. In 1881, Sir Edward Watkin formed the Submarine Continental Railway Company and moved his tunnelling operations a short way along the coast to Shakespeare Cliff (the site of the 1974-75 Channel Tunnel project and from where the present-day project will continue construction). By 1882 he had successfully bored two further tunnel headings seven feet in diameter, one nearly a mile-and-a-half in length ending some 130 feet beneath the sea-bed in the vicinity of Admiralty Pier, Dover. Work was brought to a halt, however, by the intervention of the Board of Trade against a background of military unrest of ever-increasing stridency.

In 1886 Sir Edward Watkin bought out the Channel Tunnel Company and the Submarine Continental Railway Company became the Channel Tunnel Company Limited which is the company still quoted on the London Stock Market under the title of Channel Tunnel Investments.

There was perhaps some small consolation for Sir Edward as, during his various test-bores, he discovered deposits of coal on the Shakespeare Cliff site and, in 1895, he built and operated the Dover Colliery – not a particularly rewarding venture which shut down in 1915.

The 20th-century revival of interest in a Channel Tunnel began in earnest in 1956 with the removal of Britain's military objections. In 1956 the Channel Tunnel Study Group, a consortium comprising the two 1880 British and French Channel Tunnel Companies and the Suez Canal Company, took up the challenge. Under the capable guidance of Sir Harold Harding, one of the world's leading authorities on tunnelling, and Rene Malcor, Ingenieur en Chef des Pont et Chaussees, the Group financed a series of costing, geological, engineering and traffic studies. It also sponsored independent enquiries into alternative methods of creating permanent cross-Channel links including bridges, road and rail tunnels, both bored and immersed.

These studies confirmed that the only technically feasible means of effecting a direct link with the Continent was to construct twin, seven-metre diameter, bored railway tunnels which, as the project would be financed by private capital, would be both economically and financially viable. The Channel Tunnel Study Group, encouraged by the outcome of the investigations, presented their scheme to the British and French Governments in 1960.

Shortly afterwards, Jules Moch, ex-French Ambassador to Britain, appeared on the scheme heading a group which favoured a bridge, claiming it to be superior in every respect to a tunnel. The Macmillan Government cautiously investigated the relative merits of bridge versus tunnel and, some eighteen months later, produced a White Paper which came out in favour of a tunnel, but did little to accelerate its progress.

It took a further thirteen years, a succession of five Conservative and Labour Governments, new studies, re-studies, a site investigation, selection of a financial and construction consortium to build the project, an Anglo-

French treaty, and a hybrid Parliamentary Bill before tunnelling work actually began on both sides of the Channel in 1974 – only for the project to be unilaterally abandoned by the Harold Wilson Government on January 20th 1975.

Rio Tinto Zinc Development Enterprises Limited provided the Project Management team, ably headed by Sir Alistair Frame. Construction started on the Shakespeare Cliff site at Dover in 1974. Two access tunnels were completed and the boring of a 4.5-metre diameter pilot tunnel towards France was started before work was abandoned. (These works will make a valuable contribution to the 1987 project which will continue on from where work ceased in 1975.)

In 1979, following two years joint study of 'small is beautiful', British and French Railways presented their respective Governments with plans for a single six-metre diameter running tunnel and a service tunnel for conventional rail passenger and freight traffic only.

Despite the fact that a thoroughly well-researched and investigated tunnel project, abandoned in 1975, could be reactivated, the rituals began all over again. Within months, the Minister of Transport and the news media were besieged with proposals for bridges, tunnels – both bored and immersed – and a late entry for a bridge-island-tunnel scheme. Apart from one or two refinements in design and methods of construction – none of them new – the basic concepts had all been proposed in the past.

At one stage there were a full dozen different schemes being promoted by various groups, not counting the 'do-nothing' proposal of the sea ferries who wished to cope with the future expansion of cross-Channel traffic themselves – all of which diffused the historical pre-emience of the Channel Tunnel.

It has been the subject of the closest possible scrutiny and re-examination for more years that anyone would care to recall and never more so than in its 20th-century revival. Without exception, when measured in terms of technical feasibility, cost and commercial viability, it has successfully emerged as the only practical means of establishing a fixed transport link with Continental Europe. When comparisons are made with alternative methods of achieving the same objective, such as bridges and combinations of bridges, islands and tunnels, costing escalates to at least three times as much and the tunnel becomes an even more inescapable conclusion.

It is unlikely that any of these alternative methods would ever have seen the light of day, if the Channel Tunnel had not once more raised public consciousness of the issue. Predictably, the sea ferry operators, sensing a threat to their monopoly on short sea routes to the Continent, repeated their 'anything you can do we can do better and cheaper' slogan. They said it in the 1970s and again in the 1980s, both times when the Governments of Britain and France were giving serious thought to the matter of a Channel Tunnel. Ephemeral promises of cheaper, faster, more efficient ferry services, capable of coping with all future increased traffic trends, were nailed to the masthead as they prepared to defend their position. But, alas, in the intervening years

Hector Horeau, Engineer; submarine railway between England and France, c. 1851 (*Illustrated London News*)

when the realisation of a Channel Tunnel faded, their fares remained consistently higher than the inflationary rise in the cost of living index.

The net result of the initiative of British and French Railways in 1979 was a fresh round of studies. The House of Commons Select Committee on Transport undertook a thorough examination of all the proposals for a fixed cross-Channel link. The proposers of all the schemes claimed to conform to the UK Government's economic guidelines that the British half of any project would have to be privately financed without recourse to public funds or government guarantees. This was followd by a further study of all the schemes by the Anglo/French Study Group, a committee comprising officials of the British and French Departments of Transport.

It came as no surprise, in view of the countless precedents that the two studies (published in February 1981 and June 1982 respectively) concluded in favour of a Channel Tunnel as the only scheme which could be started at once. The reports recommended seven metre diameter bored railway tunnels with roll-on, roll-off facilities for road vehicles and the possiblility of phased development. This was the scheme promoted by Channel Tunnel Developments (1981) Ltd, basically a revival of the project abandoned by the British Government in 1975, with the option of a phased construction programme.

The Anglo/French Study Group Report estimated that it would be a number of years before the more ambitious schemes for a fixed cross-Channel Link could even be started. Attractive and innovative as several were in concept, they would necessarily employ advanced construction techniques stretching present-day engineering practice beyond known and acceptable limits. This is not to say that the technical obstacles could not be overcome, but several years of technical studies would be necessary to test, prove and develop their advanced design requirements.

The principal problem faced by any scheme proposing an 'above the sea solution', is that it would create permanent navigational hazards to shipping across the busiest international waterway in the world. Upward of 500 vessels daily ply their trade up, down and across the narrow Straits of Dover, and laborious international negotiations would have to be conducted with over one hundred maritime nations to define an acceptable navigational safety system, with no guarantee that such negotiations would ever reach a successful conclusion.

Sir John Fowler/Sir Benjamin Baker, Engineers; design for a Channel Bridge, 1889

William Brown at Freeman Fox, who designed the outstandingly successful Bosphorus Bridge, proposed a conventional suspension bridge. Willie Frischmann also submitted a design for a suspension bridge for all-weather traffic use, replacing the conventional suspended road deck between the bridge piers with what can only be described as a suspended tunnel enclosing four road decks.

However, to overcome maritime objections to permanent navigational obstruction in the main international shipping lanes, all bridge schemes envisaged unsupported spans in mid-Channel, of some two-and-a-half miles in length. Currently the longest unsupported bridge span in the world is the Humber Bridge in Yorkshire, being a few feet short of one mile.

Similarly, the scheme originated by Ian MacGregor based on his association with the Chesapeake Bay Project – a seventeen mile combination of a viaduct with a short stretch of immersed tube tunnel in the centre – went to great lengths to overcome interference with the international shipping lanes. This scheme first saw the light of day when MacGregor was Chairman of British Steel; whether it originated as a promotional exercise to sell steel is not known, but it developed into being the main rival of a Channel Tunnel. Capable of taking both road and rail traffic it comprised of a road viaduct from each coastline, connected to two off shore artificial islands which were joined by an eleven kilometre immersed tube tunnel.

None of the alternative methods for constructing a fixed cross-Channel link would contribute more in terms of traffic capacity and convenience than the Channel Tunnel at a third of the capital cost.

The Channel Tunnel project, bored at an average depth some 140 feet beneath the sea bed faces none of these problems. It would provide the missing link between the British and French rail networks, enabling the unrestricted passage of conventional rail passenger and freight services throughout the Continent. The rapid railway shuttle service between terminals would transport all commercial and private road vehicles in a journey time of about half-an-hour.

Compared with these ambitious alternative options, it could be said that the Channel Tunnel is a low-technology project. This in no way detracts from it being one of the greatest 20th-century engineering projects in the Western

Model of double-decked car-carrying shuttle train, c. 1987

hemisphere. It does, however, have considerable bearing on it being a privately capitalised venture. Private investment is extremely sensitive to the technical risk element and the technical risk of the Channel Tunnel is limited by the fact that it will be constructed by tried and proven techniques, added to which it will be bored through possibly the finest tunnelling material – the lower chalk stratum – which is impervious to water and stretches between the two coastlines.

There was to be no let-up in the continuing round of studies and further joint Government investigations. Having received the Report of the Anglo-French Study Group in June 1982, the Transport Ministers of Britain and France jointly considered that it was basically a technical appraisal and that, before taking a decision, an examination of the financial implications of the various schemes was required in order to evaluate the overall position. They announced in June 1982 that they were appointing a financing group comprising the National Westminster and the Midland Banks in Britain and Banque Nationale de Paris, Credit Lyonnais and Banque Indo-Suez in France, to undertake this task.

It puzzled a number of observers that the remit of the financing group covered all the fixed-link proposals, even those which, in the opinion of the joint study group, were technically or commercially unrealistic. It therefore appeared as something of a waste of time and effort to consider their financial implications any further.

Undaunted, Channel Tunnel Developments (CTD 81) in February 1984 joined forces with two other groups who were promoting similar tunnel projects. This move brought together under the title of the Channel Tunnel Group, the five leading UK construction companies: Balfour Beatty, Costain, Tarmac, Taylor Woodrow, and Wimpey.

The report of the five UK and French Banks, under the title of the Franco/

British Channel Link Financing Group, was published in May 1984 and concluded after close examination of all competing schemes, that the tunnel scheme proposed by the Channel Tunnel Group '. . . is the only scheme that is both technically acceptable and financially viable'.

Events then began to move more positively. The National Westminster and Midland Banks joined the Channel Tunnel Group, and, in April 1985, the British and French Departments of Transport circulated a document entitled *Invitation to Promoters*. Each promoter was expected to submit in detail their respective projects by the closing date of October 31st 1985.

In July 1985 the Channel Tunnel Group merged with their French partners, France-Manche, which comprised five leading French construction companies plus the three French banks which had been members of the Franco-British Channel Link Financing Group.

On January 20th 1986 all speculation ceased when the British and French Ministers of Transport announced the award of the concession for a fixed cross-Channel link to the Channel Tunnel Group and France-Manche. This was followed by the signing of an Anglo/French Treaty by Prime Minister Margaret Thatcher and President Mitterrand at Canterbury on February 12th 1986.

The Treaty had to be ratified by rather lengthy legislative procedure through the British Parliament before work could start. Meanwhile Eurotunnel, the company joining the Channel Tunnel Group with France-Manche, came into being. Under the co-chairmanship of Alastair Morton and Andre Benard, Eurotunnel has the overall responsibility for the construction of the project and raising of private finance, and will run the project for a concessionary period of fifty-five years.

Within the short span of barely twelve months, Eurotunnel made remarkable progress: raising tranche one and two of their equity capital programme; signing a design and construct to specification contract with Transmanche-Link (comprising the ten British and French construction companies who were original shareholders in the Channel Tunnel Group and France Manche); reaching an agreement with British and French Railways for their use of the tunnel, and negotiating a one billion pound loan from the European Investment Bank.

The political climate between Britain and France has never been better and if good fortune continues to favour the project, work will restart on both sides of the Channel in the autumn of 1987 and the tunnel will be completed and operational by the spring of 1993.

Not only will its completion bring international prestige to both British and French engineering, it will also be a belated and lasting tribute to those engineers and visionaries whose perserverance and dedication kept the project alive for a span of nearly two centuries. As we pass into the 21st century, the revolutionary impact the tunnel will have on the movement of people and goods to and from the continent, will inevitably pose the question: 'How did we ever manage without it?'

Model of heavy duty shuttle wagon for long distance road transport

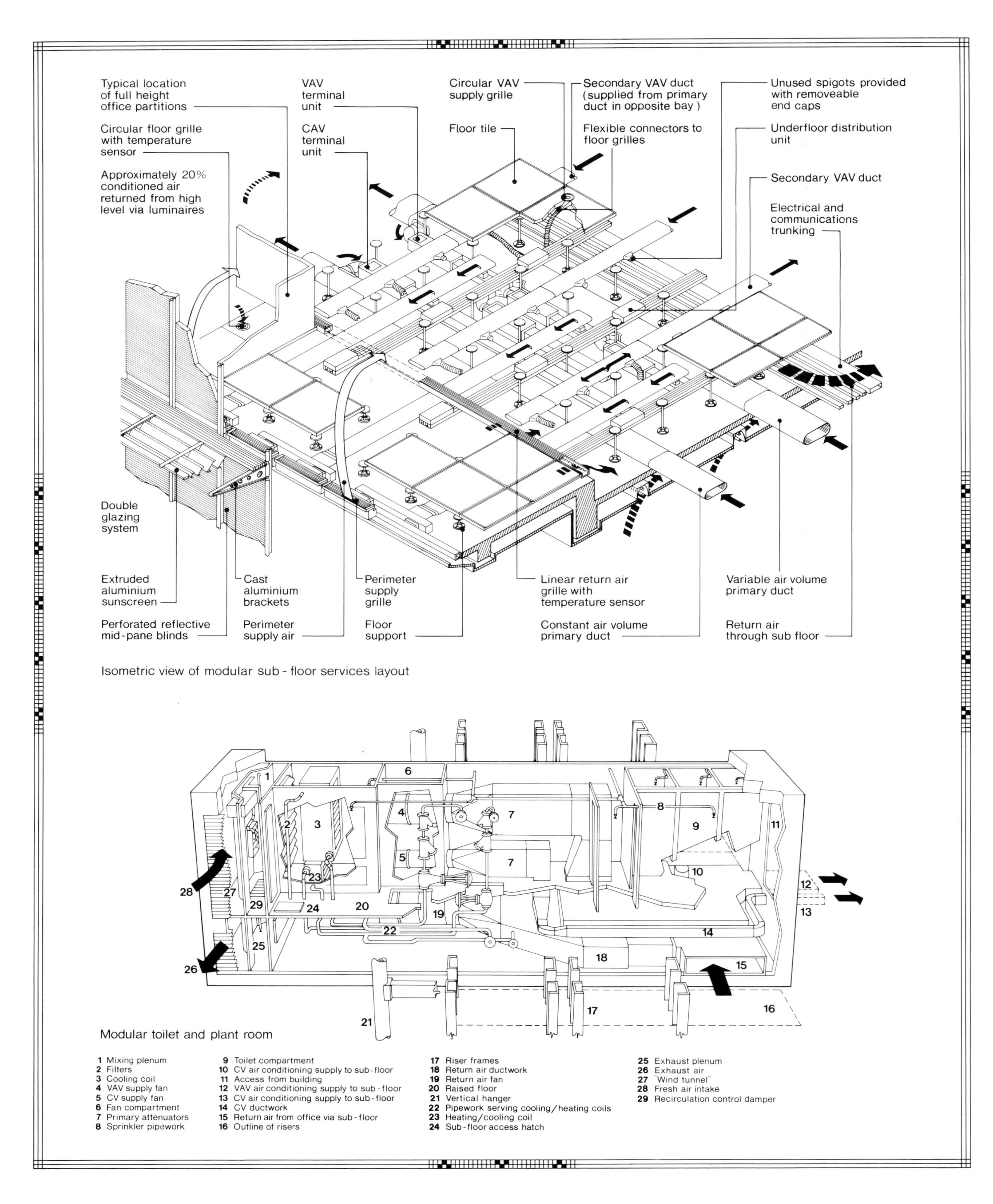

Isometric view of modular sub - floor services layout

Modular toilet and plant room

1 Mixing plenum
2 Filters
3 Cooling coil
4 VAV supply fan
5 CV supply fan
6 Fan compartment
7 Primary attenuators
8 Sprinkler pipework
9 Toilet compartment
10 CV air conditioning supply to sub-floor
11 Access from building
12 VAV air conditioning supply to sub-floor
13 CV air conditioning supply to sub-floor
14 CV ductwork
15 Return air from office via sub-floor
16 Outline of risers
17 Riser frames
18 Return air ductwork
19 Return air fan
20 Raised floor
21 Vertical hanger
22 Pipework serving cooling/heating coils
23 Heating/cooling coil
24 Sub-floor access hatch
25 Exhaust plenum
26 Exhaust air
27 "Wind tunnel"
28 Fresh air intake
29 Recirculation control damper

MAX FORDHAM

Intelligent Buildings for the Future

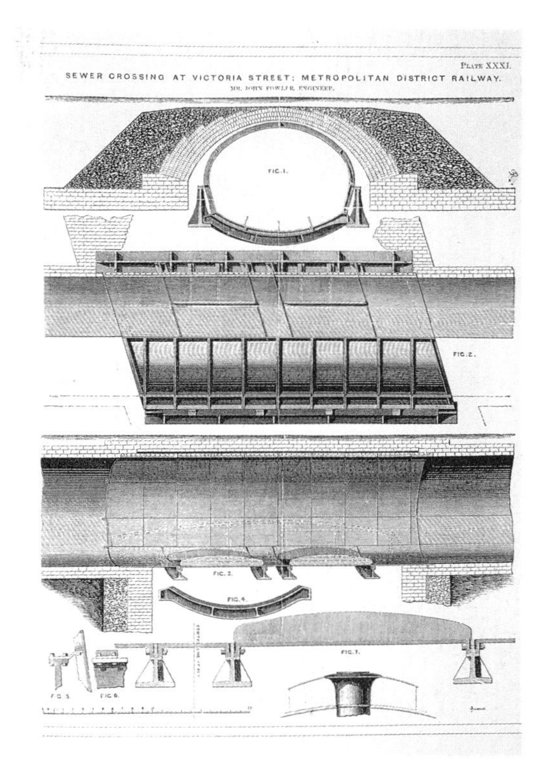

Above:
John Fowler, Engineer; sewer crossing at Victoria Street, Metropolitan District Railway (19th-century drawing)
Opposite:
Ove Arup and Partners, Engineers; Foster Associates, Architects; Hong Kong and Shanghai Bank: isometric view of modular sub floor (*top*), modular toilet and plant room (*bottom*)

Building services is concerned with the flow of energy into buildings for lighting and thermal control, and represents the major use of fuel in the world. It is imperative that we develop means of spreading the benefits of technology to the whole of the world and avoid polluting it with the effects of combustion. The efficiency of using existing energy sources must be improved and alternative sources must be developed.

Technology tends to advance very slowly through applying and modifying our ideas. Our building environment would be broadly recognisable to Mediaeval man, but our communications systems would be completely new.

During the next century new applications will be added to the artefacts we already have. Communications systems and the computer are going to increase in scope. Systems for handling energy in buildings are going to develop so as to improve comfort and degrade energy more efficiently, mainly through the application of computers to communications and control. For, the computer and communications are pretty well indivisible. In the 1950s a science fiction film predicted it all. A plant and its energy system were controlled by a computer system which could sense data emanating from the unconscious minds of the occupants of the plant. I can now see how to start putting this concept into practice but do not think it is likely to happen within the next century. The limit to the use of computers and communication lie in our own imagination.

So far as I know phased array lasers do not yet exist. But phased array radar does. This enables a completely flat emitter of very long wave light (radio waves) to behave like a movable search light under computer control without any moving parts. Phased array radar is used in defence systems, and it seems probable that something like phased array lasers will be developed.

Lighting is already controlled in special instances by digital systems. As lighting sources are developed the control of them using digital solid state devices will become general. The power handling side of solid state switching is part of the systems within buildings but the control of the systems will easily be brought under the control of communications systems.

Telephone lines are currently being used for communications between computers as well as people. Simple telephone wires cannot carry data at a very high rate but for a long time the capacity of trunk telephone lines has been increasing. The capacity of a line depends on the frequency of the signal transmitted along it and optical cables represent the highest frequency

Electric home and street lighting by arc lamps, c. 1879

available so far. I do not think that we are anticipating X-ray cables but that is the logical conclusion of this development. Michael Frayn, in a book called *A Private Kind of Place* described a life-style in which it was unnecessary for people to move outside their dwelling. Communication with other people took place by means of a holographic stereophonic telephone complete with sensations of touch and smell. This, perhaps, is a little far-fetched, for it is very difficult to imagine us being able to convey information fast enough to saturate a human being's nerve endings with data on such an ambitious scale.

At present, when a building is completed, it is usually difficult to get the services installation to behave in accordance with the original calculated design. We design and install control systems with the aim of making the building services systems adapt themselves as circumstances change. As the control systems form part of the computer calculation, the building really should be able to adapt itself to its environment. We are also able to write a special piece of software which forces the system to behave as designed so that we can then check that the installation has been properly carried out. Thus, for instance, the flows of fluids and electricity are also brought under computer control.

It is common experience that as systems grow old the sensitive manual adjustments which were never satisfactorily completed at commissioning stage slip more and more out of adjustment. With a computer monitoring system it will be possible to analyse the performance of systems based on data sent along telephone lines so that maintenance engineers will be able to repair systems before building owners even know that they have gone wrong.

When systems are designed it is assumed that the maximum load on a system is less than the sum of all the individual loads coming together. Systems have to be partly over-sized because if the load gets too big the system may fail. With computer control, systems can be more carefully designed because the computer system can decide which loads need not be served and can turn them off. Electric heating for example, can certainly be turned off to keep electric loads to within a predefined maximum. When a stand-by generator is in use the load has to be shed more drastically than during normal use and a digital control system easily controls the changed priority. Very commonly, while various pieces of plant are starting, the initial load on a stand-by generator gives way to a lower steady load and then more pieces of equipment can be added.

Ventilation represents one of the main energy loads in a well ventilated building and if the amount of ventilation can be controlled, on a room-to-room basis in accordance with the number of people and their activities, then the energy use and the standard of comfort can be improved.

At the Lloyd's Building each individual has a pad on his desk with six buttons which allow him to express his desire for more or less lighting, ventilation, or heat. His requests are used both to trigger immediate responses – turning on lights, fans, or small heat pumps – but at the same time the data is used to modify the behaviour of the central plant.

Since 1973 efforts have been made to establish energy targets for particular classes of buildings. As buildings services (energy) engineers we should be prepared to enter into contracts with building owners guaranteeing the level of energy consumption in their buildings. This cannot be made to work unless the conditions in the building are continuously monitored. Thus buildings must not be allowed to get too warm or too light or, if air conditioned, too cold. To begin with, proper digital computer control should prevent discrepancies between the energy model and reality, so that descrepancies can be detected, recorded and even become a reason for making maintenance calls.

The communication of all this data requires paths. The paths need not be simple wires. Optical and co-axial cables carry more data in less space but we need to be able to add plugs after installation instead of having to thread the manufacturer's plug through an existing building. Why shouldn't computer terminals and telephones have transmitters and receivers for ultra-sound and infra-red signals? The need to carry large quantities of data in existing buildings will make this kind of development inevitable.

We use energy in buildings to control data, to provide light, and to control our thermal environment. During the next century these benefits must be provided for the whole of the population of the world rather than just a small part of it. Most of the energy we use at present comes from burning fossil fuels and, even if there were enough fossil fuel, we are changing the chemical composition of the atmosphere in a very dangerous way. For, the atmosphere is pretty throughly mixed so that bad effects are not confined to a small part of the world in the way that for example mistakes in agriculture were confined to the dust bowl in the the United States. We are going to have to reduce the amount of energy we use that now comes from fossil fuels, although our use of energy as a whole is going to rise. About half the fuel we use in the United Kingdom is used for generating electricity but only one third of that is acutally turned into electricity. There is no possibility of increasing the proportion of electricity produced to any significant extent. But we must concentrate on using the two-thirds which is currently thrown away in cooling towers or pipes connected with the sea. New power stations should be constructed close to centres of population so that waste heat can be used in buildings. The waste heat will not be free and should be metered and charged for like any other form of energy. In the United Kingdom this would reduce our fuel consumption by nearly one half.

Nuclear energy from fusion produces no pollution unless it goes wrong. Fusion reactors should not produce the same radio-active risks. Maybe this will be our saviour. All nuclear energy produces high-grade heat as its primary output (not so high as combustion) and this heat has to be converted into electricity with a consequent by-product of large quantities of low-grade heat which we should be using to heat our buildings. There is plenty of solar energy received at the outside of the earth's atmosphere. In hot climates where heat is not required we should be able to use it for generating some

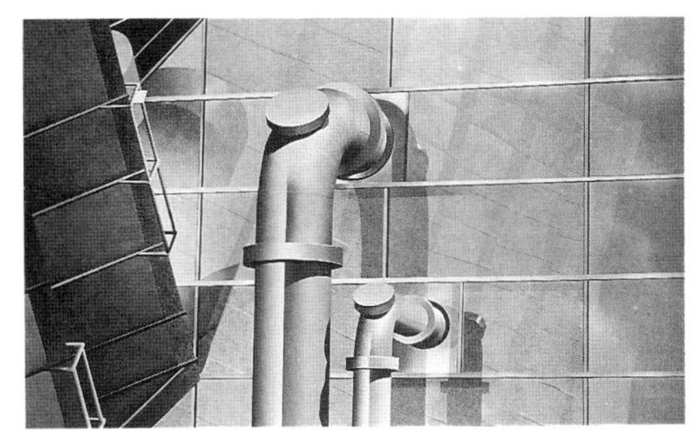

Above:
Top: Felix Samuely and Partners, Engineers; James Stirling, Architect; Cambridge History Library, c. 1964 (Ben Johnson painting). *Bottom*: Felix Samuely and Partners, Engineers; Stirling and Gowan, Architects; Leicester University Engineering Building, c. 1959-63 (Ben Johnson painting)

Ove Arup and Partners, Engineers; Richard Rogers Partnership, Architects; Lloyd's Building, c. 1986, exterior ducting

expensive electricity and also directly for air conditioning buildings. If it is to be applied in the United Kingdom it will have to be stored during the summer for use in the winter. Waves are a very diffuse form of energy and even if we could extract all the energy from waves round the United Kingdom it would not serve our energy needs. Tidal energy is on the verge of being useful; proposals for the Severn and Morecambe Bay are likely to be implemented in the next century. Photosynthesis, a lot of which takes place in the sea, is currently being used to provide our food. If we envisage this process as providing us with a significant proportion of our future energy needs, we will need to farm the sea and grow fuel in it. Wind energy has been exploited for centuries and it will be again provided we re-arrange our economic system accordingly. We are also beginning to extract geothermal energy from high temperature rock strata in the ground as a source of low grade heat.

We need to distinguish between energy sources which produce heat and energy sources which produce energy. In either case we will only exploit the more diffuse energy sources if we adjust our economic system to persuade us to do so.

Systems need to be improved so that we can develop greater control; this would make them more comfortable and should also reduce use of energy. When the whole of an installation is being controlled automatically there will be no need for commissioning and systems will work immediately they are completed. We are in fact improving the efficiency of our light sources. The least efficient incandescent sources are easily controlled but the most efficient wide spectrum light sources are more difficult to control. We are on the verge of making improvements which enable high pressure mercury vapour lamps to start immediately, even when hot, but we need additionally to be able to dim their light output. Our biological requirements for light – for seeing – are unlikely to change over the next century but, by providing more control over the direction of diffused light, we will be able to see more comfortably and with less electricity. Task lights will be controllable so that the light can be provided by three or four sources. One fixed source is not enough and one movable source is ungainly. We will be able to control our light level by talking to the light fitting.

We know that it is impossible to satisfy a group of individuals by providing one temperature for them all. Individual temperature control has to be available for people even when they are within one space. An infra-red phased array laser will be able to provide radiant heat to every individual in response to his verbal request. Alternatively, visual display units can be equipped with air supply nozzles similar to those commonly provided in motor cars. Temperature, air quantity and direction will all be controllable. Boilers as a heat source should simply disappear. In the meantime they will be improved so that more heat is extracted from the combustion process. We are currently used to the idea of clean cool products from burning gas but we are going to have to turn to coal as our primary fuel. It is bound to be wasteful converting coal into another clean fuel (gas) and we will probably have to turn to burning coal in fairly large central plants, with heat itself being supplied as hot water and metered to consumers. Even so, the size and cleanliness of coal boilers must be improved.

Heat pumps use work to take heat from a cold place and put it into a hot place. If the source of work is electricity, which has low grade heat as a by-product, then an electrically driven heat pump is not an improvement. If gas or oil are burnt in internal combustion engines to produce work and the low grade by-product is used to heat buildings then work can be used to drive a heat pump and the combination is twice as efficient as a boiler. Gas engine driven heat pumps will be developed within the next few decades.

Lloyd's Building, c. 1986, interior

If electricity is passed to a series of junctions between different metals then one set of junctions gets cold as the other heats up. At a prototype level, Honeywell have made a wall as a component of a building with junctions of appropriate metals embedded within it. In this way heat can be prevented from flowing through the wall and so the building is kept at the correct temperature.

We can easily wear clothes which enable us to be comfortable at ambient temperatures, close to those prevailing outside. We should be able to design and produce quilted, light weight flexible clothing with low emission. Television programmes certainly give us models for our imagination to work on. Heat pumps or thermo electric sections could be incorporated into our clothes to provide comfortable temperatures.

The search for a wall to act as a thermodiode allowing heat to pass through it only in the direction which we want is analogous to the alchemists' search for the philosopher's stone. It is an archetypal concept which contradicts the second law of thermodynamics. The performance of windows can be improved and we should be paying attention to the design of insulating quilted, metalised, low emissivity blinds, curtains and shutters which close with a tight seal using velcro or zips around the edge of the window. It should be possible to use insulating foams in the cavity between the two layers of glass. Computer control of insulation to windows will prevent them emitting too much heat when it is not wanted or losing too much heat when we need to conserve it.

We are currently insulating buildings thoroughly enough to provide the heat for ventilation air. We need to make buildings air tight, adding lobbies to external doors and magnetic seals of the kind used on domestic refrigerators to the doors. Low grade energy should be provided as a chargeable service, through a system of metering, like other forms of energy. Computerised communication will lead to a development of heat meters which are cheap and reliable. We must organise society to allow the benefits of circulating energy to become more widespread, using our energy resources more efficiently and relying less on combustion for the production of heat. The computer will help us to process data and use lower grade sources of energy more efficiently. The increasing order brought about by the computer will help our society to control the rate of entropy production.

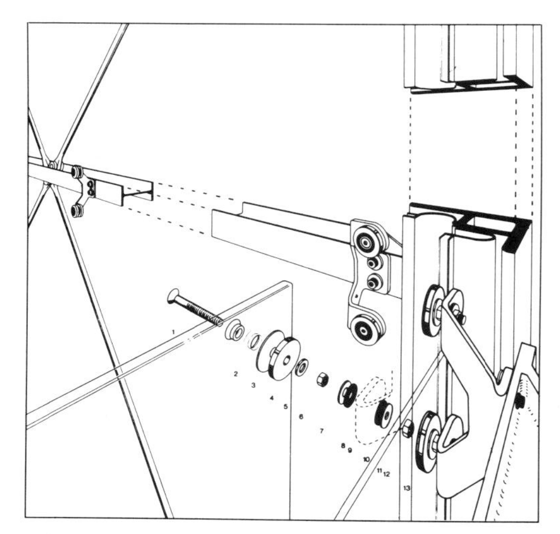

Above:
Ove Arup and Partners, Engineers; Foster Associates, Architects; glazing systems by Pilkington used for the Renault Building, Swindon, c. 1983
Opposite:
Float glass suspended assembly, 53 State Street, Boston, USA

SIR ALASTAIR PILKINGTON

Research and Product Development – Glass

I am convinced that a successfully creative company will be market led. By means of that lead from the market place its research and development will be thoroughly briefed. That brief will be supported by substantial investment in understanding the market requirements and will clearly define success. These considerations are essential for obtaining clear direction and commitment for product development work.

The field of business innovation must be defined, especially in an industry dealing with a basic product as versatile as glass. Unless effort is focused, attractive routes may be followed but no useful body of knowledge accumulated or major step forward made. The versatility of materials means, too, that attention must be paid to seeking the difficult convergence of 'what can be done with it' and 'what the market wants from it'. The glass industry has been good at getting this convergence right, with developments such as strong, light-weight milk bottles, ceramic oven and tableware, heated rear windows for cars, optical fibres for communications systems and photochromic glasses.

Our principal expertise and investment lies in the field of glass and glass-related products. This leads us into a diverse range of market applications some way removed from building and transportation, which we are generally associated with. The Pilkington Fibre Optics Factory in North Wales for example, opened in 1976, was one of the first of its kind. Since then its capability has expanded from optical cables, optical transmitters and receivers to full signal transmission sub-systems and specialised industrial inspection equipment. The research and development work which, during the late 1960s, had been mainly in the field of fibre production for short distance data transmission, resulted in the foundation of a company which has become one of the world's largest in the field of fibre optics and opto-electronics. While these high-technology activities necessarily involve us in the design and marketing of associated components, the research effort is centred on the development and marketing of our basic material – glass.

I am convinced that the means of creating a successful future for the company lies with research and development. There is nothing more important to the technical success of a company. Some mistakenly think of research and development as an overhead, a drain on profits, rather than something which creates them. Others see it as divorced from the job of production which, they believe, is concerned with making things, while

Above:
Top and bottom: external view float plant
Opposite:
Float glass installations: (*top*) Concorde, (*bottom left*) British Rail, (*bottom right*) Rolls Royce

research and development is to do with thinking about processes for the future. Wrong again. Research and development is concerned with applying science to solve problems on the production line and, using knowledge of today's production processes, conducts fundamental research from which new ideas may flow.

In a company that believes in research and development, it will be thoroughly integrated with the rest of the business; an intrinsic part of the company's strategic thinking. By its integration, it will infuse an awareness of the importance of defining the company's future. It is part of the present, part of the medium and long term futures and it will help a company think in all these time scales.

If a company has vision in its research and development, it will be able to turn its ideas into reality. It was not altogether unexpected that Pilkington should invest and develop the float glass process for, as our history prior to float shows, we already had vision as innovators.

What then is Pilkington's thinking about successful invention in a company? Let's start from the idea that man is capable of astonishing achievements once he knows what he wants to bring about. So how do we harness this capability? The first component to achieving this greatness is a conviction that a particular course of action is worthwhile. There is no greater incentive to good invention than the recognition that what you want to do is important. No greater incentive to putting ideas into effect than identifying that they are worthwhile. In a market-led company, as much time and effort will be spent identifying what is worth doing as in the doing itself. Doing the work is often relatively simple.

The second important component is the total integration of research and development into the fabric of a company. Scientists will have greater incentive to focus their efforts if they are so integrated, because they will thereby come to an understanding of, and commitment to, the needs of the market place. In industry, invention alone is not enough – it must have value in the market place. You can't afford much work that is not geared to achieving really valuable results there.

For successful invention to flourish the whole of the thinking – the setting up of objectives, the selection of people, the structure of the organisation, the methods of working – should conform to two principal requirements: 'defining success', and 'integrating the research and development activities into the business'.

Thirdly, it is important that companies are aware of their track record and capabilities; there are some activities and projects that a company has a natural ability to bring through to success and others that it has not. In Pilkington we have always seen ourselves as able to handle two or three really major research and development projects which will make big advances in various parts of the glass industry. But we have felt much less capable of creating new opportunities in the consumer markets. The further away a company gets from its own vision and from what it is already doing the more

1800TU

Hologram Production facility, St. Asaph, England

difficult it is to achieve excellence.

Now to the question of producing and handling the new idea. How do we induce its birth and nurture it? You cannot insist on invention – you cannot, as it were, say to someone 'invent' like you can say to someone 'improve'. You can, however, set the right atmosphere for invention both by the way ideas are treated and by the way people are organised to produce the sparks which may ignite. Everyone is capable of original thought and many like identifying new opportunities. No one should be frightened of making a fool of himself by suggesting something which is impracticable.

Any new idea should be treated gently at first, since it might change. A full feasibility study carried out on a new idea is dangerous. It should be sheltered like a seedling from the cold winds of criticism, put into a propagating frame and nurtured until it can be more carefully examined to see whether it is an exciting plant or just a weed. But an important stage is reached after the key questions have been defined and answered; only then should a full attack be launched against the idea to see whether it really stands up. The signals from the market play an overriding role in assessing any new venture.

Much of my thinking on the ingredients of successful invention was developed and confirmed during the float glass research and development programme. It has likewise frequently been confirmed by our successes with other major research and development projects.

Float is a revolutionary process whereby an existing market is more efficiently met. Some twenty years after the idea was conceived, the float process has completely superseded the earlier plate process in making high-quality flat glass. Plate glass was an excellent product but the grinding and polishing necessary to produce clear glass with parallel surfaces was costly. The challenge facing Pilkington was to invent a process to produce a product as good as plate but without the need for grinding and polishing.

During the 1950s, it took seven years and £7 million to transform the idea of the float process into a commercial success. Pilkington were gambling on a process which would supersede their own polished plate lines and those of its competitors, each with enormous capital already committed. If successful the process would give Britain a greater lead in flat glass manufacture than ever before. The gamble paid off in 1959 when float was announced.

In the process, a continuous ribbon of glass moves out of the melting furnace and floats along the surface of an enclosed bath of molten tin. The ribbon is held in a chemically controlled atmosphere at an high enough temperature and for a long enough time for the irregularities to melt out and for the surfaces to become flat and parallel. Because the surface of the molten tin is dead flat, the glass also becomes flat. The ribbon, while still advancing across the molten tin, is then cooled until the surfaces are hard enough to be taken out of the bath without the rollers marking the bottom surface; so a glass is produced with uniform thickness and bright fire-polished surfaces, without the need for grinding and polishing.

Some lessons are illustrated.

Big steps forward involve big risks. Much research and development work in industry is aimed at improving existing processes and products. Pilkington, for example, could have battled on with the further improvement of the polished plate process. However, we chose the more difficult route, convinced that there were big rewards to be won. The experience with float was that success was uncertain, and in many people's minds improbable, until it was actually achieved. After we proved that the technology worked it still took a long time before glass could be made more cheaply than by the plate process. Between the time when we started experiments and the time when we made the first saleable glass, we made over 100,000 tons of unsaleable glass; on the first production plant we made unusable glass for fourteen months.

I repeat, the degree of success will be directly related to the time and effort spent in defining success. Often not enough time is spent by technical and marketing people in pinpointing the area of greatest potential reward. Marketing, manufacturing, and research and development must have an equal and joint responsibility for defining and shaping the future.

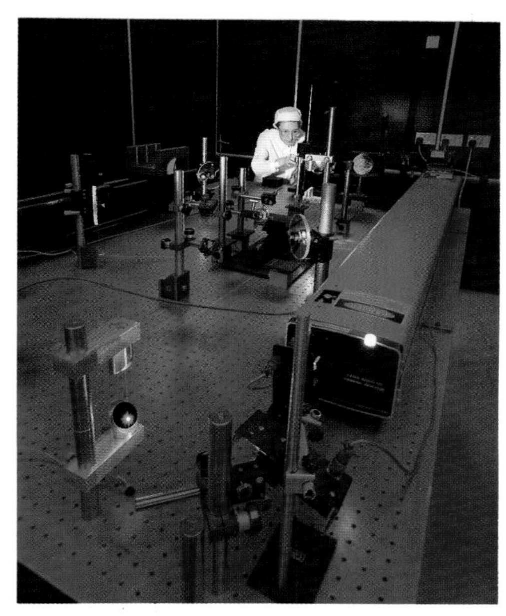

Off-line coating plant, Corby, England

Having developed float our next task was to maintain and improve product quality. The importance of quality needs to be taught and explained. Once it is understood that quality is vital then all our actions will be influenced by this conviction. I believe that it is good discipline for those at the top of a company to develop their ideas sharply enough to be able to define them in writing, although whether they should then be disseminated is another matter. Many people feel inhibited about written statements which they think sound too holy, even though I believe that quality *is* holy. Let us consider some actions and attitudes which promote quality:

- Encourage everyone to feel that they belong to a team, that they are individually important and play a vital part in the success of the team.
- Give recognition to those who achieve high performance in quality. If this recognition is personalised it is more effective even than cash, although cash may also be involved.
- Negotiate with one union only. Multi-unions can militate against the feeling of being a family working together. The Japanese make it a condition of investing anywhere that they have only one union and they consider this feeling of mutual loyalty to be critically important.
- Persuade people that quality is free – nothing is more expensive than rejects and a bad reputation – and that job security and satisfaction go with good quality.
- Employ everyone on the same basis, so that they start work at the same time and share the same canteen facilities.
- It must be absolutely clear what is expected of each person and he or she should be encouraged to make constructive criticism.
- Training and retraining are vital (General Electric in the United States quoted a figure of ten per cent of salary for training). There is no better investment in a company than well-motivated and well-trained people. Training should instil the same quality you wish to have in your product.

The concept of zero defect, getting it right the first time, is very helpful. The idea of perfection being attainable can be a real inspiration.

In the 1960s and 1970s we turned our attention to modifying the properties of float glass; 'evolutionary' research as opposed to 'revolutionary'.

In many industrialised countries energy consumption in buildings accounts for approximately fifty per cent of national energy consumption. It is estimated that in the US the heating and cooling requirements in buildings associated with the window are responsible for five per cent of national primary energy use. Subject to differences of climate and levels of industrialisation this figure is not untypical of many Western countries. The window had been regarded largely in terms of its burden on heating and cooling equipment; it was deemed a weak and capricious link in the building envelope. In response we saw as our task the development of high-insulation and solar-protection products in order to obtain better environmental control in the window.

Our first task was to research a modified float glass performance by coating the glass 'on line' in the chemically and thermally hostile environment of the float process – itself a difficult objective. We chose reflective coatings since they are more efficient in performance and economic in manufacture than absorbing, body coloured glasses. Whilst 'off line' coatings technology was being established in the 1960s, we sought economies of scale by directing our research 'on line' in the float process.

We developed the unique spectrafloat and reflectafloat solar protecting products as 'on line' processes applied in the float bath itself or in the annealing lehr. They are basic flat glass products with size and tolerance constraints generally similar to those of clear float glass. Spectrafloat is a surface modified glass, in which a coloured layer of metal ions are injected into one surface of the clear glass while it is still in semi-molten state in the float bath. Reflectafloat involves a treatment, sometimes called pyrolitic coating, applied to the hot glass during its passage through the float bath and involves the thermal decomposition of materials sprayed onto the glass in a steam of gases to form a layer which fuses to the surface. This produces a highly reflective coating.

During our 'on line' work others, outside glass manufacture, developed 'off line' solar control coating technology, giving a wide range of coating performances which were still essentially batch processes.

Our second task was to develop very high-insulation window products, following the demands of our Scandinavian market. We turned our attention to developing low-emissivity glass both 'off' and 'on line'. The resistance to heat transfer of glass is largely due to the thermal resistance at each glass surface, that of the glass body itself being fairly negligible. With multiple glazing the thermal improvement is due to the introduction of the additional glass surfaces and partly to the insulation of the enclosed space itself. The mechanism of heat transfer in the enclosed space is by conduction and convection in the enclosed gas (normally dry air) and by long-wave radiation

Above:
Top: night sight lens, military use. *Bottom*: photocaronic reactolite glass
Opposite:
Top: Felix Samuely and Partners, Engineers; MKDC (Walker, Mosscrop, Woodward), Architects; Milton Keynes Shopping Centre, c. 1973-77. *Bottom:* Anthony Hunt Associates, Engineers; Foster Associates, Architects; Willis Faber Building, Ipswich, c. 1975

J. W. Watson, Engineers; Derek Walker, Trevor Denton, David Harbord, Architects; Sewage Works, Milton Keynes, c. 1970-73

between the internal glass surfaces. By reducing the emission from the internal glass surface a large reduction in this long-wave radiation loss may be achieved and, by using a gas with improved properties (lower thermal conductivity and higher density), a reduction of the conduction and convection across the enclosed space is also obtained. Metallic surfaces, when clean, have low emission. Thus thin metallic coatings applied to glass produce surface emissivities of the order of 0.1 or less (compared with 0.85-0.9 for ordinary glass surfaces). These coatings may also increase the reflectivity of the visible and near infra-red radiation (useful for solar control glasses).

The market need, therefore, was to develop coating technology enabling surfaces to be coated with little consequent effect on light transmission, but having that low emissivity to long-wave radiation.

For the 'off line' work the challenge we set ourselves was to develop existing coating technology to produce coating performances which:

· have high reflectivity in the extreme infra-red (for thermal insulation).
· have high transparency in the visible and near infra-red (for vision, daylight and useful solar gain).
· are of acceptable high-quality appearance.
· are sufficiently durable for manufacture into multiple units.
· are stable (at least for the life of the building product).
· are affordable and cost effective.

By the end of the 1970s these criteria could be met 'off line' with a continuously operating, magnetically-enhanced sputtering plant. Pilkington's first plant was opened at Halmstad in Sweden in 1980/81 and the latest at Corby, Northants, in 1984 – both producing the low-emissivity glass product 'kappafloat'. The coatings most commonly used are based on a thin layer of copper or silver, sandwiched between protective layers of metal oxides.

In practical terms, incorporating such a low-emissivity glass can have fifty per cent improved insulation on a double-glazing unit (i.e. as effective as triple glazing) and reduces the incoming solar heat or light by only a few per cent if clear float is used. It can be combined with solar control glasses and continue to give the same high-insulation value, complete with any level of solar control desired.

By involving our kappafloat clients – the double-glazing manufacturers – and by conducting comprehensive market research the products launch was a success. I say 'success' because, while the UK is virtually the most poorly insulated country in the EEC, the public's response to kappafloat has been so remarkable that the UK market for low-emissivity glass has now become the largest in Europe!

To achieve this our marketing was supported by research projects in real buildings and by careful co-operation with Government and research bodies in the amendment of building regulations and institutional codes to pave the way for the acceptance of this innovation. This market 'preparation' is an essential element in new product development and has to start well before the product launch, in some cases at the initiation of research and development. Multiple glazing with low-emissivity coating is rapidly becoming the standard form of conventional glazing. Thus our current research and development objective is to produce a high-quality, low-emissivity product 'on line' in the float process. Additionally, we are planning new investments to determine future market needs for window technology. In the Fenestration 2000 project, with assistance and co-operation from the Department of Energy and Libby Owens Ford – our flat glass company in North America – we are involved in defining future performance requirements for the window in order to more clearly define research and development objectives. This involves close collaboration with experts outside our company specialising in economics and building construction materials technology, as well as others in political, social and technical aspects, so that we can clearly understand the context of the future building industry and its probable effects on detailed performance specifications. The definition of future markets must involve Government, universities and other specialist bodies outside our own company. From this work we will obtain a better definition of the variable performance window as part of the future 'intelligent building'.

Felix Samuely and Partners, Engineers; Derek Walker, Stuart Mosscrop, Christopher Woodward, Architects; Milton Keynes Shopping Building, 1972-77

Let me summarise my main points. Company effort for product development must be led by market needs. Investment in defining those needs is well spent. It is also important that 'success' be defined, for there is always a temptation to *do* rather than identify what is worth doing. Both process improvement and new process development are necessary for industrial success, the former provides a greater guaranteed reward but both the risks and rewards of the latter will be greater – taking big steps will involve big risks. Nevertheless, process improvement is vital as the search for major advances has to be paid for. The field of innovation must be defined and that must focus people's thinking in order to concentrate their efforts. Attention must also be paid to seeking convergence between 'what can be done with a material' – such as glass – and 'what the market wants from it'. A company must not pursue too many major projects although there may be many ideas that show potential. The atmosphere must be set right to encourage innovation. New ideas should be treated gently at first and only later subjected to rigorous testing when key questions have been defined and answered. New processes and process improvements should be designed with people in mind, not just as technical exercises. The contribution of all those with ideas should be sought both inside and outside the company; knowledge is not a prerogative of senior people. Industry must plan its use of human resources and aim to make the best use of all the talents they possess. It is important to differentiate between project direction and project management; it is important that the Board should believe in research and development and it is encumbent on research and development to generate this support. Finally and most importantly, research and development should give, and be seen to give, value for money.

PROFESSOR DEREK WALKER

Today's Engineers – The Legacy Lives On

Above:
Top: Büro Happold, Engineers; 'Pink Floyd' umbrellas, Stuttgart concert, c. 1979. *Bottom*: Ian Liddell testing the rig for the umbrellas
Opposite:
Anthony Hunt Associates, Engineers; Foster Associates, Architects; The Sainsbury Centre, University of East Anglia, c. 1978, interior view

In describing today's elite group of structural engineers, one is aware that architectural prejudice colours the selection process and this prejudice becomes frustrating when one discovers qualities common to them all.

High-quality structural engineering is a small world and the atelier process which develops the invention and elegance of technical solutions is, alas, confined to very few practices. The oldest established group is Freeman Fox, founded in 1857 by Sir Charles Fox, fresh from the dissolution of Fox Henderson of Smethwick. Fox and his partner were the real heroes of the Crystal Palace in carrying out for Paxton detailed design, fabrication and erection of the Exhibition Building in the unbelievably short space of eight months. Paxton was not Fox's only link with the heroics of the first half of the century; he also fulfilled every schoolboy's ambition by acting as an engine driver on the Liverpool-Manchester Railway for a short spell just prior to becoming a pupil of Robert Stephenson and, later, assistant engineer on the London-Birmingham Railway. The Freeman Fox practice has always stood for invention and innovation, from the first railway era and the Crystal Palace, to the Sydney Harbour Bridge and the Festival of Britain. They have been the outstanding bridge builders of the 20th century and their consistency and quality has been the yardstick for today's consultant engineers.

Unlike Fox, Baker, Bazalgette and Williams, the giants of the post-war years, were of continental extraction. Ove Arup was from Denmark and Felix Samuely from Austria. Their paths briefly crossed in 1933 during Samuely's first post in England with J. L. Kier – the contractors in an engineering office under Ove Arup's direction.

Many brilliant engineers passed through the office, including Frank Newby who has worn the mantle of the great man with such distinction, Sven Rindl whose Tatlin Tower will be remembered by all those who revelled in the Constructivist show at the Hayward in the early 1970s and Anthony Hunt, who subsequently formed his own practice and engineered most of Team 4 and Norman Foster's earliest works with such panache: Reliance Controls at Swindon, Willis Faber at Ipswich and the Sainsbury Centre at the University of East Anglia. Hunt's work and philosophy remind me so much of Konrad Wachsman whose pursuasive homage to the perfect joint also inspired Newby, Happold and others who completed their early apprenticeship in the United States.

Ove Arup, like Samuely, gathered around him exceptional designers; he

Right: Sainsbury Centre, typical cross section
Opposite:
Top: Sainsbury Centre, interior view. *Bottom*: exterior cladding

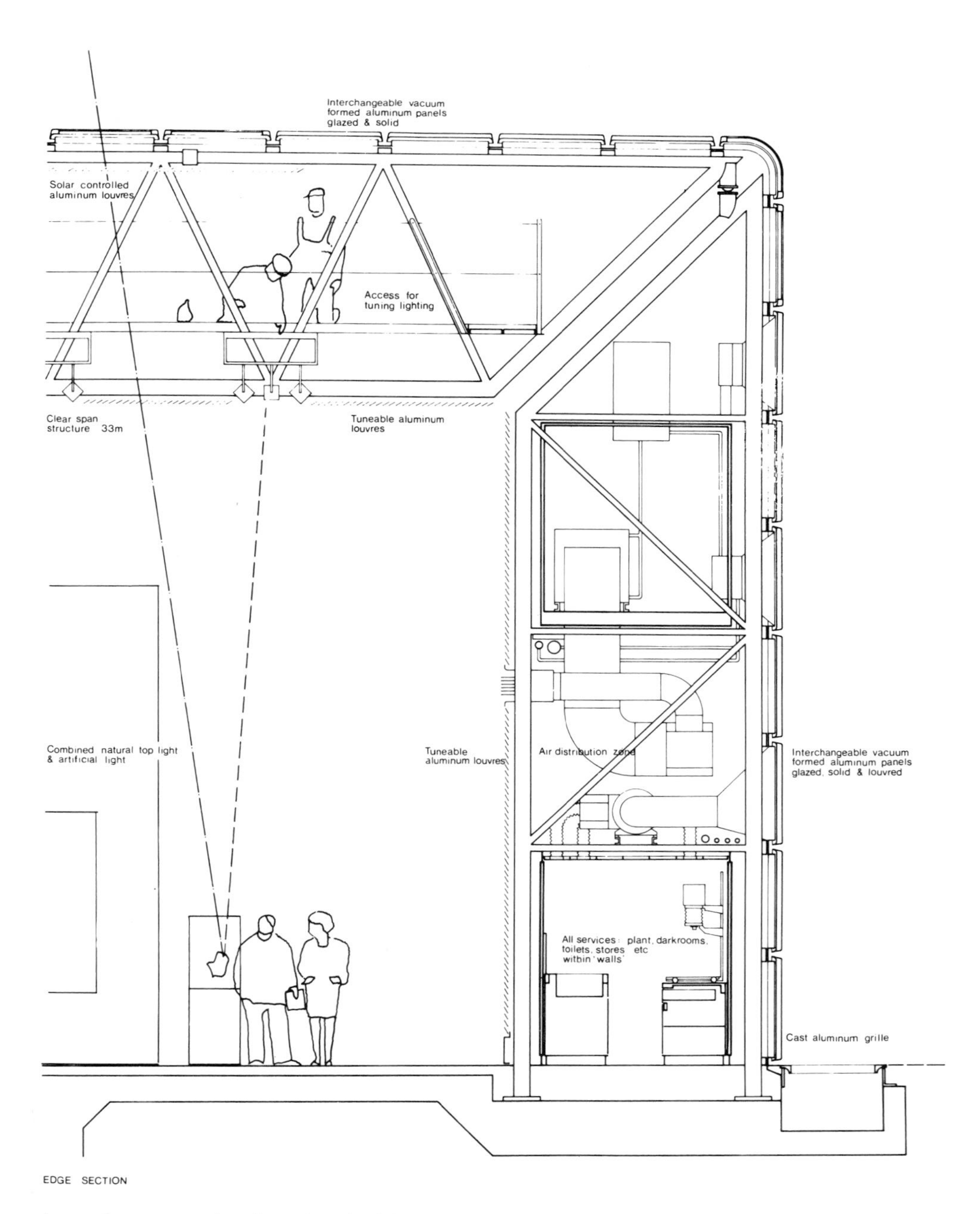

has always praised and upheld the notion of multi-disciplinary work. He has been the Robert Stephenson of the 20th century: a designer who could organise and inspire simultaneously, capable of generating loyalty, dictating quality and delegating responsibility with the certainty and infallibility of a papal edict. Jenkins, Hobbs, Zunz, Beckmann, Michael, Rice, Smythe, Lewis, Henkel and Barker are all superb designers who have continued to develop in depth the unique quality of the Arup organisation. Their great strength has always been an incessant zeal for research and invention, an attitude attracting brilliant graduates like bees to honey.

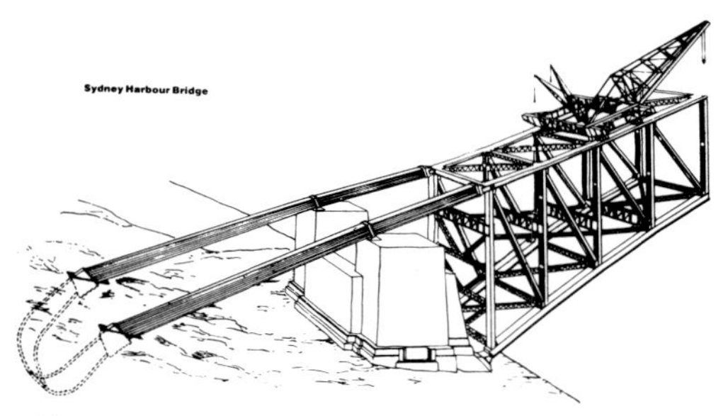

Above:
Ralph Freeman, Architect; Sydney Harbour Bridge, Australia, c. 1924, arch erection
Opposite:
Top: Forth Road Bridge, c. 1964, main span 1,004 metres. *Bottom*: Bosphorus Bridge, c. 1983, 1,074 metre main span, dual three-lane highway
Page 222:
Top: Sydney Harbour Bridge, general view today. *Bottom*: The Humber Bridge, 1,410 metres
Page 223:
The Victoria Falls Bridge, c. 1906, plan, general elevation and structural details
Page 224:
The Humber Bridge: erecting a box in the main span (*top*), Barton access jetty for pier site (*centre left*), cables being painted with red lead (*centre right*), erecting the Barton access (*bottom left*), tower foundation diagram (*bottom right*)
Page 225:
The Humber Bridge: viewed from Hessle Tower (*top*), construction shot (*bottom*)

Freeman Fox and Partners

Freeman Fox is one of the oldest firms of consulting engineers in the world, and has, since it was founded by Sir Charles Fox in 1857, been responsible for a wide range of major engineering works in over thirty countries. During the early years its work was largely concerned with railway projects including bridges, in Britain and abroad. Over the last fifty years activity has extended to other fields, including bridges, roads, expressways, tunnels, dams, hydro-electric and thermal power stations, buildings, special structures, radio and optical telescopes, mass freight and passenger transport systems.

From 1969, when the group carried out further studies and set up the network for the Hong Kong Metro, Freeman Fox have developed an in-depth know-how of mass and light transit railways, participating in projects in Iraq, Kuwait, Taiwan, Greece, the United Kingdom, Mexico and Venezuela as well as continuing an involvement in Hong Kong. This combined experience has placed the firm in the forefront of dealing with passenger transport systems.

The practice has also had a tradition of engineering works underground with the design and constructions of tunnels, particularly immersed tube tunnels. Douglas Fox was knighted for the Mersey Tunnel and many years later the practice completed the 6,000-foot cross harbour Hong Kong Tunnel. Of the two thousand bridges designed by Freeman Fox, there are many memorable structures and a plethora of innovation and controversy.

Baptism by fire has always been a feature of the practice. Thus, when Ralph Freeman joined the firm in 1902 almost his first job was to make the calculations for the Victoria Falls Bridge. The project, completed in 1904 with a span of 500 feet, was the largest bridge of its type in the world. The Fox connection was maintained with Charles Beresford Fox, son of Francis, working on the team.

The tradition continued with Sydney, the new Thames Bridge, the Forth Road Bridge, and the Severn and Wye Bridges on the English/Welsh border. In recent years the Bosphorus Bridges and the Humber Bridge consolidated the large span development work pioneered by the practice. The Humber project in particular – the world's largest single-span bridge with a main span of 1,410 metres – is a masterpiece of design, economy and elegance.

In 130 years, Freeman Fox have maintained a reputation in all six continents, with railways from Cape Town to Rhodesia, the Snowdon Mountain Railway, the Argentine Central Railway, Victoria Falls Bridge, a definitive Report on the Channel Tunnel (in 1906), St. Paul's Cathedral, Otto Beit Bridge over the Zambesi River, Auckland Harbour Road Bridge, High Marnham Power Station, Algonquin radio telescope, Ontario, Erskine Bridge, River Indus Suspension Bridge, Bangkok Expressway, Simon Bolivar Bridge over the Panama Canal and the Baghdad Metro – just a few from a portfolio in which tradition is equated with quality performance. Innovation and invention is the life blood which will extend the firm's dynasty into the 21st century.

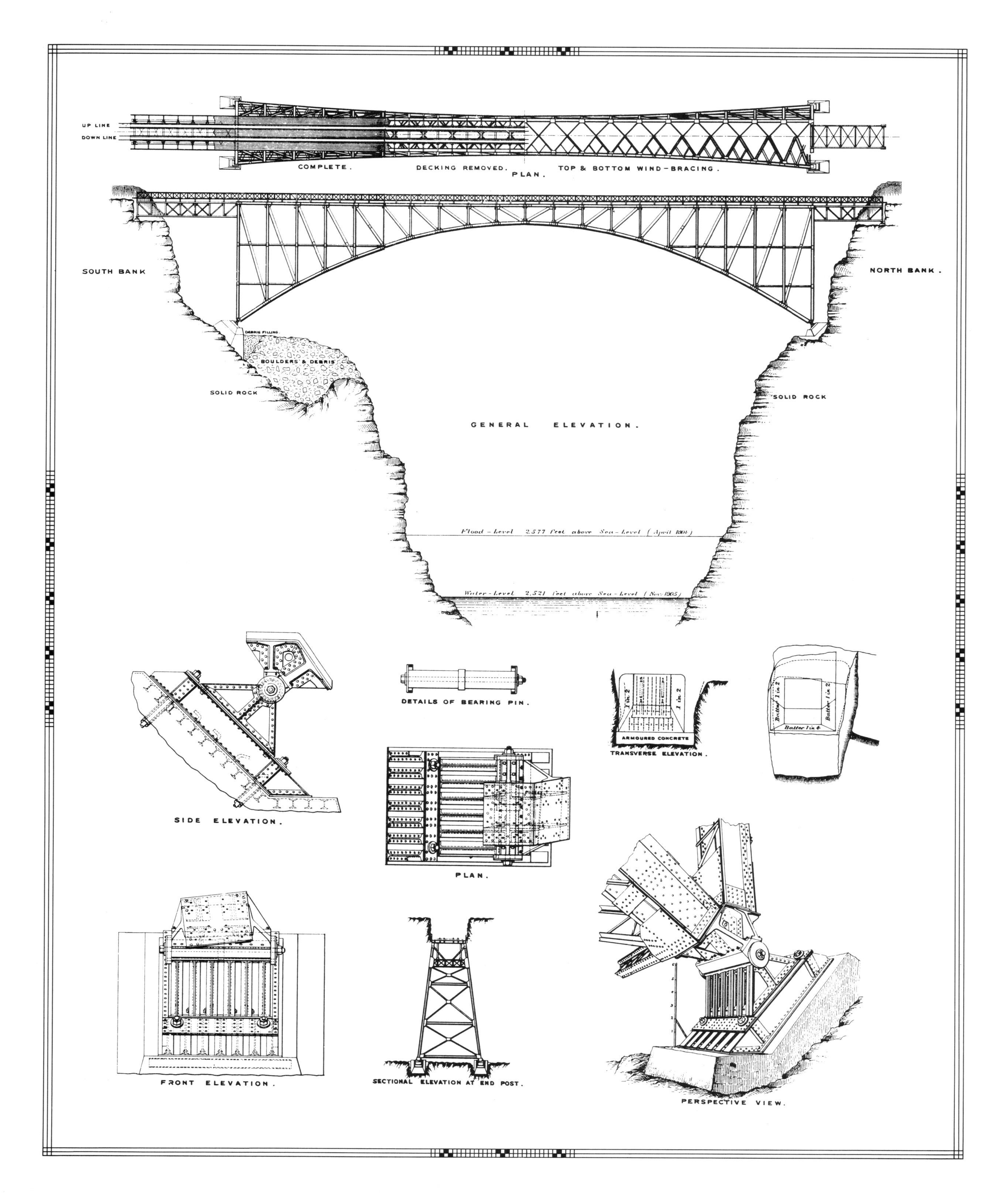
UP LINE
DOWN LINE
COMPLETE.
DECKING REMOVED.
PLAN.
TOP & BOTTOM WIND-BRACING.
SOUTH BANK
NORTH BANK.
DEBRIS FILLING.
BOULDERS & DEBRIS
SOLID ROCK
SOLID ROCK
GENERAL ELEVATION.
Flood - Level 2,577 feet above Sea - Level (April 1904)
Water - Level 2,521 feet above Sea - Level (Nov. 1905)
DETAILS OF BEARING PIN.
1 in 2
1 in 2
ARMOURED CONCRETE
TRANSVERSE ELEVATION.
Batter 1 in 2
Batter 1 in 2
Batter 1 in 4
SIDE ELEVATION.
PLAN.
FRONT ELEVATION.
SECTIONAL ELEVATION AT END POST.
PERSPECTIVE VIEW.

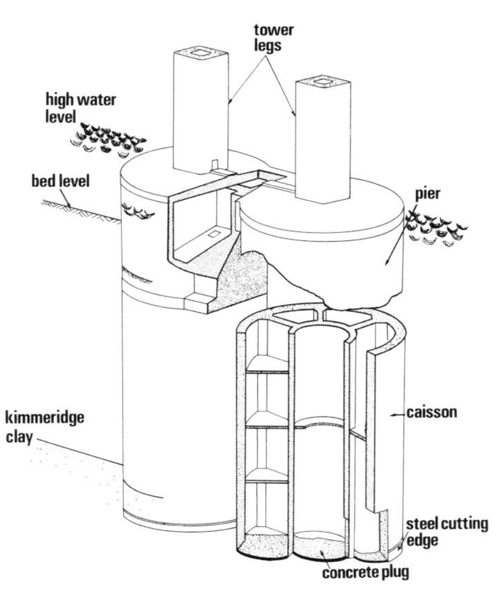
tower legs
high water level
bed level
pier
kimmeridge clay
caisson
steel cutting edge
concrete plug

Felix Samuely and Partners

Felix Samuely was born in Vienna in 1902. He studied in Berlin, where he obtained his engineering degree in 1923 and, apart from a year in an architect's office in Vienna, worked with contractors up to 1929 when he started his own practice with Stephen Berger. This lasted until 1931 when he went to Russia. During those two years Samuely designed the first commercial welded-steel building in Berlin, worked with Erich Mendlesohn and built a factory with Arthur Korn. On welding he writes:

> I have worked on the problem of welding of steel construction since 1928 when I had occasion to design an experimental, welded structure for a water cooler for Siemens Schukert. This tower was constructed as an imitation of a riveted building and it struck me forcibly that this was the wrong method of procedure and that welded steelwork should have its own design!

After two years in Russia working on the design of a steelworks and researching steel construction, Samuely returned to Berlin via China and came to England in 1933. He obtained work with J. L. Kier contractors, his first job being the calculations for the Penguin Pool and, on winning the Bexhill competition, Mendlesohn and Chermayeff invited him to act as engineer. He set up practice with Cyril Helsby and was later joined by Conrad Hammam.

The practice prospered and Samuely was the engineer for many of the Modern Movement buildings of the late 1930s. He collaborated with such architects as Wells Coates, Lasdun and Connell, Emberton, Goodhart-Rendel, Pilichowski, and Ward and Lucas on both steel and concrete structures. He was also an active member of the MARS group and in 1942, with Arthur Korn, became joint author of the *MARS Plan of London*. He had a passionate interest in transport and independently proposed his own rail plan for London.

After Samuely's death at the age of fifty-seven in 1959, the practice has been carried on by his brilliant protégé, Frank Newby, a Yorkshireman whose considerable skills have further extended the scope of the practice. Newby has also continued the tradition of collaboration on innovative building types, such as Leicester University Engineering Building with Stirling and Gowan; the Aviary at London Zoo with Snowdon and Price; Cambridge Library, St. Andrew's Halls of Residence, Olivetti and the Tate Extension all with James Stirling; W. D. and H. O. Wills with SOM; in addition to projects as diverse as Bristol Cathedral, the US Embassy in Grosvenor Square and the Shopping Building at Milton Keynes.

Newby's work and scholarship was acknowledged in 1986 by the award of a gold medal by the Institution of Structural Engineers.

Many of Britain's most gifted architects owe a great deal to Samuely and Newby; their teaching contributions at the Architectural Association and within their own practice have enriched engineering in Britain and influenced the pioneering instincts of many of their collaborators and colleagues.

Above and Opposite:
Stirling and Gowan, Architects; Leicester University Engineering Building, c. 1959-63: flue seen through structure (*top*), glazing at the corner (*bottom*), general view (*opposite*)
Page 228:
Powell and Moya, Architects; 'The Skylon', Festival of Britain, c. 1951: erection of the Skylon – main vertical 250 feet, maximum diameter of thirteen feet built up by welded frames cradled by three pylons seventy feet each, held together by prestressed cables
Page 229:
Top: Barry Gasson, Architect; The Burrell Collection, Glasgow, 1971-84, laminated timber trusses are supported on concrete columns.
Bottom: Brown Daltas Associates, Architects; SAMA Bank, Jeddah, c. 1978, joint detail to transmit up to 1,000 tons (*left*); office building, Jeddah, c. 1980, reinforcement details (*right*)
Pages 230-1:
Lord Snowdon, Cedric Price, Frank Newby, Architects; Snowdon Aviary, London Zoo, c. 1962: general view and structural details

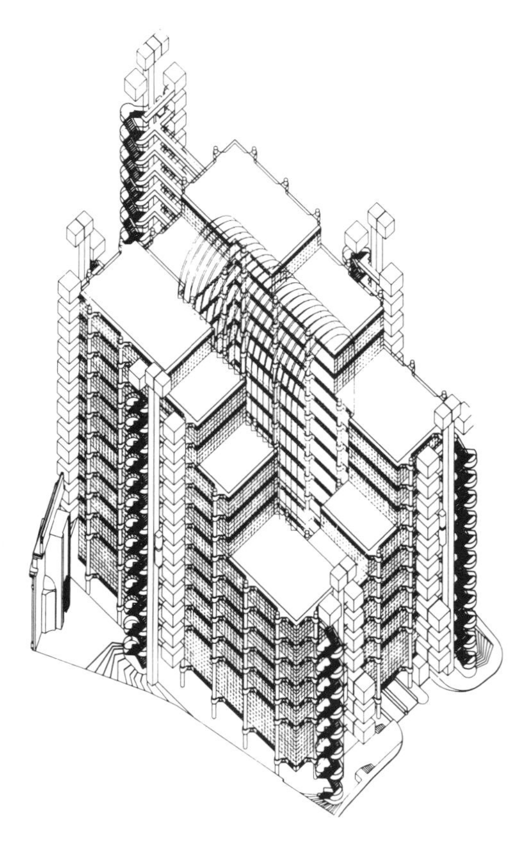

Above:
Richard Rogers Partnership, Architects; Lloyd's Building, London, c. 1986, axonometric
Opposite:
Top: Kingsgate Footbridge, University of Durham, c. 1961-4. *Bottom*: Lloyd's Building, main entrance
Pages 234-236:
Piano and Rogers, Architects; Centre Georges Pompidou, Paris, c. 1971-6: up view of service ducts (*page 234*), details of service elevation, gerberettes and escalators (*page 235*), detail section through escalators, view to piazza and upper walkway (*page 236*)
Page 237:
Frei Otto, Architect, with Edmund Happold and Ian Liddell; Garden Centre, Mannheim, c. 1975, construction of timber lattice

Ove Arup and Partners

The work of Ove Arup and his Partners is legendary. They have managed the impossible – the combination of high-quality work in a massive dispersed organisation. They have engineered many of the seminal buildings of the last fifty years – the Sydney Opera House, Pompidou Centre, Hong Kong and Shanghai Bank, Stuttgart Staatsgalerie – and they still find time for little gems like the Menil Collection Museum, IBM's Travelling Museum,the Patscenter and the Durham Footbridge. They have provided a finishing academy for bright young graduates and have generously made their research facilities available to most architects interested in the extended learning curve of new construction techniques and environmental and material analysis. Their philosophy is adequately summed up by Ove himself in a paper given at the Building Services Inaugural Speech at the Institution of Civil Engineers in 1972:

> Architects and engineers both see themselves as designers. And although the majority of engineers and a great number of architects can hardly be called that, it's the designers I am concerned with here. For the design, as I use the word, is the key to what is built; it is the record of all the decisions which have a bearing on the shape and all other aspects of the object constructed. These decisions are unfortunately not all taken by the designer but they must be known to him and integrated into a total design.
>
> We must distinguish between routine design, which does not require any creative thinking, and what may be labelled original, innovative, conceptual or creative design. Creative design must of course build on previous experience and contains and employs predesigned parts, and it may even consist almost entirely in assembling such parts to create an entity. But building is always tied to locality and to the people one builds for, and they vary from case to case. The synthesis required to create an entity, a whole which economises in means yet fulfils the aims, is an artistic process.
>
> Art is solving problems which cannot be formulated before they have been solved. The search goes on, until a solution is found which is deemed to be satisfactory. There are always many possible solutions, the search is for the best – but there is no best – just more or less good. Quality is produced if the search doesn't stop at a second-rate solution but continues until no better solution can be found.
>
> An engineer who doesn't care a damn what his design looks like as long as it works and is cheap, who doesn't care for elegance, neatness, order and simplicity for its own sake, is not a good engineer. This needs to be stressed. The distinctive features of engineering are mainly matters of content – the nature of the parts and the aims. The success of the whole undertaking depends on the right allocation of priorities and whether the resulting entity has this quality of wholeness and obvious rightness which is the mark of a work of art.

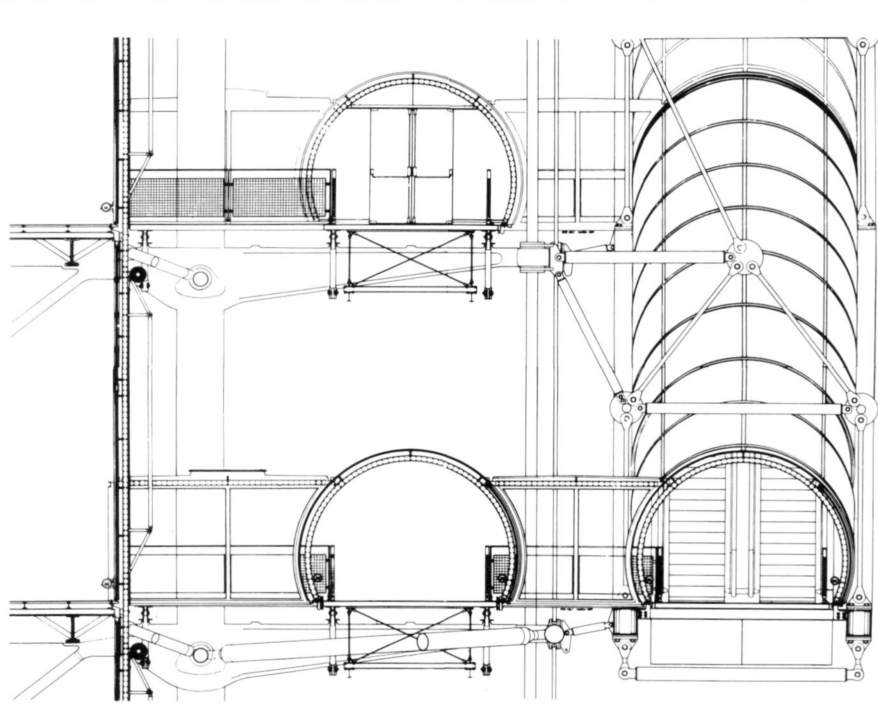

Above:
The Partners, 1987
Opposite:
Derek Walker Associates, Architects; Energy Pavilion, WonderWorld, Corby, 1986: energy playground model (*top*), cross section (*bottom*)
Page 240
The 'Pink Floyd' umbrellas, Stuttgart, c. 1979
Pages 241-2
Rolf Gutbrod/Frei Otto, Architects; Sports Stadium, Jeddah, c. 1980: detail of cable nets (*page 241*), long view and construction of cable nets (*page 242*)
Page 243
Derek Walker Associates; Kowloon Leisure Park, Hong Kong, 1987
Page 244
Top: Frei Otto, Architects; Garden Centre, Mannheim, c. 1986. *Bottom*: Arni Fullerton, Architect; 58° North Project, Canada, c. 1983, development model, transparent air-supported roof
Page 245
Aviary, Munich, c. 1980, details and general shots during construction and on completion

Büro Happold

For eighteen years Edmund Happold served Ove Arup and Partners as a design engineer and as executive partner for the innovative group of Structures 3 within the practice. Mannheim, Arctic City, the Council of Ministers in Riyadh, Pompidou and the hotels at Mecca and Riyadh came under his design initiative and, during that period, he developed a relationship with Frei Otto that has been outstanding in its consistently inventive development of light-weight structures.

Within Structures 3 a group of friends – Buckthorpe, Liddell, Dickson and Macdonald – started to develop the philosophy which was to become the rationale behind Büro Happold, which was formed when Happold took the chair of building engineering at the University of Bath.

Since 1976 Büro Happold has practiced working in a bewildering variety of international environments: long-span structures, air-supported structures for Arctic conditions, airship frame design, bicycle design, buildings, tunnels and pneumatic flood control systems. If Arup is the Stephenson of the current period, Happold is the John Smeaton – versatile, passionate, a science-based artist with strong entrepreneurial talents. As current President of the Institution of Structural Engineers, he is using this platform to reactivate and renew the values of invention and leadership within the engineering profession. The sad reflection I have as an architect is that there are still too few engineers who can span the chasm that separates inventive design from humdrum mediocrity. This is not a factor in assessing the Happold group for they have assumed a personality well-adapted to the complexities of many of today's building types; looking at constructional problems in a multi-disciplinary way means that different members of the team assume leadership of a project at different times during its conception and implementation. For, in the final analysis, we should be producing buildings which reflect the times and we will never do this without an orchestra which can play in tune.

Perhaps it is appropriate to describe the unique quality of today's engineering by summarising Happold's personality – entrepreneurial, inventive, analytical, literate and lively. This is precisely how it was in 1837 and long may the legacy be maintained.

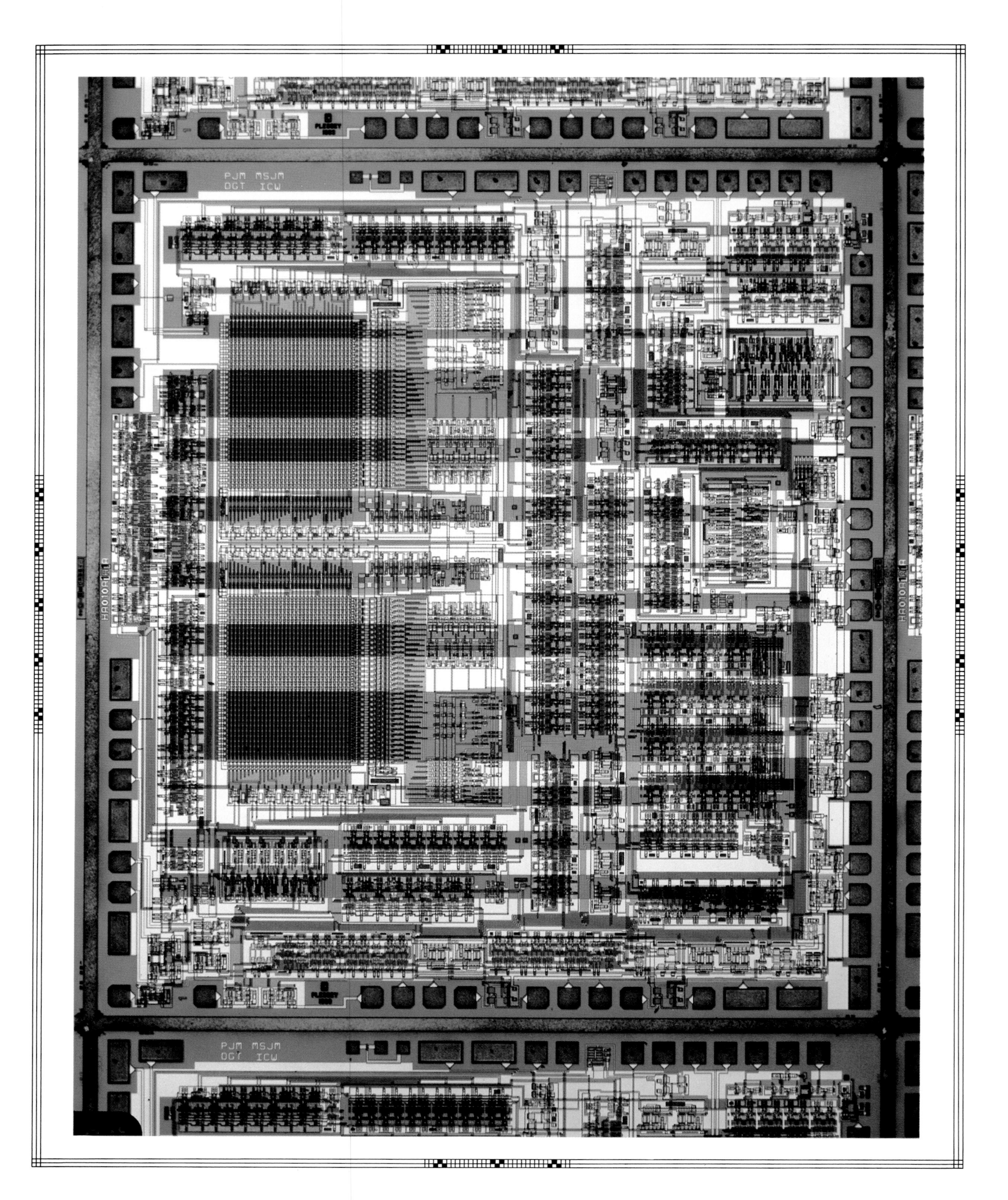
PJM MSJM
DGT ICW

DR. JOHN C. BASS

Solid State Technology – The Stimulus of Electronics

We live in times of ever-expanding electronic communications and information processing systems – between people and between electronic processing and computing machines. Information of all kinds is processed within the machines and transmitted across networks of cables and optical fibres, or broadcast by radio, microwave beams or satellites, with instantaneous worldwide communication. The world has been called an 'electronic village' since time and distance are no longer problematic factors within communications.

This great expansion over the past twenty years has been due to the growth of microelectronics – the silicon microchip – and, more recently, optoelectronics. These are the major fields of what is known as 'solid state technology' and they are derived from applied research and development in the physics and chemistry of materials called semiconductors. Semiconductors are weak conductors of electricity, unlike metals which conduct currents strongly.

Solid state technology, and the resulting worldwide microelectronics industry, has been perhaps the biggest ever technological thrust, involving billions of pounds worth of international research. It could all be said to have started in 1948 with the invention of the transistor at Bell Telephone Laboratories. This came about through research into the ill-conceived physics of detection devices used in early radio and radar systems. These employed metallic point contacts – 'cat's whiskers' as they were called – on silicon crystals or other semiconductors.

If we look back into the 19th century, electrical systems and the early telegraphic communication systems required only readily available solid state materials, such as iron to produce magnetic fields, copper to conduct electricity and glass or natural fibres to provide insulation. Telegraphic communications were performed manually by the operator. With the advent of the electronic valve came electronic control of electrical signals and more sophisticated materials were introduced, such as tungsten and thorium to provide the electronic emission and current flow in valves. Valves also operated much faster than manual switches and, more importantly, control grid electrodes within them allowed weak electrical signals carried by the current to be amplified. As valves developed further, electrodes were introduced to improve their signal amplification characteristics and so the electronic valve became the foundation of radio communications and electronics until the early 1960s. But, despite all these improvements, the valve had the great disadvantage of requiring considerable electrical power

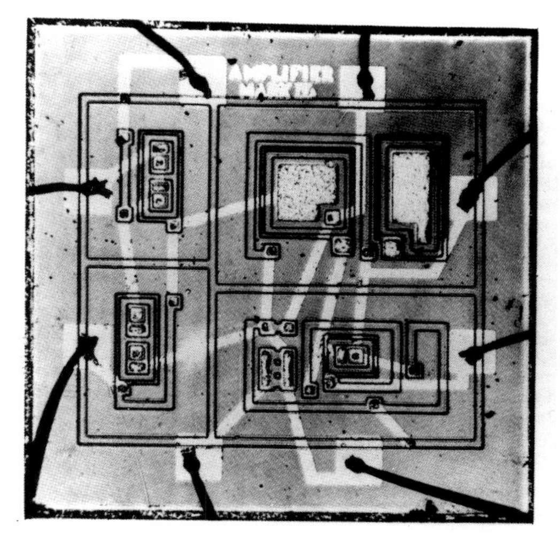

Above:
A very early silicon microchip of 1963, the Plessey Mark IV A amplifier. It contains six transistors, seven resistors and two capacitors. It operates up to 5 MHz. The smallest features are 30 *um* across
Opposite:
A modern silicon microchip developed by Plessey for 565 Mbytes per second fibre optics communications

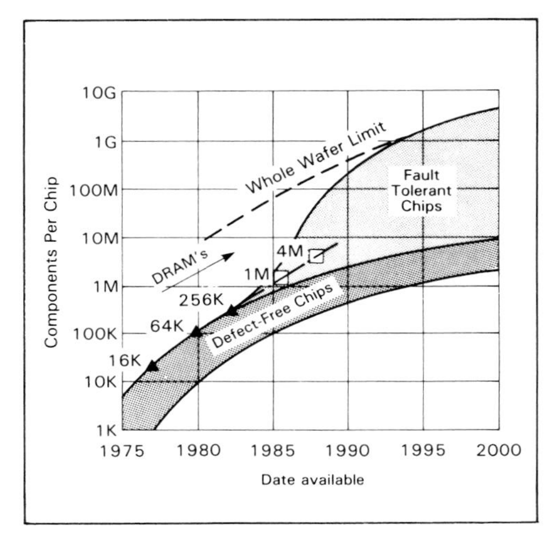

Above:
The increasing complexity of very large scale integration of silicon chips.The figure also shows the benefit fault tolerance as a practical solution to the design of large chips
Opposite:
Top: A typical silicon wafer also showing some test elements among the chips. *Bottom:* a modern gallium arsenide microwave amplifier chip for 6000 MHz operation. Notice the microwave transistor lines and the square spiral inductors

simply to heat its cathode to sufficiently high temperatures to produce electronic emission in the valve. This high temperature was the major cause of their restricted operating life. So early electronic systems which required large numbers of valves – like the first computers of the 1940s and 1950s – were impracticable and unreliable.

The 1948 invention of the transistor gave the breakthrough which led to the possibility of really complex electronic systems. The transistor introduced first germanium and then silicon to the range of solid state electronic materials. Current flows in such semiconductors without the need for heating power and, as a consequence, they proved much more reliable than valves, as well as being faster to operate because they were much smaller.

It was soon realised, both in the United States and in this country, that it should be possible to build integrated assemblies of the tiny transistors and their connections within the surface of a silicon layer. This would give building blocks for applications in systems like the new computers. Likewise, the possibilities of complex but reliable miniature circuit blocks, consuming very small amounts of power, were recognised for their potential military value in such areas as space or missile systems, or light-weight portable service radios. From these early beginnings of microelectronics we now have an all-pervasive technology.

The greatest impact of solid state technology so far has been produced by silicon microelectronics – the silicon chip. We can now place a million transistors or other components on a silicon chip about a centimetre square. The linewidth – the smallest feature size of a circuit on a chip – has decreased dramatically over the past quarter-century. A typical linewidth in the first integrated circuits in the early 1960s was thirty micrometres. Today, linewidths are commonly of the order of one micrometre (which is about one eightieth the diameter of a human hair). But how much smaller can we go? The answer is that extrapolations based on historical trends suggest that linewidths around 0.1 micrometres will be reached by the mid-1990s. However, it is around 0.1 micrometres that theoreticians believe that current flow in devices will no longer behave classically. At such small geometries the motion of the current-carrying electrons may change dramatically. This is because impurities and defects within the silicon crystal have separations which are of the same order, so that whereas at larger geometries electrons see many such locations as they travel across a transistor, whose scattering and slowing effects on the electrons are averaged, this is no longer the case at very small geometries. Thus electrons are expected to travel, unimpeded or ballistically rather as they do in valves, and this may have profound effects on transistor operation and their electronic noise generating processes.

The complexity of today's silicon chips is produced by very sophisticated manufacturing processes. These borrow techniques from the printing industry, using replication by photolithography, to produce maybe two hundred identical silicon chips on a disc-shaped wafer of single crystal silicon which, in today's industry, is generally six inches in diameter. There are very

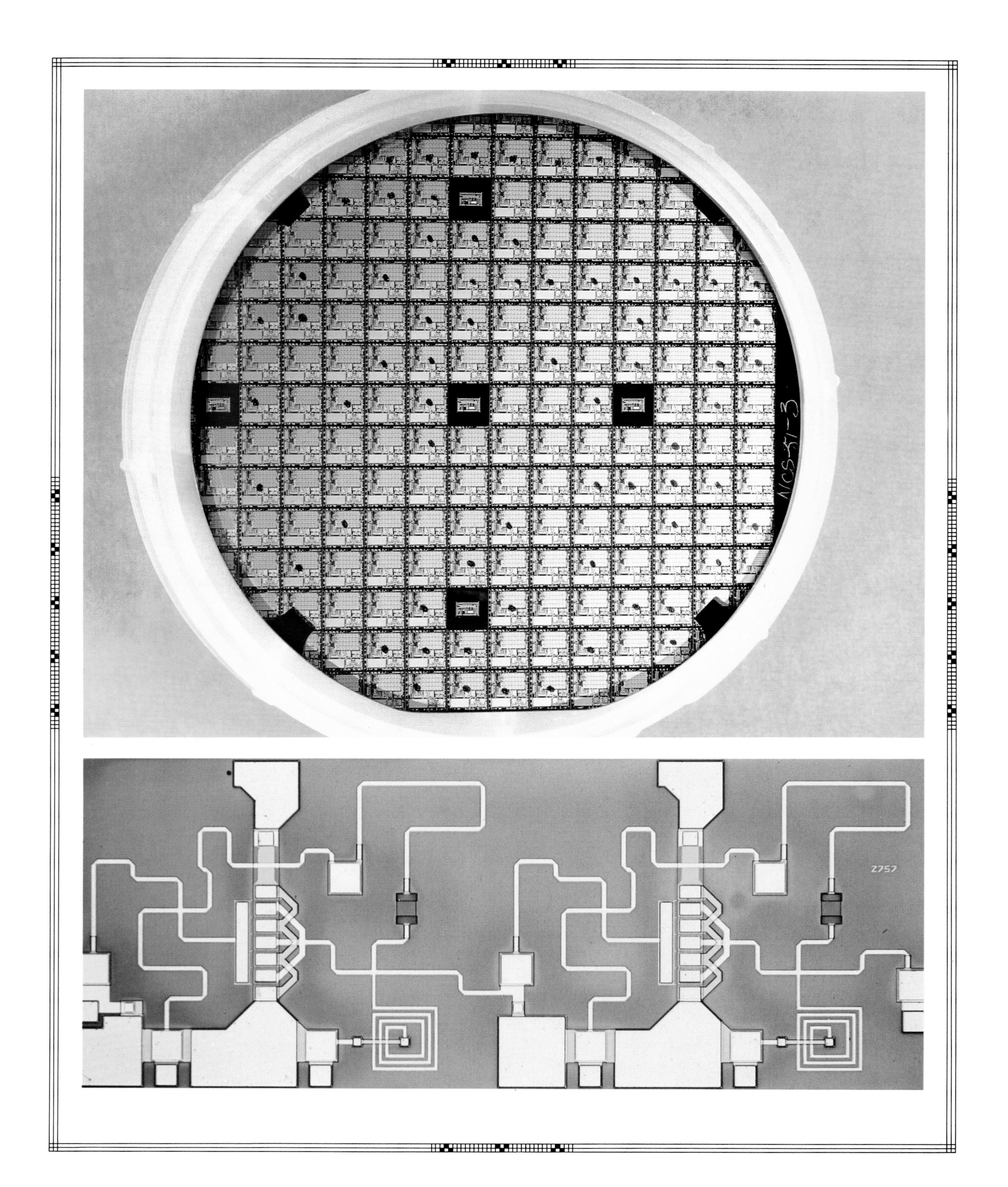

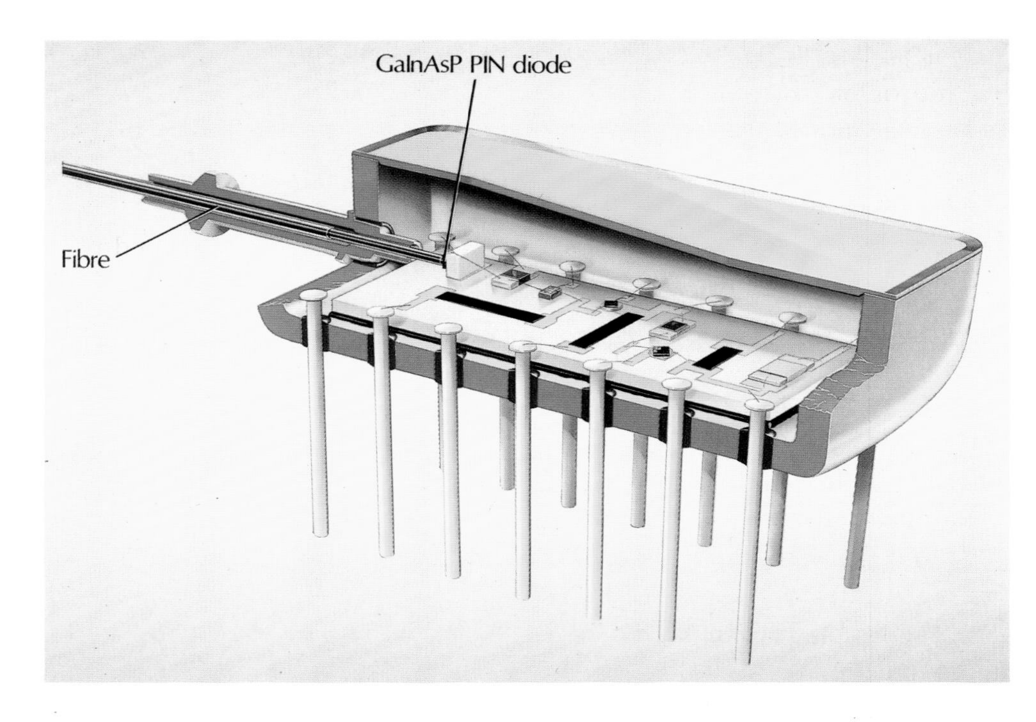

Detectors: once the optical signal is received it has to be detected. Semiconductor devices are used, and for long wavelength systems Plessey has developed a hybrid package containing a PIN photodiode aligned to a fibre pigtail and incorporating an FET transistor and associated circuitry. A cut-away diagram of the package is shown above. This receiver is stable, requires a low bias voltage and shows excellent sensitivities out to data rates of 1 Gbit/s or more. Along with the light source, it can be incorporated into a complete regenerator unit.

many steps in the sequence of building circuits on silicon chips. The transistors on them are usually made by oxidation of the silicon wafer to give electrical isolation between areas, and also through the introduction of atomic 'doping' impurities of different types to give surface regions of electronic properties different from that of the original silicon wafer. Connecting holes and areas are delineated by photosensitive resists, exposed through photographic masks of micron tolerance. Annealing procedures must also be carried out to restore the crystal to its original condition following the introduction of the impurities. Metals and alloys are deposited to give the pattern of contacts between transistors and circuit elements, using techniques such as vacuum evaporation or sputtering from the plasma of gas discharge. After the chip pattern on the wafer is built up, it is cut into its separate chips, each of which might contain as many as a million transistors for a memory, or twenty thousand gates of electronic logic for a communication circuit. The chips are then mounted on to packages and fine gold wires connect their electrodes to the pins of the packages.

These processing steps and techniques offer great flexibility and have given rise to a great range of different processes using the basic families of transistors and circuit components. The two major families are the bipolar transistor and the metal-oxide-silicon (MOS) transistor. The former have higher switching speed characteristics and somewhat higher drive powers than the latter, and are often used in signal processing applications. MOS transistor circuits are often used in a complementary configuration (CMOS) of balanced pairs of devices. These are slower in operation than bipolar circuits but consume very little power. These are often used in memory or processing applications where minimum power is required, such as in watches or the great mass of telecommunication circuits and computers.

It is apparent that manufacture of such microscopic chips brings with it great problems in yield with which the industry has constantly struggled to achieve economic and profitable production. Dirt particles must be eliminated by carrying out processing in clear air rooms in which by today's standards there are less than ten particles (and these must be below one micrometre in size) per cubic foot of air. Highly purified water and pure chemicals, with the order of parts per billions of impurity, must also be used. Furthermore, no chip circuit can be better than the photographic mask set used to define it. It has been found over the years that the yield of chips is highly dependent on their area because it is difficult to make circuits much larger than the distance between blemishes of various kinds. As a consequence, chip sizes have remained generally around six-to-ten millimetres square, although the individual devices on them have shrunk a hundredfold over the years. New attempts are being made to overcome these limits through 'wafer scale integration' which provides very large chips for massive processing applications. But, for these to succeed, fault-tolerance, through redundancy in the many individual circuits on the wafer-size chip, will be necessary to avoid any areas of physical imperfections.

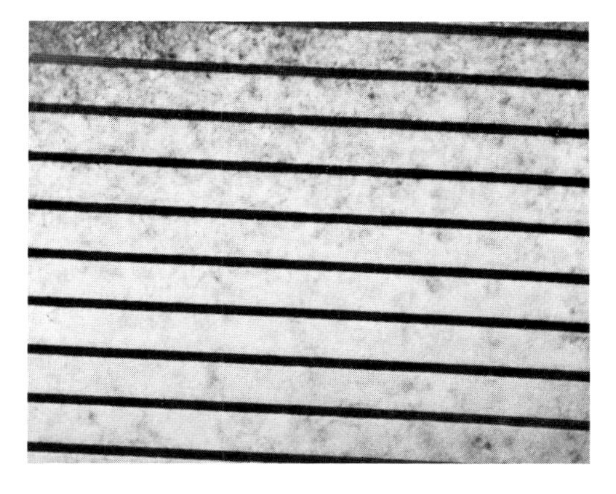

Top: a 'super lattice' or multiple quantum-well structure in gallium arsenide and gallium aluminium arsenide taken by electron microscopy. *Bottom*: a 0.3 micron wide feature (the Y shape) in a gallium arsenide device produced by electron beam lithography

Progress towards ever smaller features has been the continuous preoccupation of the industry. As size diminishes, circuits on chips become faster and consume less power since the current carriers have less distance to travel. However, the relatively long wavelength of the ultra-violet light needed for photolithography is, at today's limits of around one micrometre, barely able to cut the extremely sharp features into the photoresist covering the wafer. To overcome this and advance to even smaller geometries, much effort is now devoted to electron beam lithography. The wave length associated with the particle-wave duality of electrons is a fraction of an atom in diameter which provides no basic limitations. The research effort on electron beam lithography is therefore largely concerned with such matters as developing photoresists of adequate sensitivity and resolution to electron exposure. Also there are the engineering practicalities of constructing highly stable electron beams with sufficiently intense current and control to allow lithography at economically high production rates on silicon wafers. An alternative technique under study is the use of very short-wave length X-rays for lithography through sets of closely aligned masks. These X-rays would be produced by synchrotrons but, unfortunately, present synchrotrons, like particle accelerators generally, are quite large facilities. Considerable effort will be necessary to produce the table-top size models that the microelectronics industry would need.

In considering the solid state technology of silicon chips we should not forget the parallel development of techniques for the design of complex circuits. With twenty-thousand gates on a typical logic circuit or several hundred thousand bits of memory as design requirements, only the most sophisticated computer-aided design methods could cope with translating a specification for a system requirement into a physical layout of micron-sized

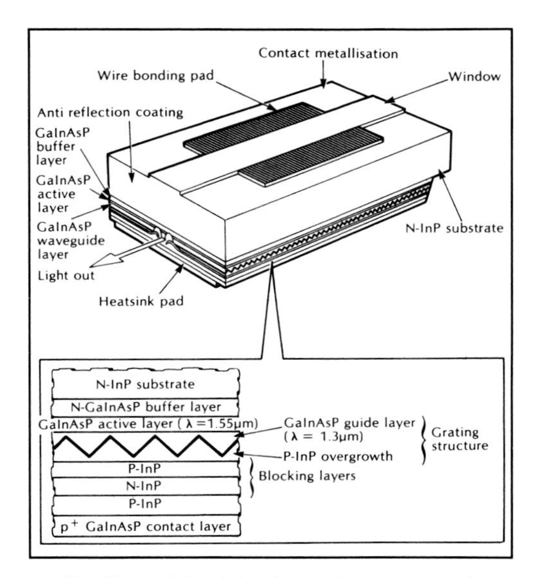

A distributed feed-back semi-conductor laser for coherent optical communications

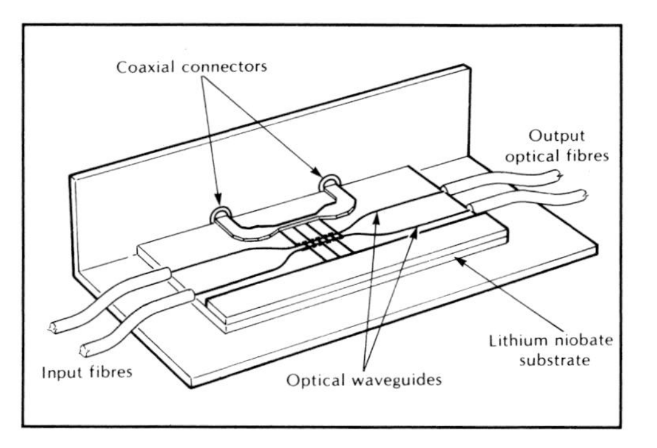

An integrated optic switch in which a travelling wave of electric field is used to switch light between optical waveguides

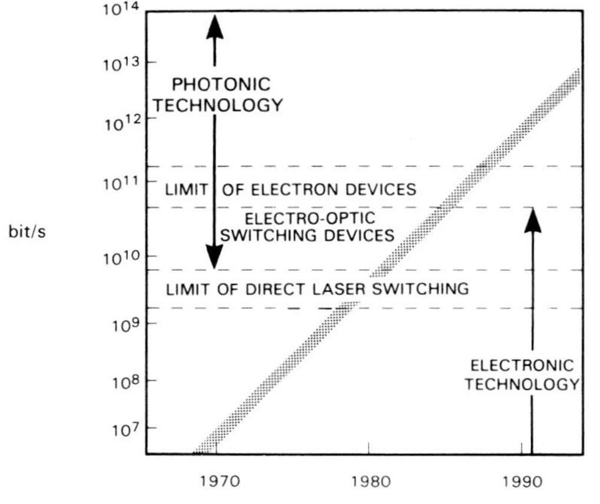

The realm of photonic technology

diffusions, insulations or connections on a silicon chip.

Silicon chips must be packaged and so the techniques of interconnecting the silicon chip into its equipment through large numbers of fine connecting wires, in sophisticated plastic or ceramic packages, continue to present their own challenges. There has also been much innovation of new packaging techniques, such as surface-mounting packages, as well as in techniques for improving heat dissipation from them.

The operating speed of solid state devices depends on their speed of electron transport. On an atomic scale electrons behave as if they were waves – a result of the particle-wave duality of the quantum theory of matter. When electrons pass through the crystalline lattice of a semiconductor they behave as if their mass were much smaller than that of an electron travelling in free space. In the case of silicon, electrons move as if they had one fifth of the mass of free electrons. But, since the early 1960s, it was known that in the compound semiconductor, gallium arsenide electrons behaved as if they had only one fifteenth of the mass of the free electron. This suggested that transistors made of gallium arsenide could have about three times the switching speed of silicon transistors. As a result, since that time we have seen the rise of a new solid state technology based on a gallium arsenide. This finds its application in the highest frequencies, at the microwave region of the spectrum where radar or satellite communications are major fields. Potentially too, gallium arsenide will also be used for very high speed digital applications which silicon may not attain. However, such is the massive technological base of silicon, that it is a real and ever-extending competitor to gallium arsenide in digital performance.

Gallium arsenide is just one example of a semiconductor that can be made from elements of the so-called Groups III and Group V of the periodic table. The III refers to three valence electrons possessed by elements such as gallium, indium or aluminium; the V refers to five valence electrons which, for example, antimony, arsenic and phosphorus possess. Because eight electrons forming a closed shell around each atom is a primary building block in compound chemistry, many combinations of III-V compounds allowing this are thus possible. Moreover, the proportions of each element may be variable. Thus we have the prospect of tailoring the properties of these materials to suit specific electronic and optoelectronic requirements.

A process known as 'epitaxy' is of great importance in the manufacture of III-V devices and a great variety of epitaxial growth techniques have been developed. The purpose of epitaxy is to grow thin, perfect layers a few microns thick on the surface of a bulk crystalline substrate wafer so that their crystalline structure exactly matches that of the substrate. In hetero-epitaxy, substrate and layer are of different compositions and this is proving an essential technique in the manufacture of the most sophisticated devices. There are three general ways of growing epitaxial layers; liquid phase epitaxy, in which a substrate is placed in a liquid melt which is slowly cooled to give an epitaxial surface layer; vapour phase epitaxy, in which surface deposition takes place from a gaseous stream over the substrate. Finally there is molecular beam epitaxy, in which the surface layer is grown on the substrate atom-by-atom in an ultra-high vacuum.

The first gallium arsenide circuits were made in 1976 by Plessey and, independently, by Hewlett Packard. The major activity in gallium arsenide today is the development of an integrated, or monolithic, circuit technology. Today's circuits use field effect transistors as device switches or amplifying elements. Unlike silicon, there is a high resistivity form of gallium arsenide which allows isolation between the devices of a circuit and the design of microwave transmissions lines on the circuits.

To illustrate just one application of gallium arsenide circuits in the radar field we might consider their possible use in phased antenna arrays. The phased array radar has its aperture filled with an array of low power gallium arsenide circuits. Each of these will contribute to the total power output of the radar in its transmitting mode. Each will receive signals in the receiving mode, and these can be added together. By varying the relative operating phases across the face of the arrays the beam pattern can be electronically steered in space or even altered in beam shape. Thus we should be able to replace the familiar cumbersome rotation of radar dishes by electronically steered fixed dishes. These will allow extremely rapid response in air traffic control for example, or in the interception of missiles from many different directions. The concept of phased arrays has been established for over twenty years but it has proved too expensive to implement except in one or two large installations. Development of gallium arsenide integrated circuits should be the breakthrough we need for these revolutionary systems.

The flexibility of III-V compounds has introduced new types of devices to improve on the base capabilities of allium arsenide high frequency devices. Heterostructure devices incorporating gallium aluminium arsenide and gallium arsenide together has led to bipolar devices which have greater power handling capabilities than FETs. The other type of device is high electron mobility transistor (HEMT). In one form of this there is a sandwich of epitaxial layers. One layer may have few defects or impurities and is thus relatively perfect but it contains few current carriers. Adjacent to this are layers which are less perfect but contain more carriers. By careful design of the layer configurations, carriers from the imperfect layer enter the perfect layer where they are able to travel, unimpeded by defects, along its length. In this way such sandwich structures make it possible to have high electron concentrations with high mobility. This then leads to enhanced device performance as well as lower noise characteristics than the original FET devices in gallium arsenide. These sandwich structures are simple examples of 'super-lattice' structures. These consist of many alternate layers each of different materials from the III-V families of just a few atoms thickness. These synthesised semiconductor structures should lead to many new devices which have entirely new parameters for high speed applications.

Gallium arsenide integrated circuit technology is, in may ways, where

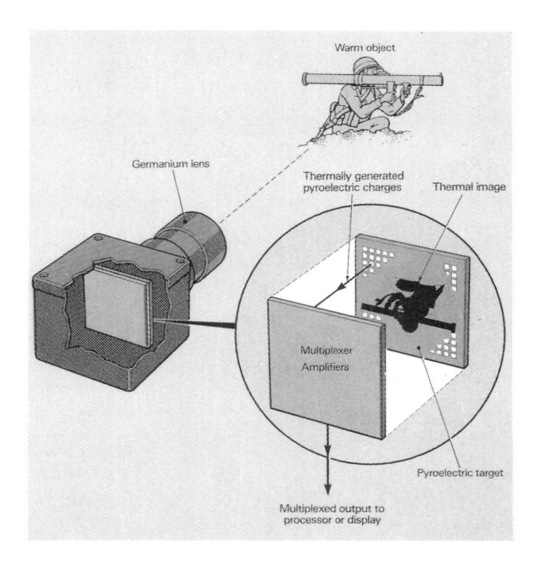

Top: an application of pyro-electric sensors in the military field. *Bottom*: a pyro-electric thermograph

Right:
Schematics and biosensor operation
Opposite:
A biosensor chip operating by detection of changes in the surface conductivity of enzymes

Biosensor Operation

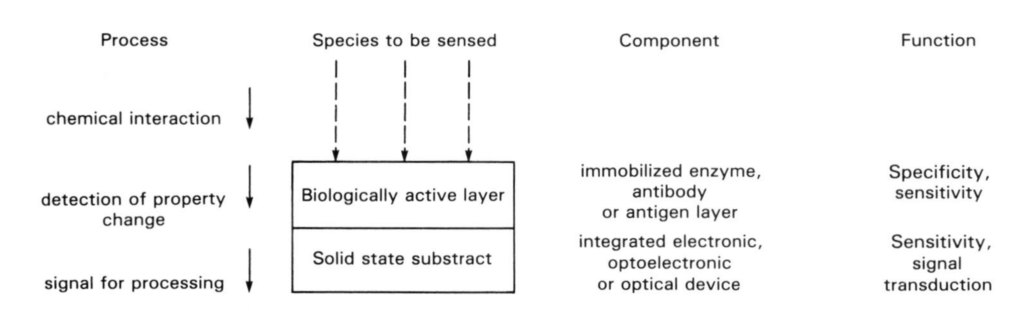

silicon was some twenty years ago. The size of the future commercial market and its take-off point are uncertain. However it is clear that gallium arsenide will hold an indispensable role in high speed electronics from microwaves to the millimetre-wave end of the spectrum, as well as in ultra-high speed digital applications.

Solid state technology forms the kernel of optical information technology through the ability of semiconductor devices to generate and detect light. This takes place through the interaction between light and electrical current as electron transitions occur across the band structures of semiconductors. Light is already the most important medium by which tehnological man communicates with his environment, whether through the printed page or through pictures on a television screen, or as characters on a liquid crystal computer display. However, optical information technology uses light, not just visually, but through its properties as an electro-magnetic wave of extremely high frequency.

The impact of solid state technology shows nowhere more clearly than in the rapid development of fibre optic communications systems over the past few years. The breakthrough occurred in the 1970s when glass manufacturers adopted technology similar to that used in the microelectronics industry to ensure the highest purity in their glasses. By eliminating the chemical impurities – mainly iron and water vapour – which produced large attenuation losses along lengths of glass fibre, they were able to realise what had, until then, been simply a theoretical concept of long distance communications along fibres. This was originated at Standard Telecommunications Laboratories in England.

We now have fibres installed in practical cables whose attenuations are as low as 0.3 to 0.5 dbs/km. These represent the most transparent materials ever made which, if the seas were as transparent, would allow us to see to the bottom of the deepest oceans.

The further contribution of solid state technology in establishing fibre optics is the successful development of lasers and optical detectors from the III-V family of semiconductors as transmitting and receiving devices. Nevertheless, despite today's apparent sophistication, fibre optic systems are

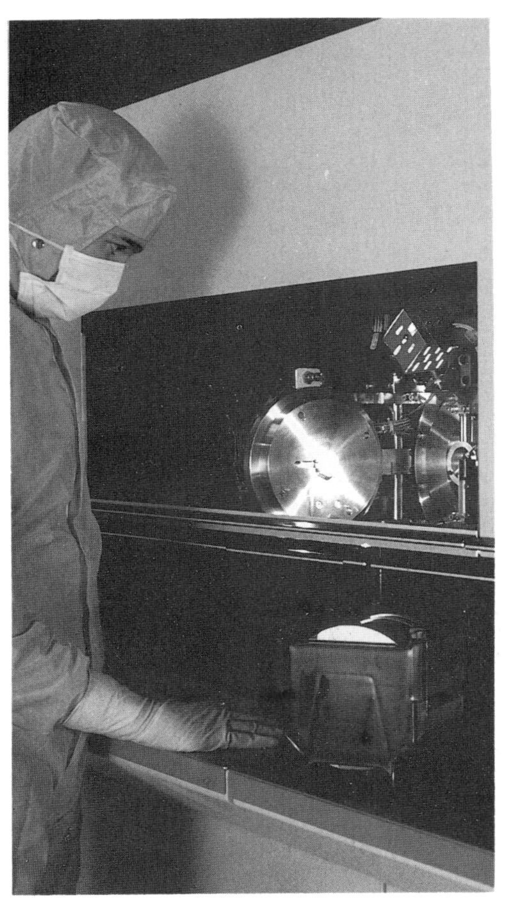

Sputterer for coating silicon wafers with an aluminium layer that is etched to provide electrical connections

essentially simple in that the laser emits a pulse burst of optical power which is detected in the simplest way without the use of any of the sophisticated frequency sensitive detection techniques of radio. Solid state devices to give coherent optical techniques are now the research challenge which will both improve system sensitivity and conserve optical band width. To do this, the most advanced semiconductor lasers will be necessary, emitting as closely as possible a pure (i.e. monochromatic) optical wavelength. This can be mixed, by analogy with the radio technique of super-heterodyning, with a similar laser local oscillator. Such semiconductor lasers represent a *tour de force* of today's technology in the complexities of their multiple semiconductor layer structures and the optical grating filters which are embodied in them to produce emission at a pure optical frequency.

Just as the discrete electron device of the transistor led to electronic integrated circuits, recent research has seen discrete optoelectronic devices being incorporated into optoelectronic integrated circuits (OEIC's) as well as the development of integrated optics techniques.

Integrated optics uses optical waveguide techniques to steer light beams in circuits that modulate, multiplex, switch or carry out other functions which have analogues in electronics. These new techniques will have major influences on future fibre optic communication systems. But, as well as that, we can look towards 'photonics' or digital optics which may take us towards undreamed of bit rates of information processing capacity. Here we shall be dealing with entirely optical systems without electronic circuits in them. They will operate through the physics of interactions between light and solid state materials. Most of these take between 10^{-12} and 10^{-14} seconds to occur. At such operating speeds the techniques of electronic wiring to transport charges and build-up signals will not be adequate. Instead, lightwaves interacting with each other through a responsive solid state medium will be used to transfer information around such systems. The optical computer may be one of the final outcomes of this research, although we are only at the start of a long haul. So far, research on the fundamental logic elements of such a computer is giving much encouragement. Optical computers and optical switching matrices, using the potential which the small wavelength of light offers, could process vast arrays of information points in parallel, perhaps at the level of 10^8 bits per square centimetre. We are then also reminded of similar problems of wafer scale integration in silicon for large processing applications which, equally, would need such packing densities and also present major technological questions of power dissipation.

Today, the field of optical information technology presents the same scope for solid state physics and materials that its part in evolving silicon chips has given to electronics.

Sensors are the means by which electronic or optoelectronic systems observe and measure the world. They have applications as diverse as monitoring the processes in an oil refinery, detecting the heat emitted from cancers in the human body, observing craft flying hundreds of miles away and measuring the changes in the local magnetic field as a nuclear submarine travels under the ocean. In all of these, and many others, solid state technology provides the key sensing and measuring devices. Such a broad field is best understood by some specific examples of different types of sensors. Here we shall also be noting some of the most advanced research today in the solid state technology field.

A region of the electromagnetic spectrum where there are many requirements for sensors is the infrared region between about five and twenty microns. It is in this region that objects around room temperature emit their maximum thermal radiation. This immediately suggests many applications such as remote sensing, thermal imaging techniques, medical thermography, energy monitoring, fire detection and so on.

Significant advances are being made in the field of infrared detectors using the pyroelectric effect. Pyroelectric materials operate by releasing electrical charge as a result of absorbing heat. Some of the most useful materials are ferroelectric ceramics of the lead zirconate family. These have an inherent electrical polarisation, the magnitude of which changes with heat absorption and thus with increase of temperature. This is the source of the charge release which can then be detected and amplified by following circuits. At the present stage of development, pyroelectric sensors can, broadly, measure temperature differentials in a scene of around 0·1°C. They can be built as arrays with a back plane of silicon circuits to provide an electronic readout from a thermal image. The great range of microelectronic assembly and packaging techniques are now being deployed to give two-dimensional sensor arrays of high picture resolution.

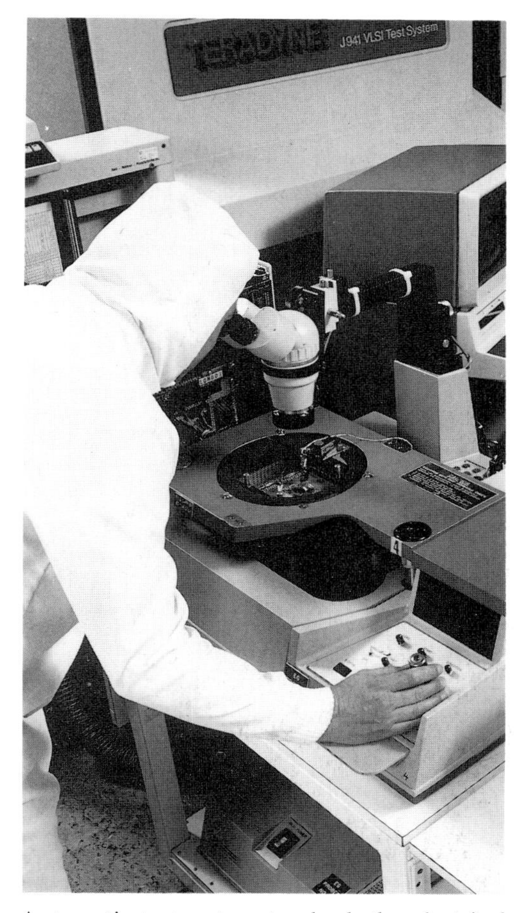

Automatic test system to check the electrical specification of finished integrated circuits

There are often situations in which, apart from sensing radiation in the electromagnetic spectrum, it would be very useful to sense and measure electronically. The materials used so far in solid state technology have been largely inorganic materials of relatively simple structures or, indeed, elements themselves; silicon, gallium arsenide and lead zirconate tantalite, for example. If we could learn to use more complex molecular materials in electronics and optoelectronics – such as organic compounds, polymers and biological materials – we might be able to produce much wider ranges and subtleties of sensing functions. The emerging technology of molecular electronics and its application of biosensors is one such new field. The basic sciences which will allow the development of biosensors are fourfold. In the biological sciences we have molecular biology and genetic engineering and, in the physical sciences, solid state physics and molecular science. These sciences, in turn, have given rise to two technologies which are expected to dominate the future – solid state technology, through microelectronics, and biotechnology. Biosensors will depend heavily on both these technologies. At their simplest, biosensors consist of a solid state substrate device and an overlaying biological layer. The function of the biological layer is to react in some way to sense or analyse a substance, possibly as a chemical species in a liquid environment. The reaction then causes new products or molecular configuration changes which result in changes in the electrical or optical properties of the biological layer. If the underlying solid- state device is sensitive to surface charge,like a field effect transistor or, alternatively, carries an integrated optic waveguide on it, then the changes in the biological layer can be sensed electronically or optoelectronically.

Such biosensors are still largely devices of the future but they offer a very wide field of exciting possibilities. Perhaps one of the most important may be the chemical analysis of body fluids. Such real-time analysis could in turn lead to controlled medicinal drug release systems in patients. Biosensors will also find industrial applications, as in the control of fermentation or in pollution monitoring and in future bioengineering processes themselves. So too, they may be used in many security applications from the detection of nerve gases to the detection of concealed explosives. And, ultimately in the next century, molecular electronic information systems may mimic the functions of the brain itself in capacity and intelligence.

The technologies of today are the results of applied research into, and development engineering of, the scientific knowledge and discoveries of several decades ago. The history of solid state microelectronics, optoelectrtonics, sensors and electronics materials demonstrates that well. In the debate on the relevance of science to industrial progress, it has been wisely said that there are only two sorts of science – applied science and not-yet-applied science – and it is surely true that there are many not-yet-applied science fields which will lead to future solid state technologies. What, for example, will be the applications in high speed electronics and sensors of the newly-discovered elevated temperature superconductors? What will be the devices that may come from research into organic carbon-based semiconductors and optically active materials?

It is certain that solid state technology will, for many years to come, continue to expand in its scope and applications as the most important enabling technology of electronics.

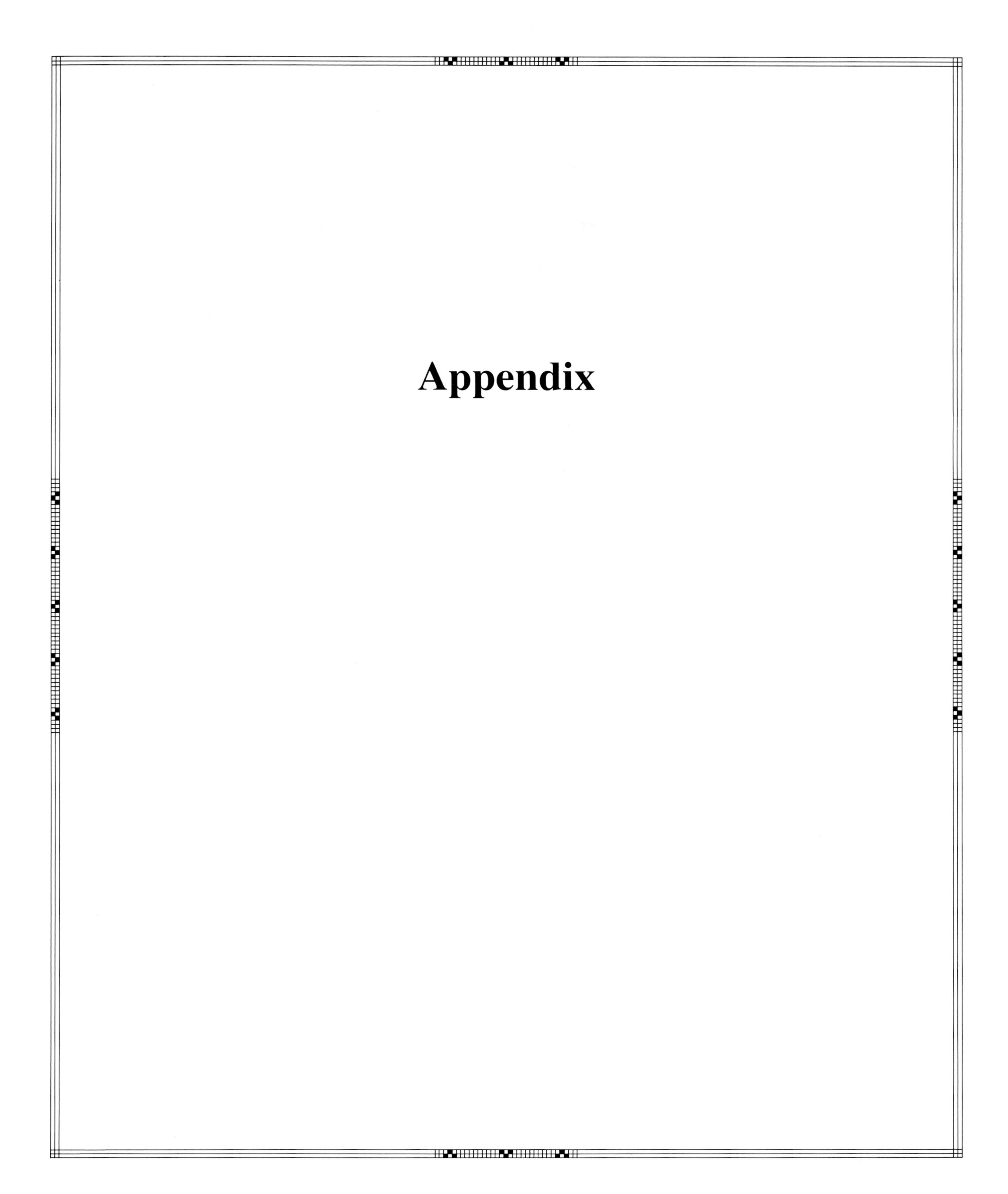

Appendix

Ove Arup and Partners: Centre Pompidou

Biographies

Ove Arup & Partners: Centre Pompidou

AHM, POVL (1926-)
Povl Ahm was educated in Denmark and graduated from Aarhus University. He joined Ove Arup & Partners in 1953. From the start, Ahm was concerned with the design of a number of well-known buidings, including St. Catherine's College Oxford and Coventry Cathedral. Made an associate partner in 1956 and taken into partnership in 1965 he continued to be involved with a number of innovative structural engineering projects. The development of the partnership's civil engineering activities in the United Kingdom owes much to Ahm's drive, perseverance and dedication. Under his leadership a number of imaginative bridges have been constructed, many of which have won awards. Important transportation projects have included major elements of the Tyne and Wear Metro and the rapid transit system in Hong Kong. Ahm has also been deeply involved in the firm's activities in the Far East and is currently a member of British Railways' Board Design Panel. Ahm became Chairman of Ove Arup & Partners in 1984.
See Ove Arup.

Ove Arup & Partners: Sydney Opera House

ARUP, SIR OVE (1895-)
Ove Arup founded the consulting engineering practice in 1946 and formed the partnership three years later. At first the principal strengths of the firm lay in civil and structural engineering, but in the following four decades the partnership has diversified. Today Ove Arup & Partners is an international design consultancy providing engineering services which embrace all aspects of the construction industry. In the formative years its success was largely due to the initiative, drive, enthusiasm and philosophy of Arup himself. His unswerving dedication to innovation, excellence and integrated working enthused all around him. Born in Newcastle upon Tyne in 1895, Arup was educated in Denmark where he studied philosophy before obtaining a degree in engineering. Returning to the United Kingdom in 1923, he was at first interested mainly in marine and general civil engineering work. It was during the early 1930s that he became deeply interested in the Modern Movement in architecture and began his life-long association with the leading young architects of the period. Reinforced concrete was the construction material that gave designers the opportunity to create new forms. It also gave Arup the inspiration to design structures that were reference points for the future; the thin, entwined, reinforced concrete ramps of the Penguin Pool at London Zoo constructed in 1934, Highpoint I and II at Highgate, and his work on underground shelters were all significant contributions to engineering. Many honours have been bestowed on him, including Gold Medals from the Institution of Structural Engineers and Royal Institute of British Architects, several Honorary Doctorates and a knighthood in 1971. When he started in private practice he selected a number of collaborators who were to make their own contribution to the development of the practice in the following years.

ARKWRIGHT, SIR RICHARD (1732-1792)
English textile industrialist and inventor of a revolutionary mechanical spinning machine. In 1771 Arkwright – who had made and lost a fortune as a wigmaker and hair dyer – constructed a water-powered machine which spun cotton yarn. The thread produced was stronger than that made on James Hargreaves's spinning jenny – invented in the late 1760s – and Arkwright introduced his invention in mills in the north of England. It formed the basis of mass production in the cotton industry.

ASPDIN, JOSEPH (1799-1855)
An English bricklayer and stonemason credited with the invention of Portland cement. In 1824 Aspdin patented his process for grinding and burning clay and limestone to produce a substance that hardened when mixed with water. He named it 'Portland cement' due to its resemblance to the stone of that name quarried in Dorset. Portland cement is a basic constituent of concrete and indispensable to modern masonry construction.

ATKINS, WILLIAM (1902-)
After World War II, Atkins saw the opportunities offered by the new steelworks. He established the 'Lambda' system of precast concrete framework and became manager of major engineering projects worldwide, including the rapid transit rail system and London Docklands projects in the UK. Additionally, he was a project manager for the French Channel Tunnel authorities, oversaw the treatment and daily distribution of 14,000 million litres of drinking water to Baghdad from the River Tigris; headed production of sugar-cane power alcohol in Mali in a bid to reduce imports of fuel; was design consultant for the King Fahad Medical City in Riyadh; was a project manager for the Annaba Steelworks in Algeria and designed off-shore platforms in Indonesia.

BAIRD, JOHN LOGIE (1888-1946)
Scottish engineer, inventor and television pioneer, Baird gave the first public demonstration in his attic workshop of what he called 'seeing by wireless' in London's Soho in January 1926. His audience consisted of members of the Royal Institution and their wives, and overnight he became a much sought-after celebrity: the first person to have successfully televised moving objects, which included a ventriloquist's dummy. Two years later he demonstrated colour tele-

vision, and, in 1929, began a daily black-and-white service using a BBC transmitter. In 1936, however, the BBC became interested in a system developed by EMI and Marconi. For three months the opposing systems were transmitted to the public and, to Baird's annoyance, the rival system was favoured as giving a sharper picture and was exclusively adopted.

BAKER, SIR BENJAMIN (1840-1907)
Sir Benjamin received his early training in a South Wales ironworks and then became Sir John Fowler's chief assistant in London, constructing the Metropolitan and District railways. He designed the cylindrical vessel in which Cleopatra's needle was brought to England in 1877-78 and wrote a number of papers on the construction of bridges from 1867 onwards. He was in charge of the design and building of the Forth Bridge in 1890 in association with Sir John Fowler. Later he was commissioned by many government and municipal authorities and his services were of great value to the Institution of Civil Engineers. He received a KCMG in 1890 and a KCB in 1902.

BARLOW, WILLIAM HENRY (1812-1902)
As a civil engineer he patented the saddle-back rail in 1832. He was sent by Maudslay and Field to Constantinople for six years to manufacture ordnance for the Turkish Government. In 1844 he became chief engineer of the Mildland Railway and, between 1844-86, took out several patents. Between 1862-69 he worked on the Bedford to London Railway and was involved with the roof at St. Pancras Station – then the largest span in Britain at 240 feet. In 1860, with Hawkshaw, he completed Brunel's suspension bridge near Bristol and in 1879 appeared on the court of enquiry into the Tay Bridge disaster. Between 1882-87 he was design consultant on the Tay Bridge replacement scheme and, during 1879-80, became President of the Institution of Civil Engineers.

BAZALGETTE, SIR JOSEPH (1819-1891)
Bazalgette was chief engineer on the creation of the Metropolitan Board of Works in 1855 and was concerned with the construction of the main drainage system and the Thames Embankment. His drainage plans were delayed until 1858 because of official obstruction, but in 1865 the system was opened. The section of the Thames Embankment between Westminster and Vauxhall was built between 1860 and 1869. The length between Westminster and Blackfriars was opened by the Prince of Wales in 1870. Chelsea Embankment followed in 1871-74, and, in 1876, the Northumberland Avenue section was built. Bazalgette designed new bridges at Putney and Battersea and the steam ferry between North and South Woolwich. He died before his schemes for the Blackwall Tunnel and a bridge at the Tower were realised.

BELL, ALEXANDER GRAHAM (1847-1922)
Scottish-American teacher of the deaf and inventor of the telephone, Bell was Professor of vocal physiology at Boston University when, in 1876, he patented his 'electrical speech' machine. He gave its first public demonstration later that year at the Centennial Exposition at Philadelphia to mark the hundredth anniversary of the Declaration of Independence. The transmitter and receiver were set some 500 feet apart and Bell declaimed into the handset: 'To be, or not to be, that is the question . . . ' Later that day, Emperor Pedro II of Brazil, who was on a state visit, sat at the receiver and heard the quotation from *Hamlet* come over the wire. He jumped up and shouted excitedly: 'I hear! I hear!' Bell later retired to an island home in Nova Scotia. 'I have become so detached from the telephone,' he stated, 'that I often wonder if I really invented it – or was it someone else I had read about?'

BESSEMER, SIR HENRY (1813-1898)
Born in Hertfordshire, Bessemer was a prolific inventor, but is chiefly known for the Bessemer process, used in the production of steel. This had an enormous effect on the industry by cheapening production costs. He decarbonised cast iron by forcing a blast of air through the mass of metal when in a molten condition and was able to convert melted cast iron into a malleable, perfectly fluid state. He erected steel works in Sheffield to develop this process.

BOBROWSKI, JAN (1925-)
Since 1986 Bobrowski has been President of the Concrete Society where he chairs debates on the 'Fire Resistance of Concrete Structures' and sits on committees at the Institution of Structural Engineering. He is also involved with a Department of Environment study group on 'Construction Times for Industrial Building' and is author of numerous engineering and technical papers. He was awarded an FIP Medal in 1978 for his work with prestressed concrete and the Oscar Faber Award for a paper entitled 'The Design and Analysis of Grandstand Structures'.

BRAMAH, JOSEPH (1748-1814)
One of the most versatile and inventive 18th-century engineers, he invented the hydraulic press and, in 1784, patented the famous Bramah Lock, still made today. In forty years of working life he was granted eighteen patents including: locks, hydraulic presses, water closets, fountain pens, fire engines, printing machines, carriage brakes and suspension. One of the earliest exponents of machine tools and the principle of interchangeable components, he was founder of the first School of Machine-Tool Makers. Described as the 'father of fluid power', Bramah's invention of hydraulic power was the greatest contribution to food production since Jethro Tull's seed drill of 1701. Without hydraulic power many of the engineering feats of the 19th century such as the launching of Brunel's 'Great Eastern' and the erection of Stephenson's tubular bridge across the Menai Straits would have been impossible. Bramah talked of 'rowing or forcing' (sailing) vessels in calm weather by means of rotary steam engines, propulsion with a stern paddle wheel-driven by a worm drive, and a 'wheel with inclined fans like the vertical sails of a windmill'. This was one of the first known suggestions for screw propulsion.

BRASSEY, THOMAS (1805-1870)
Born near Chester, at the age of sixteen he was apprenticed to a surveyor, becoming a partner and eventually manager of the business. He was contracted to build a portion of the Grand Junction Railway in 1895 for Joseph Locke, and later completed the London and Southampton railway line which involved a contract of £4 million and the employment of 3,000 men. He worked on lines in Scotland and Northern England. With his partner, Mackenzie, he took on the construction of a railway from Paris to Rouen, of which Joseph Locke was the engineer and was also engaged on five other French lines with his partner and, on his own account, worked on several lines in England. During the following years he was consulted on the construction of railways in Holland, France, Spain and Italy. One of his largest undertakings was the Grand Trunk Railway of Canada – 1,100 miles in length with its fine bridge over the St. Lawrence River. In this work he was associated with Sir M. Peto and E. L. Betts. Besides railway works, he also had subsidiary coal and iron works, dockyards etc.

BRUNEL, ISAMBARD KINGDOM (1806-1859)
Born at Portsmouth, he was the son of Sir Marc Isambard Brunel. At fourteen he was sent to study at the College Henri Quatre in Paris and, in 1823, became assistant engineer on the project of the Thames Tunnel which started in 1825 but was terminated in 1828. He designed the suspension bridge over the Avon at Clifton in 1864 and the Hungerford Bridge in London. At the age of twenty-seven Brunel became engineer on the Great Western Railway. His last and greatest railway works was the Royal Albert Bridge at Saltash. He also took the lead in designing the first 'Great Britain' steamship which was built at Bristol, launched in 1858, and became the earliest vessel to make regular

I.K. Brunel: Launching the Great Eastern

transatlantic voyages. He then designed the 'Great Eastern' which was the largest ship afloat at the time and the first to use the screw propeller. The first voyage was made from Liverpool to New York City in 1845. He had also been involved in the construction of many docks, piers and hospitals. Overcome with worry and overwork, Brunel died on September 15th 1859.

CARTWRIGHT, EDMUND (1743-1823)
Cartwright was an English clergyman who won fame as inventor of the power loom. In 1784 Cartwright visited Richard Arkwright's cotton spinning mills at Cromford, Derbyshire, and felt that a similar machine could be constructed for weaving. A year later he built and patented the power loom, and opened a spinning and weaving factory in Doncaster, Yorkshire. In 1789 he also devised a wool-combing machine and a rope-making machine three years later. He spent some £40,000 of his own money on perfecting his inventions and soon ran out of funds. In 1793 he gave up his works, assigning his patent rights to his brothers Charles and John. But compensation came in 1809 when Parliament made him a grant of £10,000 'for the good service he has rendered the public by his invention of weaving'.

CAYLEY, SIR GEORGE (1773-1857)
English designer and builder of the first man-carrying glider and the inventor of what became known as 'caterpillar tracks'. In September 1853 Sir George instructed his coachman to pilot the glider across a small valley near Brompton in Yorkshire. The monoplane flew for some 500 yards and came down with a thump. The coachman – John Appleby – hauled himself clear and reported back to his master. 'Please, Sir George,' he stated 'I wish to give notice. I was hired to drive, not to fly!'

COCKCROFT, SIR JOHN (1897-1967)
In 1932 Cockcroft designed the Cockcroft-Walton generator with E.T. Walton to disintegrate lithium atoms by bombarding them with protons. Research on the splitting of atoms established the importance of accelerators in nuclear research. During the Second World War, Cockcroft became the director of Atomic Energy for the National Research Council of Canada and, in 1946, director of Atomic Energy Research at Harwell. In 1948 he was knighted and in 1951 jointly won the Nobel Prize for physics with Ernest Walton for their pioneering use of particle accelerators in studying the atomic nucleus. Cockcroft became Master of the newly-founded Churchill College Cambridge in 1960.

COCKERELL, SIR CHRISTOPHER (1910-)
English boat-builder and inventor of the hovercraft. In 1950 Cockerell bought a boatyard on the Norfolk Broads and set about improving his boat's performance. He felt that an air cushion on which a craft could ride would reduce water friction and wave resistance, and experimented with a model built out of household bits and pieces. Five years later, after overcoming numerous technical difficulties, Cockerell patented the 'hovercraft' as he named it. In 1959 the first full-scale experimental model was launched at Cowes, on the Isle of Wight, and that summer the first Channel crossing by hovercraft was made from Dover to Calais. A regular cross-Channel hovercraft service was started in 1968, with passengers and cars being carried at speeds of up to sixty-five knots.

DAVY, SIR HUMPHRY (1778-1829)
English chemist and inventor of the Davy miners' lamp whose naked flame would not ignite fire damp – an explosive mixture of methane gas and air which forms in coal mines. Davy devised an oil lamp in which the flame was enclosed by a cylinder of wire gauze. The fine mesh absorbed the heat of the flame before it could contact the gas and thus greatly reduced the number of fatal explosions. So that the lamp could be widely adopted, Davy refused to patent his invention. In 1818 he was knighted for his humanitarianism and services to mining.

DICKSON, MICHAEL (1944-)
Dickson was educated at Eton College, Windsor, Cambridge University and Cornell University USA. As well as being a partner in Büro Happold, he is a tutor at the Royal College of Art and an Honorary Research Fellow at the Department of Building Engineering and Architecture, University of Bath. He is a member of the British Standards Committee for ISC/NFE8: Corrosion of Metals and Alloys, and convenor for a study group on long-span structures. He also chairs an *ad hoc* committee on the appraisal of sports stadia at the Institution of Structural Engineers. He is involved in

Michael Dickson / Büro Happold: Munich Aviary

many civil and structural commissions all over the world, including Kowloon Park, Hong Kong; the Energy Pavilion at WonderWorld, Corby; Aston University Conference Centre; the Aviary in Munich and the 58° North Air Supported Structure.

DUNICAN, PETER (1918-)
Peter Dunican joined Arup in 1943. During the war he was concerned with the design and construction of defence structures. The post-war reconstruction programme engendered his strong technical and social interest in low-cost housing. Later Dunican was to become a leading advocate of high-rise construction and the industrialisation of the building process. He was appointed Chairman of the National Building Agency in 1978 and became President of the Institution of Structural Engineers for 1977-78, a body which he served with enthusiasm and dedication throughout his professional career. A full partner in 1956, Dunican became the first Chairman of Ove Arup Partnership when the constituent practices were incorporated with unlimited liability in 1977. He retired from the post in 1984 when Ronald Hobbs and Jack Zunz became co-Chairmen. Peter Dunican was awarded a CBE in 1978.
See Ove Arup.

DUNLOP, JOHN (1840-1921)
Scottish veterinary surgeon who invented the first pneumatic bicycle tyre. In 1887 Dunlop, who had a practice in Belfast, Northern Ireland, bought his nine-year-old son John a tricycle fitted with solid rubber tyres. The boy complained of being jarred as he rode along the unevenly paved streets, and in 1888 his father constructed tyres with rubber air-tubes which passed smoothly over the roughest surfaces. In the same year Dunlop patented the invention, which was produced commercially by a Belfast firm in 1890. The firm, in which Dunlop had 1,500 shares, grew into the Dunlop Rubber Company and its pneumatic tyres were largely responsible for the popularity of cycling. It also helped make possible the growth of the automobile industry. Unknown to Dunlop, a patent

for a pneumatic tyre had been taken out in 1846 by a Scottish engineer, Robert William Thomson. But Thomson had not developed his invention and the Dunlop company was able to establish its rights as it had introduced and patented such improvements as rims and valves.

FABER, OSCAR (1886-1956)
The son of the Danish Commissioner of Agriculture in London, from 1911 onwards Faber played a prominent part in the development and use of reinforced concrete in Britain at a time when many engineers distrusted it. From simple deflection load tests he developed his theory of 'Plastic Yield in Concrete' and the resistance of reinforced concrete beams to shear. Among his notable works were the rebuilding of the Bank of England and House of Commons, including heating, ventilation and air-conditioning, as well as Africa House, India House and many factories. During World War II, Dr. Faber flew to America to advise Sir Winston Churchill on the Mulberry Harbour project. In 1922 he co-wrote 'Reinforced Concrete Design' with P. G. Bowie, which was to become a standard work and in 1951 Faber was awarded a CBE for his work on the House of Commons.

FAIRBAIRN, SIR WILLIAM (1789-1874)
A Scottish engineer, in 1817 Fairbairn went into partnership with James Lille. They set up a lathe and began a business in Manchester, which became famous throughout the world. Fairbairn investigated the use of iron in ship building and made many experiments on its strength as a building material. In 1835 he established a ship-building yard at Millwall, London. Unfortunately, however, other matters took up too much of his attention and the business was sold at a loss. In 1845 he was employed with Robert Stephenson on the construction of the tubular railway bridge across the Conway and Menai Straights. He declined a knighthood in 1861, but accepted a baronetcy in 1869.

FARADAY, MICHAEL (1791-1867)
An English chemist and physicist, Faraday was largely self-educated. In 1825 he was appointed director of the laboratory at the Royal Institute and in 1833 was made Fullerian Professor of chemistry for life. He was an outstanding lecturer, but had a poor memory so made it a habit to write everything down. His 'diary' is in fact published in seven volumes (1932-36). He is best remembered for his invention in 1831 of a dynamo which generated its current by mechanical means. That same year he also invented a transformer. He succeeded in liquifying several gases, investigated steel alloys and produced several kinds of optical glass – one of which became important as the substance in which, when placed in a magnetic field, Faraday first detected the rotation of the plane of polarization of light. It was also the first substance to be repelled by the pulse of a magnet. In 1825 he discovered benzene. He succeeded in greatly improving methods used in laboratories and was buried in Highgate cemetery.

FERRANTI, SEBASTIAN (1864-1930)
Ferranti was the pioneer of the generation and transmission of electricity at high voltages, and the father of modern electrical engineering and supply industries. In 1881 he equipped his school in Ramsgate with electric bells and the following year patented an a.c. generator. It was known as the 'Ferranti Alternator' and was used at Cannon Street Railway Station. He also patented an arc lamp and a meter to measure electric current. 1882 was also the year in which he worked for the Grosvenor Gallery Electric Supply Corporation, supplying it with a generator. Between 1888-90 Ferranti was chief electrician to the London Electric Supply Corporation, where he designed and built the Deptford Power Station. The site for this had access to both coal fuel and cooling water which had been used thirty years earlier by Brunel to launch the 'Great Eastern'. Deptford Power Station was designed to be capable of lighting the whole of London at that time. Overhead mains at 10,000 volts were transformed to 2,400 volts at local sub-stations and then to one hundred volts in the domestic cellars. Public concern at the 10,000 volts supply resulted in the adoption of a low-power d.c. system for lighting London. The high-power system was abandoned at Deptford, but Ferranti's ideas were eventually internationally adopted. In 1892 he returned to research, manufacturing 176 patents in forty-five years and promoting the conservation of coal, urging instead the use of electricity for light, heat and power. The Electrical Age had begun.

FOX, SIR CHARLES (1810-1874)
A civil engineer and contractor, Fox was appointed engineer on the London and Birmingham Railway by Robert Stephenson in 1837. Fox Henderson and Company was responsible for many important station roofs, including Liverpool Tithebarn Street in 1849, as well as Bradford Exchange, Paddington and Birmingham New Street in 1850. In 1851 Fox Henderson and Company erected the Crystal Palace in London's Hyde Park for the Great Exhibition, later re-erecting it with additions on Sydenham Hill. Fox was knighted for this along with Joseph Paxton and William Cubitt. In conjunction with G. Berkley he built the first narrow-gauge railway in India. Among his many railway projects are: the Medway Bridge, Rochester; three bridges over the Thames; railways in Lyons, Geneva, Wiesbaden, Zealand, Queensland and Cape Town; the Toronto three-and-an-half foot gauge line and the approach to Victoria Station in London.

FOX, SIR DOUGLAS (1840-1921)
The son of Sir Charles Fox, in 1874 he became senior partner in Sir Douglas Fox and Partners where he was later joined by Ralph Freeman. The firm acted as consulting engineers in many railway projects including the Mersey Tunnel in conjunction with Sir James Brunless – for which both men were knighted in 1886. Fox's projects included the Snowdon Mountain Railway, the Euston and Hampstead Tube (northern line), the Central Argentine Railway, the British South African Chartered Company, in association with Sir Charles Metcalfe, and the South India Railway Company. He also presented three papers to the Institution of Civil Engineers: 'Light Railways in Norway, India and Queensland', 'Widening Victoria Bridge and Approaches' and 'Description of Excavating Machine on the W. Lancs. Railway'.

FOX, FRANCIS (1818-1914)
In 1846 Fox joined the staff of I. K. Brunel as assistant engineer on the South Wales Railway and in 1854 became engineer on the Bristol and Exeter Railway until it amalgamated with the Great Western Railway in 1876. Fox built the Chard and Cheddar Valley Branch Railway and it was on his recommendation that the Bristol and Exeter Railway adopted the 'block system'. In 1883 he built the Weston Super Mare New Station and Loop Line, in 1884 the Tiverton and N. Devon Branch Line and, the following year, the Exeter Valley Line.

FOWLER, SIR JOHN (1817-1898)
Fowler was a civil engineer born near Sheffield, where his father was a land surveyor. He was involved in the railway construction business from a very early age and set up in business for himself in 1844. He was engaged to lay out the railway system which eventually was amalgamated under the title of Manchester, Sheffield and Yorkshire Railway. He was the engineer on the London Metropolitan Railways and the pioneer of the underground railway which was noteworthy in that it was built not by tunnelling, but by excavating from the surface and then covering in. In 1865 he was elected President of the Institution of Civil Engineers. He was strongly opposed to the project of a Channel Tunnel to France and, in 1872, tried without success to obtain consent to a Channel ferry

scheme. He was created a baronet in 1890 on completion of the Forth Bridge on which, with his partner Sir Benjamin Baker, he was joint engineer.

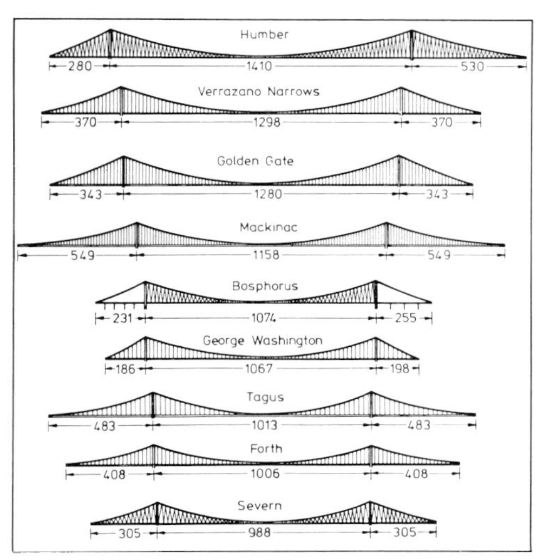

Freeman Fox: Spans of Freeman Fox Bridges

FREEMAN, SIR RALPH (1880-1950)
Best known for his work on the Sydney Harbour Bridge, Freeman joined the Birmingham offices of Messrs. Fox Henderson in 1902 where he calculated how to take the railway over the Zambesi by means of the Victoria Falls Bridge. While surveying, Charles Beresford Fox fell into the gorge but was saved by the branch of a small fig tree. In 1906 Freeman was involved in the report on the Channel Tunnel and with Dorman Long submitted a successful tender for the design and construction of the Sydney Harbour Bridge. In 1928 he was responsible for the arch bridge over the River Tyne in Newcastle and in 1951, with the aid of a computer, designed a startling 365-foot aluminium shell 'Dome of Discovery' for the South Bank Festival of Britain Exhibition. Before his death he was also responsible for three major road bridges – the Firth of Forth, the Severn and the Humber.

FRIESE-GREENE, WILLIAM (1855-1921)
English portrait photographer and a pioneer of the motion-picture camera, Friese-Greene is said to have given the first 'public' demonstration of his 'magic box' in 1889 – when he dragged a London policeman into his studio and sat him in front of a screen. The startled constable watched scenes of cabs, pedestrians and horses, which the inventor had filmed earlier that day in Hyde Park. Soon afterwards, Friese-Greene patented his invention, but in 1891 his photographic business failed and he became bankrupt. Thirty years later he attended a meeting of the Cinematograph Exhibitors' Association. He made an emotional speech, appealing to the industry to settle its differences for the sake of 'the universal language'. A few minutes later he suffered a fatal heart attack and was taken from the meeting by handcart to a mortuary – where it was found that he had exactly 1s. 10d. on him – just enough to buy a good ticket for the cinema.

GOOCH, SIR DANIEL (1816-1889)
Notable as chairman of the Great Western Railway and for laying the first transatlantic cable. In 1837 Brunel appointed Gooch company chairman of the Great Western Railway. When the seven-foot gauge engines which Brunel had ordered were causing difficulties Gooch replaced these with reliable ones like Stephenson's 2-2-2 Northstar engine. The first new engine to arrive in 1840 was the 'Firefly'. In 1840 Gooch sited the Great Western Railway works at Swindon before moving on, in 1864, to join the Telegraph Construction Company which charted Brunel's great iron ship – the 'Great Eastern' – to lay the first Atlantic cable in 1866; the same year in which he became a baronet. Gooch had returned to the Great Western Railway as chairman in 1865 where he oversaw the completion of the Severn Tunnel in 1887. Earlier, in 1873, the second transatlantic cable was installed, again using the 'Great Eastern'.

Edmund Happold: Mannheim Garden Centre

HAPPOLD, EDMUND (1930-)
Professor Edmund Happold is Professor of building engineering in the School of Architecture and Building Engineering at the University of Bath, and is principal partner of Büro Happold. He is a member of the Penney Committee, advising the British Government on structural safety, and member of several British Standard and other committees covering civil engineering and building services. He joined Ove Arup & Partners, London, in 1957 as a design engineer working on the structure of Sydney Opera House, built in 1959, and Coventry Cathedral. After spending two years as structural design engineer with Severud, Elstad and Krueger Associates in New York, he returned to Ove Arup & Partners, London, becoming executive partner for Structures Division 3. He was responsible for the structural engineering of Exeter University Science Buildings, Bootham School (RIBA NE Medal 1967), the Hyde Park Cavalry Barracks, the British Embassy at Rome, the hotel and conference centres at Riyadh and Mecca (UAI competition winners), the Pompidou Centre, Paris (winner of an international open competition), the Garden Centre in Mannheim and many other works. He received the Guthrie Brown Award 1970-71, the Oscar Faber Medal 1974, and the Henry Adams Award 1976. He was elected to the chair at Bath in 1976, and became President of the Institution of Structural Engineers in 1987.

HOBBS, RONALD (1923-)
Ronald Hobbs joined Ove Arup & Partners in 1948. He collaborated with Jenkins on many projects including the prestressed concrete arches at the Bank of England Printing Works at Debden, and was taken into partnership in 1961. Following his work with Jenkins he became involved in what was to become one of the the most significant developments of the Partnership – that of integrated design. Together with Philip Dowson, Hobbs more than anyone was responsible for the early development and establishment of Arup Associates as one of the foremost multi-disciplinary practices in the United Kingdom. He also initiated the use of management contracts in the industry. He was Chairman of Arup Associates and is now co-Chairman of Ove Arup Partnership. Ronald Hobbs' concern with team work between client, designer and contractor to achieve the best possible design with fast construction has been extended to his work with the Joint Contracts Tribunal, where he chaired the working party for the JCT Form for management contracts. See Ove Arup.

Anthony Hunt: Sainsbury Centre, model

HUNT, ANTHONY JAMES (1932-)
Hunt is a structural engineer with an interest in articulate, exposed structures which are forward looking and light-weight. He has worked both with Felix Samuely and Partners and Hancock Associates where he oversaw structures. He started his own practice in 1962, then worked on the Reliance Controls Factory with Team 4.

In 1978 he worked with Foster Associates on the Sainsbury Centre of Visual Arts at the University of East Anglia where he developed an extraordinary extendible frame. He had earlier worked with Foster Associates in 1975 on Willis Faber Dumas in Ipswich, the internal structure of which was revealed at night through its smooth skin of highly reflective glass. In 1980, in collaboration with the architect Michael Hopkins, he produced a minimal, lightweight, patera steel system and in 1982 collaborated with Richard Rogers on the Inmos Microelectronics Factory, in which the uncluttered production area was achieved through incorporating structure and sevices on the outside. In 1984 Hunt worked on the Schlumberger Cambridge Research Limited Building, again with Michael Hopkins, which had a suspended steel membrane structure.

JENKINS, RONALD (1907-1975)
Ronald Jenkins, a graduate of Imperial College, first worked with Arup in 1943. During the early years Arup encouraged him to initiate design methods to solve the problems of thin reinforced concrete slabs or plates. He went on to develop techniques using matrices for the analysis of plates and particularly shells. Jenkins introduced an elegance and rigour into analytical techniques completely absent in the United Kingdom at that time. His work on the design of the shell roofs for the Brynmawr Rubber Factory was well in advance of its time. He inspired younger colleagues as well as other designers who used his techniques in their work at universities. Taken into partnership in 1949 Jenkins played a major role in the advancement of the firm's technology during the early years. He retired in 1973 to become a consultant to the practice. Ronald Jenkins died after a long illness two years later.
See Ove Arup.

KERENSKY, OLEG ALEXANDER (1904-1984)
Educated at Northampton Engineering College, before joining Freeman Fox and Partners in 1946, principal projects had been as chief engineer on construction of Wandsworth Bridge and Avonmouth Oil Jetty. Engaged by Freeman Fox and Partners as their principal bridge designer he worked on designs for the Severn and Forth bridges and Auckland Harbour. Kerensky was made a partner in 1956 and took charge of projects including the Medway Bridge, Grosvenor Railway Bridge and the M2 Motorway. Once President of the Institution of Structural Engineers, he was a member of the permanent committee of the International Association of Bridge and Structural Engineering, and of the court and council of the City University.

LEWIS, MICHAEL (1927-)
Michael Lewis joined Ove Arup & Partners in 1951 and, with Zunz, led the firm's activities in Southern Africa from 1954 to 1962. He went to Sydney in 1963 to co-ordinate the design, management and construction of the Sydney Opera House. Lewis was the founder of the Australian branch of the firm which is now a flourishing partnership with offices in six major cities. On his return to London in 1974 he was appointed an Executive Partner of the Civil Engineering Division and Director of Ove Arup Partnership in 1977. Lewis is a council member of the British Consultants Bureau and is its Chairman for 1986-7. Currently he is Chairman of the partnership's Civil Engineering Board.
See Ove Arup.

Ian Liddell / Büro Happold: Pink Floyd Umbrellas

LIDDELL W. I. (1938-)
Liddell was educated at Fettes College, Edinburgh, St. John's College, Cambridge and Imperial College, London. His discipline is civil and structural engineering. Projects include the Garden Centre at Mannheim; the British Embassy in Riyadh; the Sports Stadium for the Air Defence Corps in Jeddah; the Demountable Theatre for Royal National Eistedfod; the Exhibition Tent for 'Challenge to British Genius Exhibition'; and the 12,000m^2 'Christ for All Nations' Mission Tent.

LOCKE, JOSEPH (1805-1860)
Locke was a civil engineer noted for economic construction of railways and his aversion to tunnels. Between 1835-37 he set up on his own to build the Grand Junction Railway, running from Birmingham to Warrington. Other lines constructed by him in the late 1830s and 1840s included London to Southampton, Sheffield to Manchester (taking over the work from C. B. Vignoles), Paris to Rouen, Barcelona to Mattaro and the Caledonian Railway running from Carlisle to Edinburgh via Glasgow (in collaboration with J. E. Errington). Although Locke's lines were economical and avoided tunnels, difficulties were experienced on steep gradients between Lancaster and Glasgow, particularly in wintry conditions. In 1847 Locke became Liberal MP for Honiton in Devon.

MARCONI, GUGLIELMO (1874-1937)
Italian physicist who invented the first practical system of wireless telegraphy. In 1985, after experimenting for a year in two attic rooms in his family villa near Bologna, Marconi sent a wireless signal from his farm to a field some two miles away. The following year he offered his invention to the Italian Ministry of Posts and Telegraphs, which felt it was no great improvement over the electric telegraph, which carried tapped-out messages along wires. In frustration, Marconi went to London, where, in June 1896, he applied for the world's first radio patent. The next month he successfully demonstrated the system to London Post Office officials. This was quickly followed by demonstrations on Salisbury Plain. Next year Marconi formed the Wireless Telegraph and Signal Company Limited – afterwards renamed Marconi's Wireless Telegraph Company Limited. By 1897 wireless telegraph had a range of twelve miles and, by 1901, the first wireless signal was sent across the Atlantic. Marconi took little interest in the development of popular radio broadcasting, which began in America before the First World War. Three years later he shared the Nobel Prize for physics with the German physicist Karl Ferdinand Braun, who had greatly increased the range of Marconi's transmitter. Marconi later worked on the development of short-wave wireless communication, which largely provided the foundation for long-range radio broadcasting in the modern world.

MARTIN, JOHN (1923-)
John Martin joined Ove Arup & Partners in 1957, became an Executive Partner in 1968 and a Director of Ove Arup Partnership in 1977. He worked with Arup on the Kingsgate Footbridge at Durham University and has been responsible for several of the firm's major structural engineering projects. Martin has been particularly concerned with the development of a number of multi-disciplinary engineering groups to realise well-integrated and co-ordinated building services and structural engineering designs for buildings. Significant projects carried out under his direction include Habitat, Wallingford, and the National Exhibition Centre, Birmingham. He is currently the chairman of Ove Arup & Partners Building Engineering Board.
See Ove Arup.

MAUDSLAY, HENRY (1771-1831)
Principally remembered for inventing the slide-rest lathe and mechanised production line, Maudslay was to iron craftsmanship what Chippendale was to furniture. Having perfected the Bramah lock and the hydraulic press, Maudslay invented the slide-rest lathe which transformed the crude hand-held tool into a precision instrument. A screw thread produced by this lathe resulted in Maudslay's rise to fame and fortune with the screw-cutting lathe; producing screw threads of any diameter or pitch with speed and accuracy. Maudslay also produced a ship's block-making machine for Marc Brunel, capable of outputting 160,000 ship's blocks a year using a vastly reduced labour force: the world's first fully-mechanised production line. Other inventions included a patent calico printing and coin minting plant, adopted by the Royal Mint and exported to nearly every country in the world.

Ove Arup & Partners: Sydney Opera House

MICHAEL, DUNCAN (1937-)
Duncan Michael joined Ove Arup & Partners in 1962 to become an Executive Partner in 1970 and a Director of Ove Arup Partnership in 1977. He leads the structural engineering groups in the UK practice, has responsibility for the partnership's offices in the USA and Hong Kong as well as for projects in the Republic of China. Michael is particularly interested in high-rise structures and was responsible for the Hopewell Centre in Hong Kong – the tallest reinforced concrete building in the Far East. At present he is active in the affairs of the Institution of Structural Engineers.
See Ove Arup.

NEWBY, FRANK (1926-)
Frank Newby has been senior partner of F. J. Samuely and Partners since 1959. He joined the firm in 1949, later becoming Felix Samuely's research assistant. In 1952 he spent an influential year in the USA working with Wachsmann, Eames, Saarinen and Severud. He is noted for his close collaboration with architects and follows Samuely in the search for new and appropriate structural forms and techniques.

Frank Newby: Snowdon Aviary, London Zoo

His work with leading architects includes the US Embassy in London, Leicester University Engineering Laboratories, the Snowdon Aviary, Clifton Cathedral, the Burrell Collection and SAMA banks in Saudi Arabia. He was also commissioned with Cedric Price in 1972 to prepare proposals for a 'Draft for Development Code of Practice for Air Supported Structures'. Frank Newby's interests lie in the history of engineering and in architectural education on which he writes and widely lectures. He became an Honorary Fellow of the RIBA in 1979 and was awarded the Gold Medal of the Institution of Structural Engineers in 1985.
See Felix Samuely.

PARSONS, SIR CHARLES (1854-1931)
English engineer who invented the practical steam turbine. In 1884 Parsons joined a firm in Gateshead, County Durham, which made electric dynamos driven by a reciprocating (to-and-fro) steam engine. He felt that their efficiency could be greatly increased if they could be driven direct by a spinning shaft. That year he constructed a steam-powered turbine engine. Within four years some 200 of the machines had been built and were mainly used to supply lighting for ships. In 1889 Parsons founded his own engineering firm in Newcastle upon Tyne and the following year installed two of his machines in the city's power station. Five years later he started the Parsons Marine Steam Company and fitted a turbine engine aboard a forty-four ton experimental vessel, named the 'Turbinia'. At first the British Admiralty scoffed at such a means of propulsion. To show the authorities that his engine really worked Parsons gate-crashed a naval review held at Spithead in 1897. The hundred-foot 'Turbinia', with Parsons at the helm, raced between the lines of warships at a speed of 34.5 knots – more than seven knots faster than the swiftest destroyer afloat in the British fleet. Following this success Parsons was commissioned to build two turbine-driven destroyers – the 'Viper' and the 'Cobra' – and in 1905 the Admiralty decreed that all its warships should be fitted with turbines. The Cunard Shipping Company followed suit by adopting the engines for its 30,000-ton liner 'Carmania'. Two years later the newly-launched 'Mauretania' was fitted with turbines, and on her second voyage – from New York to Queenstown, Southern Ireland – she established a record time and held the Blue Riband of the Atlantic from 1907 to 1931. Altogether, Parsons – who was knighted in 1911 – took out more than 300 patents. Motorists are still grateful for his other significant invention – non-skid car chains.

PAXTON, SIR JOSEPH (1801-1865)
Apart from being an English architect, Paxton was also an ornamental gardener. In 1825 he was employed in the arboretum of the Horticultural Society's gardens at Chiswick and, in the following year, became superintendent of the Duke of Devonshire's gardens at Chatsworth and manager of his Derbyshire estates. In 1836 he began the building of a grand conservatory, 300-feet in length, which was finished in 1840 and formed the model for the Great Exhibition Building of 1851. Paxton's most interesting design perhaps was that done for the mansion of Baron James de Rothschild at Ferrieres in France, but he also designed many other important buildings. He was Member of Parliament for Coventry from 1854 until his death.

PILKINGTON, SIR ALASTAIR (1920-)
Pilkington was the English engineer who invented float glass. In 1952, while working for his family glass-making firm, Pilkington had the idea of producing plate-glass of a uniform thickness and without flaws. For the next few years he headed a team which, in 1958, turned out its first square foot of float glass – so called because, on leaving the melting furnace, the liquid glass floats along the surface of a bath of molten tin while cooling. The flat surface of the tin imparts a smooth, bright finish to the surface of the glass. The following year the process was officially announced, and there was a great demand for the perfect, quarter-inch thick glass. Pilkington, who was knighted in 1970, revealed that the glass had originally been made on 'old and tattered' equipment. But when the worn out parts were replaced the quality was greatly impaired. It was only by reproducing similar conditions to those of the decrepit equipment that they were able to make consistently good glass.

RICE, PETER (1935-)
Peter Rice, a graduate of Queen's University Belfast, joined Ove Arup & Partners in 1956.

Ove Arup & Partners: Centre Pompidou

He immediately became involved in the design of the Sydney Opera House, later becoming resident engineer on site. On his return to the United Kingdom he developed an interest in light-weight structures and was responsible for the design of the Federal Garden Show Pavilions, Mannheim and structures in Riyadh in collaboration with Frei Otto. Rice has been responsible for notable buildings such as the Pompidou Centre in Paris with Piano and Rogers, Lloyd's of London with the latter and the Menil Gallery, Houston, Texas, with the former. See Ove Arup.

ROBERTS, SIR GILBERT (1899-1978)
Roberts was educated at the City and Guilds College before working as an assistant under Ralph Freeman of Sir Douglas Fox and Partners on the design of the Sydney Harbour Bridge. Between 1936-43 he was engineer in charge of development and construction for Messrs. Sir William Arrol and Company Limited, and then became director and chief engineer of the same firm, taking charge of all design. He extended the application of welding and high-tension steel to many important structures, including Croydon Power Station, and invented a new jointing method for dock gates. In 1949 Roberts became a partner of Freeman Fox and Partners, a position he was to retain for twenty years. In this capacity he took charge of design work on the Severn Bridge, the Auckland Harbour Bridge, the Volta River Bridge and the Forth Road Bridge. He was knighted in 1965 for his work on the Forth Bridge, and a special function grip bolt used on the Forth, Severn, Humber and other bridges of the period is known as the 'Roberts Bolt'. Roberts also designed a radio telescope, various crane structures and invented the suspension bridge development for the Severn Bridge.

SAMUELY, FELIX JAMES (1902-1959)
In 1929 Samuely became a partner of Berger and Samuely, consulting engineers in Berlin where, between 1931-33, he acted as consultant to the Russian Government in Moscow on the welding of structural steel. In 1933 Samuely came to England where he started a consulting engineering practice. He designed the first all-welded steel structure for a department store in Berlin and the first in Britain for the De la Warr Pavilion in Bexhill. Other notable early works were Simpsons of Piccadily and a block of flats at Palace Gate, London. Between 1942-44 he developed welded tubular steel construction, working on aircraft hangars, an Ealing film studio and the Truvox Factory. At the 1951 Festival of Britain he was consulting engineer for the Transport Pavilion, the Power and Production Building and the 'Skylon'. He pioneered space structures and folded slab construction in concrete, steel and timber. His later work included the Government Pavilion and British Industries Pavilion at the Brussels Exhibition, and the American Embassy in London. Always revolutionary in approach, he reduced the effects of railway noise and vibration by building on a bed of cork at the W. A. Gilbey office at Camden in collaboration with the architects Mendelsohn and Chermayeff.

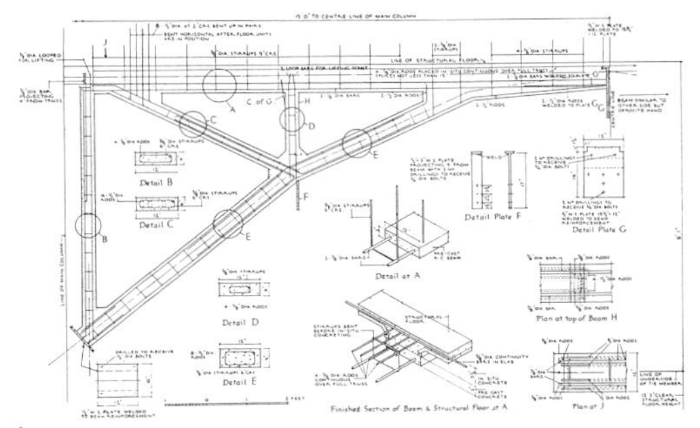
Felix Samuely: Factory at Malago, Bristol

SMEATON, JOHN (1724-1792)
The first self-styled consulting engineer, Smeaton was born near Leeds, the son of an attorney. He was employed in his father's office, but in 1742 became apprenticed to an instrument maker, later setting up his own business. Smeaton made improvements to various mathematical instruments used in navigation and astronomy, rebuilt Eddystone Lighthouse in 1759 and prepared many designs for engineering projects, including the construction of canals and harbours. He designed a series of windmills and pumps, as well as building and repairing bridges.

STEPHENSON, GEORGE (1781-1848)
On September 27th 1825, Stephenson's 'Locomotion' steamed away along ten miles of track between Darlington and Stockton, in the north-east of England. The occasion was the opening of the world's first public steam railway. Although the line was intended for freight, the engine drew some 450 people on its inaugural run and hauled the thirty-eight open carriages at speeds of between twelve and sixteen m.p.h. Twenty-two years earlier the Cornish engineer Richard Trevithick had designed a locomotive

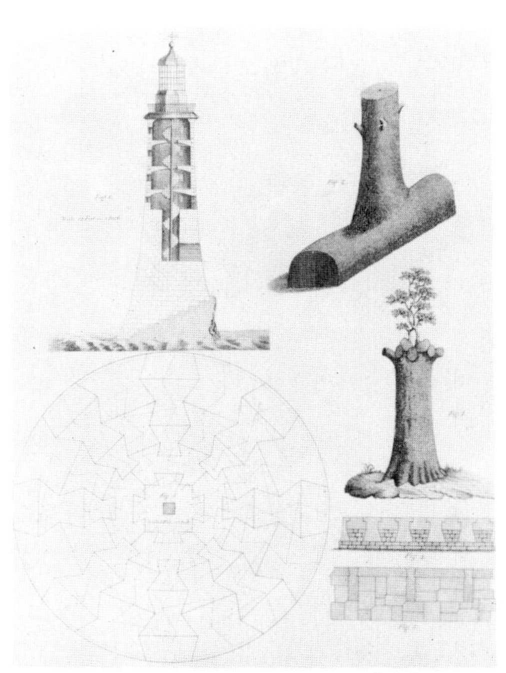
John Smeaton: Ideas for Eddystone Lighthouse

built at the Coalbrookdale Ironworks in Shropshire. However, his engines were plagued with mechanical problems and it was the Stephenson family who brought about the invention's progress and success. The son of a colliery fireman in Wylam, Northumberland, George Stephenson built engines which hauled as much as thirty tons of coal over distances of up to eight miles from the mines to the wharves. At eighteen he'd been to night school to learn to read and write, and insisted that his son Robert – who later helped him in his work on locomotives – received the kind of formal education which he had lacked. In 1829 speed tests were held to choose a locomotive for the recently completed Liverpool-Manchester Railway. Robert Stephenson's 'Rocket' won the trial, and the following year it and seven other Stephenson-built engines pulled passenger coaches along the forty miles of track at speeds of around thirty m.p.h. Stephenson refused most of the honours that were offered to him – including a knighthood and a 'safe' seat in the House of Commons – and retired in 1845 to devote himself to gardening.

STEPHENSON, ROBERT (1803-1859)
Only son of George Stephenson, he assisted his father in surveying the Stockton and Darlington, and Liverpool and Manchester railways. In 1824 he took charge of engineering operations in South America for the Colombian Mining Association of London. After resigning in 1827, he returned to England and managed his father's factory in Newcastle, helping with the improvements of his father's locomotive. He specialised in the construction of tubular girder type railway bridges, including the Royal Border

Robert Stephenson: Britannia Bridge, section

Bridge at Berwick-upon-Tweed, the Britannia Bridge over the Menai Straights and the Conway Bridge. In 1847 he took a seat in the House of Commons as member for Whitby, and retained it until his death in 1859. He is buried in Westminster Abbey.

SWAN, SIR JOSEPH (1828-1914)
Swan was an English physicist and chemist who pioneered electric lighting. In 1860 Swan devised his first primitive electric light, but the bulbs were short-lived and the light itself was not strong enough. A painstaking worker, he took almost twenty years to perfect his invention, and it was not until February 1879 that he was able to light the drawing-room in his house in Newcastle upon Tyne with electric lamps. It was the first domestic room in Britain to be illuminated in that way. The Swan lamps proved a great success and two years later were installed in the House of Commons. At around the same time the American inventor Thomas Alva Edison was demonstrating *his* model of the electric light bulb and patenting a number of refinements and improvements which were vital to the invention's reliability. Rather than take their rival claims to court, the two men formed the London-based Edison and Swan United Electric Company in 1883 to manufacture the bulbs. Swan – who worked for some years in a firm of photographic plate-makers in Newcastle – had earlier produced his first successful major invention; in 1871, dissatisfied with the wet plates then in use, he perfected a dry plate which greatly simplified the processing of photographs. Seven years later he invented photographic bromide paper – still frequently used for printing from negatives.

TELFORD, THOMAS (1757-1834)
Notable civil engineer and designer of the Menai Suspension Bridge. Telford was originally a mason before teaching himself architecture. In 1786 he was appointed surveyor to Shropshire where his tasks included bridge construction; the most notable of these being three over the River Severn. In 1793 he became engineer to the Ellesmere Canal Company, building two giant aqueducts to carry the canal over the Cerioy and Dee valleys in Wales at Chirk and Pont-y-Cysyllte. His novel use of cast-iron plates fixed as troughs to the masonry brought Telford national acclaim. In 1803 he was appointed engineer of the Caledonian Canal and built over 900 miles of road. Between 1819-25 Telford built two famous suspension bridges over the Menai Straits and the River Conway, making masterly use of wrought-iron links to suspend the roadway deck. Other major works included the Göta Canal in Sweden and St. Katherine's Dock in London. Telford was made first President of the Institution of Civil Engineers when it was founded in 1828.

TREVITHICK, RICHARD (1771-1833)
Richard was the only son of the manager of several Cornish mines and was a precursor of George Stephenson in the construction of locomotive engines. On Christmas Eve 1801 his road locomotive carried the first load of passengers ever transported by steam and, in 1803, a steam-driven vehicle made by him ran in London from Leather Lane along Oxford Street to Paddington. In 1805 he constructed a circular railway in London, near Euston Square, which carried the public at between twelve to fifteen miles per hour. He also constructed a high-pressure steam threshing engine in 1812. In 1814 he agreed to supervise the construction of engines used in mine workings in Peru, eventually returning to England in 1827 to die penniless and without recognition for his inventions at Dartford.

TURING, ALAN MATHISON (1912-1954)
Turing was a mathematician who pioneered the evolution and programming of modern computers. In his 1937 Ph.D. paper he proved that a universal machine could be built that, with suitable programming, would be capable of doing the work of any machine designed for special purpose problem solving. Turing championed the theory that computers had thought potential and, in 1952, conducted a study of morphogenesis – the development of pattern and form in living organisms. Turing's death was the result of accidental potassium cyanide poisoning during an electrolysis experiment.

VIGNOLES, CHARLES (1793-1875)
Vignoles was an eminent civil engineer credited with the invention of the flat-bottomed railway line during the 1830s – the same period at which R. L. Stevens was developing his version in the USA. In 1830 Vignoles, in collaboration with John Ericsson, patented a method of ascending steep inclines by a centre rail gripped by two horizontal wheels operated by a lever. It was 1837 when Vignoles introduced the flat-bottomed rail – an inverted T-section very similar to present-day rails although these latter are hardened to reduce wear and controlled cooling during manufacture reduces internal cracking (one of the main causes of broken rails and derailment). Vignoles also laid the foundations for the Liverpool and Manchester Railway, and the Sheffield and Manchester Railway. His main work in later years, between 1853-55, was the suspension bridge over the Dnieper at Kiev – then the longest suspension bridge in the world.

WALLIS, SIR BARNES NEVILLE (1887-1979)
English inventor and engineer, Wallis won fame as the designer of the 'bouncing bombs' which, in 1943, burst the Eder and Möhne dams in the industrial Ruhr Valley in Germany. Having trained as a marine engineer he became a designer in the airship department of Vickers Limited, where, between 1916-22, he was the firm's chief designer and was responsible for the successful Government airship, R.100. In the Second World War he also devised 'Grand Slam' penetration bombs, which bored through the earth to blow up such targets as railway tunnels. He also invented the geodetic, or criss-cross construction which both strengthened and lessened the weight of the RAF's Wellington and Wellesley bombers. He helped in the development of Concorde – the world's first supersonic airliner to go into service – and pioneered the 'swing-wing' supersonic aircraft which were taken up by the Americans. At the time of his death he was still working on a hypersonic plane which would fly to Australia in four hours without refuelling.

WATSON-WATT, SIR ROBERT (1892-)
An eminent physicist, author and adviser for radio and radar research, in 1937 Watson-Watt, in collaboration with A. P. Rowe, launched the first operational studies of radar and claimed to have given the discipline its name. Parallel research into radar took place in Australia, America, Canada and among the Free French Forces, but it was not until Watson-Watt visited

the USA in 1942 that operations research to introduce radar into the war effort really began and radar's extraordinary potential was fully understood.

WEX, BERNARD PATRICK (1922-)
During 1942-47 Wex was attached to the Royal Armoured Corps and in 1950 joined Freeman, Fox and Partners where his projects included the Severn and Forth Road bridges, the Auckland Harbour Bridge, the Ganga Bridge in India and a number of large power stations. His bridges in Pakistan and India were all-welded and Wex gathered valuable experience of foundation problems and lattice tower structures. In 1969 Wex was made a partner of Freeman, Fox and Partners and became involved with the construction of a number of steel and concrete bridges, including those over the Avonmouth and Humber. Wex has extensive overseas experience, having taken part in bridgework in South Africa, Pakistan, Bangladesh, Kuwait and worked on the submerged tube steel structure for the Hong Kong Tunnel.

WHEATSTONE, SIR CHARLES (1802-1875)
Wheatstone was an English physicist who devised the first electric telegraph to go into regular public use. Although Wheatstone's first creative interest was in musical instruments – in 1829 he invented the concertina – he later became fascinated by sound and how it could be transmitted. He discussed the problem with the American physicist Joseph Henry, who also advised his fellow countryman Samuel Morse on his experimental work. Wheatstone later teamed up with the electrician Sir William Cooke, and in 1837 they devised and patented the first practical telegraph – similar to the one on which Morse afterwards sent his first message.

WHITTLE, SIR FRANK (1907-)
English aviator, engineer and pioneer of the jet engine. As a fighter pilot with the peacetime Royal Air Force, Whittle felt that the propeller and conventional piston engine were too heavy and unreliable for the high-speed and high-altitude flying of the future. In 1930 he patented his first jet engine and, while still a serving officer, formed a company called Power Jets Limited at Rugby, Warwickshire, to develop and manufacture his invention. Seven years later a Whittle W1 jet engine was fitted to a specially designed Gloster E28/39 airframe and a series of test-bench trials were held. In May 1941 the experimental plane made its successful maiden flight with Whittle looking on. But it was not the world's first jet-powered flight; that honour had already gone to the Heinkel He-178 designed by a young German aeronautics wizard, Hans von Ohain, in the late 1930s. Although British and American air experts realised the importance of Whittle's invention – made quite independently of the German's – the Gloster Meteor jet fighter did not go into action until towards the end of the Second World War. Meanwhile, von Ohain devised a German jet fighter, the Messerschmitt Me-262, which entered combat service in 1944.

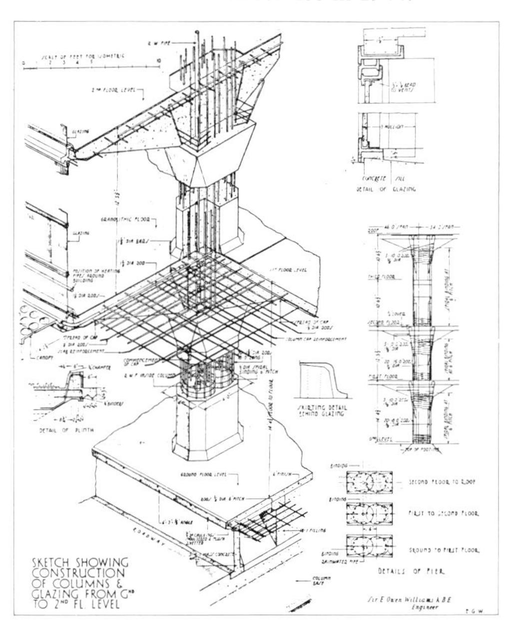

Owen Williams: Boots Factory

WILLIAMS, OWEN (1890-1969)
Williams' best-known works date from his 1930s Modernist phase: the Boots factories; the Dorchester Hotel and the Daily Express buildings. He is also notable for his involvement in 1958 with Britain's first motorway – the M1. In 1924, at the age of thirty-four, Williams was knighted. In 1932 he condemned the newly erected Sheffield City Hall – executed in classical stone – inclining to the view that a glistening all-steel structure would have more accurately reflected the city's manufacturing heritage. Williams adored the weight and mass of concrete and the Boots 'Wets' Building in Nottingham clearly demonstrated his mastery of reinforced concrete, patent glazing and top lighting. His Daily Express buildings in London and Manchester heralded a new age with their innovative use of glass and black vitrolite – at Manchester the rolling presses can be seen at street level through the curved glass walls. The smooth curtain walls used on these buildings were years ahead of the celebrated Peter Jones store in Sloane Square. His later career was largely dominated by motorway construction, although the cantilevered reinforced concrete structures at the Heathrow BOAC aircraft hangars possessed an impressive brutality.

WOOD, GEOFFREY (1911-)
Geoffrey Wood, a graduate of City and Guilds College, joined Arup in 1947 after serving with distinction in the Royal Engineers. He was responsible for many interesting structures built of aluminium, some of which were constructed for the 1951 Festival of Britain. Much of the partnership's work in West Africa, particularly Nigeria, was initiated by Wood who was also involved in the West African schools and universities programmes. Subsequently he helped establish the firm's reputation as highway engineers with the world's leading funding agencies. Taken into partnership in 1949, he was involved with the design of industrial projects which were later to lead to the establishment of Arup Associates as a totally integrated design practice. Wood retired in 1977 and now lives in New Zealand. He was awarded the CBE in 1980.
See Ove Arup.

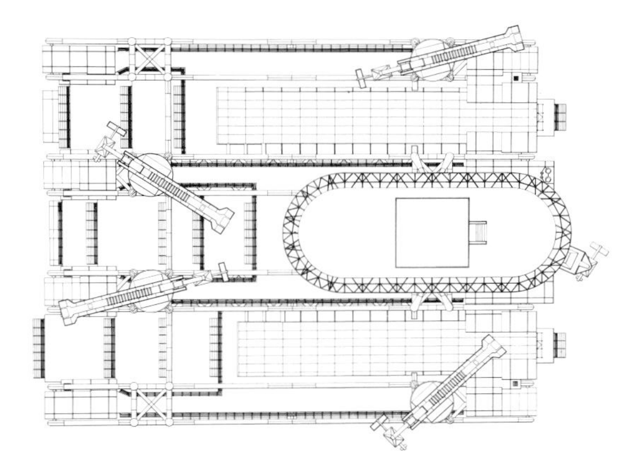

Ove Arup & Partners: Hongkong Bank

ZUNZ, JACK (1923-)
Jack Zunz first joined Ove Arup & Partners in 1950. After working in the London office for four years, he returned to Southern Africa to lead the firm's activities in that area. He returned to the United Kingdom in 1961 and was taken into partnership in 1965. During this time Zunz was involved in the design and construction of tall structures – buildings, towers and masts. He was subsequently responsible for the design of the roofs of the Sydney Opera House as well as other projects. These include several aircraft hangars, the Carlsberg Brewery at Northampton and the Renault Centre, Swindon. In 1977 he was appointed Chairman of Ove Arup & Partners, but was able to maintain an active involvement in a number of the partnership's innovative projects, the most notable of these being the realisation of the Hongkong Bank. He was appointed co-Chairman of Ove Arup Partnership in 1984.
See Ove Arup.

Great British Inventions in their Time

1700 ▷ 1850

1701 **Seed drill.** First practical seed drill invented by the agricultural experimenter Jethro Tull of Wallingford, Berkshire.

1709 **Iron Smelting.** Coke-burning process introduced by Abraham Darby at Coalbrookdale, Shropshire.
Piano. Built by Bartolommeo Cristofori, Florence, Italy.

1712 **Steam engine.** Piston-operated engine worked by atmospheric pressure, invented by Thomas Newcomen, erected at Coneygre Colliery, Tipton, Staffordshire.

1733 **Flying shuttle.** Device for speeding up cotton weaving introduced by John Kay of Bury, Lancashire.

1743 **Silver plate.** Process discovered by the cutler Thomas Boulsover of Sheffield, Yorkshire. Originally used for button manufacture.

1754 **Rope-making machine.** Patented by the English inventor Richard Marsh.

1757 **Sextant.** Devised by Captain James Campbell, RN.

1761 **Marine chronometer.** First instrument sufficiently accurate for ocean navigation made by John Harrison of Yorkshire.

1767 **Spinning machine.** Spinning jenny, first successful spinning machine, invented by an illiterate weaver, James Hargreaves, of Standhill, Lancashire. It produced thread much faster than was possible with a spinning wheel, enabling cotton-thread manufacture to catch up with weaving, mechanised by John Kay in 1733.

1768 **Threshing machine.** First practical thresher invented by the Scottish agriculturist Andrew Meikle.

1769 **Spinning machine.** Water-powered machine patented by the Lancashire industrialist Richard Arkwright. Built in 1771.

1777 **Iron boat.** Craft, 12ft long, sailed on River Foss, Yorkshire.
Steam-engine. James Watt's separate-condenser engine starts work in British mines. The separate condenser allowed steam to be cooled away from the cylinder. Low-pressure steam reinforced atmospheric pressure in driving the piston.

1778 **Duplicating machine.** James Watt of Birmingham made a flat-bed press that produced a stencil from absorbent paper treated with a fixing solution.

1779 **Spinning machine.** 'Mule' invented by Samuel Crompton of Bolton, Lancashire. Combined characteristics of the spinning jenny (1770) and Arkwright's water frame (1771).
Steam-engine. Rotative steam-engine built for the Birmingham button manufacturer James Pickard. Pickard's factory was the first with machinery driven by steam power.

1779 **Bridge.** First bridge constructed wholly of iron built by Abraham Darby across the River Severn at Coalbrookdale, Shropshire. Span 100ft.

1785 **Power loom.** Patented by the Rev. Edmund Cartwright, rector of Goadby-Marwood, Leicestershire, following a visit to Arkwright's cotton mills.

1786 **Lifeboat.** Coble (fishing boat) converted according to the specification of Lionel Lukin; introduced at Bamburgh, Northumberland.
Clutch. Invented by John Rennie, a Scottish engineer, for regulating machinery used in a flour mill.

1792 **Gas Lighting.** William Murdoch lit his office at Redruth, Cornwall, with coal gas.

1794 **Planing machine.** Introduced by Horace Miller at his works in Preston, Lancashire.

1796 **Hydraulic press.** Invented by the English engineer Joseph Bramah.
Iron-framed building. Benyon, Marshall & Bage flax mill constructed with iron girders at Shrewsbury, Shropshire.

1803 **Railway locomotive.** First used at the Coalbrookdale Ironworks, Shropshire, by the Cornish engineer Richard Trevithick.

1807 **Gas lighting.** Street lighting introduced in Pall Mall and Golden Lane, London. Domestic gas lighting installed by William Murdock in house of cotton-mill owner George Lee at Salford, Lancashire.

1808 **Mass production.** Manufacturing system first applied by Marc Isambard Brunel and Henry Maudslay, to the production of pulley blocks at Portsmouth for the Royal Navy.

1810 **Electric lamp.** Arc lamp demonstrated before Royal Institution, London, by Sir Humphry Davy.

1812 **Jack.** Hydraulic jack patented by the English engineer Joseph Bramah.

1815 **Safety lamp.** Invented by Humphry Davy; first used at Hebburn Colliery, Co. Durham. Invented independently by George Stephenson.

1818 **Tunnelling machine.** Invention of tunnelling shield by British engineer Marc Isambard Brunel makes possible deep tunnelling in soft soil.

1824 **Imperial measures.** Imperial standards of weights and measures given legal force by British Parliament.

1826 **Reaping machine.** First practical machine, invented by the Scottish engineer Patrick Bell.

1829 **Elevator (lift).** Passenger elevator installed at the Coliseum, a panorama building in Regent's Park, London, by William George Horner.

1831 **Dynamo.** Device for generating electric current by mechanical means demonstrated in London by Michael Faraday.
Transformer. Invented by Michael Faraday in London.

1832 **Linear induction motor.** Patented by William Sturgeon, London.
Corrugated iron. Made by John Walker of Rotherhithe, Kent.

1834 **Railway signals.** Fixed signals introduced on Liverpool & Manchester Railway.

1839 **Steam hammer.** Designed by the Scottish engineer James Nasmyth; first to be erected was at the Le Creusot ironworks in France, 1840.
Bicycle. Crank-driven machine built by Kirkpatrick Macmillan, blacksmith of Courthill, Dumfries.
Electric telegraph. First in regular commercial use installed by Professor Charles Wheatstone and W.F. Cooke over 13 mile stretch of Great Western Railway between Paddington and Hanwell.

1840 **Postage stamp.** Adhesive postage stamp – 'Penny Black' – introduced by British GPO.

1843 **Christmas card.** Designed by John Calcott Horsley, RA, for Henry Cole, writer, art critic and founder of public lavatories, who was too busy that year to write to all his friends.

1845 **Pneumatic tyre.** Patented by R.W. Thomson of Edinburgh and fitted to the wheels of a brougham.
Tarmacadam. Mixture of stones (macadam) and tar first laid in London Road, Nottingham.

1850 ▷ 1900

1851 **Modular construction.** First large building constructed from prefabricated parts for the Great Exhibition in London.

1853 **Glider.** First manned flight was by a coachman, John Appleby, in an aircraft built by Sir George Cayley and flown 500 yds across a valley at Brompton Hall, Yorkshire.

1856 **Bessemer process.** Method of mass-producing mild steel economically announced by Henry Bessemer before British Association, Cheltenham.

1860 **Electric light.** Sir Joseph Swann devised his first primitive version.

1857* **Oil rig.** Erected by G. C. Hunaus at Wietz, Hanover, Germany, to drill for crude oil.

1862 **Plastics.** First plastic articles, moulded from Alexander Parkes' 'Parkesine' (celluloid), were exhibited at the International Exhibition, South Kensington, London. Plastics manufacture was commenced by the Parkesine Company, 1866.

1869 **Bridge.** First concrete bridge, a temporary structure, built by John Fowler at Cromwell Road, London.

1873 Maxwell's *Treatise on Electricity and Magnetism.*

1874 **Electric car.** Three-wheeler built by Sir David Salomons, Tunbridge Wells, Kent.

1876 **Cast steel.** Process for making cast steel introduced by Sir Henry Bessemer, England.

1877 **Differential gear.** First practical application for road vehicles, on the Starley Wonder four-wheeled sociable cycle, manufactured at Coventry, Warwickshire.

1878 **Microphone.** Demonstrated by Professor D. E. Hughes at the Submarine Telegraph Company offices, London.

1881 **Power station.** Inaugurated at Godalming, Surrey, using hydroelectric generating equipment supplied by Siemens. Generated electricity for industry, street lighting and domestic use.

1884 **Steam turbine.** Earliest practical apparatus designed by Charles Parsons of Gateshead, Co. Durham.
Time zones. Decision of the International Meridian Conference divides world into zones, in each of which clocks show a standard time.

1888 **Carburettor.** Spray carburettor of modern type designed by Edward Butler of Erith, Kent, and fitted to his Petrocycle motor tricycle. Invented independently by William Maybach, Germany, in 1893.
Pneumatic tyre. First air-filled rubber cycle made by John Boyd Dunlop, who applied it to his son's tricycle, Belfast Northern Ireland.
Duplicating machine. Cyclostyle machine adapted for wax typewriter stencil by the Hungarian immigrant David Gestetner of London.

1890 **Compression ignition engine.** Patented by Herbert Akroyd Stuart; manufactured by Richard Hornsby & Sons, Grantham, Lincolnshire, 1982.
Railway electrification. First electric locomotive in regular public service on underground, City & South London Railway.

1891 **Stoney** invents the word 'electron' to describe the elementary particle of electricty.

1892 **Vacuum flask.** Invented by the Cambridge physicist Sir James Dewar for use in laboratory work. Adapted for domestic use in 1904 and named 'Thermos' (Greek for 'hot').

1894 **Turbine-driven vessel.** Turbinia, built by Charles Parsons at Newcastle upon Tyne, Co. Durham.

1895 **Radio.** Demonstrated by 20-year-old Guglielmo Marconi at the Villa Grifone, near Bologna, Italy.

1897 The 'discovery' of the electron by J. J. Thompson (April 30).

1900 ▷ THE PRESENT DAY

1901 **First wireless transmission.** Across the Atlantic between Newfoundland and Cornwall, achieved by Marconi.

1902 **Disc brake.** Fitted to a car by the pioneer British motor manufacturer Frederick Lanchester.

1903 The theory of thermionic emission (Richardson).

1904 **Thermionic valve.** Diode valve patented by John Ambrose Fleming of London.

1905 **Safety glass.** Patented in Swindon, Wiltshire, by solicitor John Crewe Wood and fitted to his own Peugeot Bébé.

1906 **The triode.** Dunwood discovers the detecting capacities of crystals.

1910 **Nuclear physics.** Rutherford discovers the atomic nucleus and the proton.

1913 **Stainless Steel.** First made by Harry Brearley, Sheffield, Yorkshire.

1916 **Tank.** First experimental model built by Foster & Company of Lincoln to the design of William Tritton.

1930 **Atomic particle accelerator.** John Cockcroft and Ernest Walton build a machine at Cambridge University with which to bombard atoms.

1936 **C. S. F.'s resonating cavity magnetron.** Electronic television in several countries. The B.B.C. opens the Alexandra Palace studios. Publication of Alan Turing's paper 'on computable numbers'.

1937 **Jet engine.** First test run of experimental Whittle unit, designed by Frank Whittle of British Power Jets Ltd.

1945* **Atomic bomb. Detonated at Alamogordo Air Base, New Mexico, July 16. Hiroshima, Japan, devastated by atomic bomb, August 6.**

1947 **The 33 rpm microgroove.** The point contact transistor.

1948 **The hologram.** (Dennis Gabor), Mark I (Manchester University).

1956 **Telephone.** First transatlantic telephone cable, Scotland to Newfoundland, begins operation.

1959 **Hovercraft.** First full-sized model, the SR-N1, launched at Cowes, Isle of Wight. Principle patented by Christopher Cockerell in 1955.
Fuel cell. First hydrogen-oxygen fuel cell suitable for practical application demonstrated by Francis Bacon of Cambridge after 27 years' research. Used for generating current for radio transmitters in US space satellites.

1961* **Spacecraft.** First manned spacecraft, *Vostock I*, piloted by Major Yuri Gargarin, launched from Baikonur, Western Siberia.

1962 **Satellite broadcast.** First satellite broadcast (Telstar) between the US and Europe.

1969* **Man on the moon.** July 21 Apollo lands on the moon, 238, 857 miles away from earth.

1973 **Videotex.** British televison stations demonstrate transmission of Oracle and Ceefax information on home receivers.

1974 **Videotex.** Prestel system of transmitting information developed by Sam Fedida of the British GPO Research Centre. Public service launched on television in 1979.

* Non-British Invention

Bibliography

ELECTRONICS

ACKERMAN, OTTO, AND BECK, EDWARD. 'Electronic Oscillograph', *Westinghouse Engineer*, November 1944.

AGRAIN, P. AND NOZIÈRES, P. (ED). *De la thermodynamique à la géophysique: Hommage au Pr. Rocard*. Edition du C.N.R.S., 1977.

AISBERG, E. *La radio et la télévision. . . mais c'est très simple*. Ed. Radio, 1972.

D'ALBE, EDMUND. *The Life of Sir William Crookes*. Fisher Unwin, London, 1923.

ALEKSEEV, N.F. AND MALYAROV, D.E. *The Generation of Powerful Oscillations with a Magnetron in the Centimeter Band*. Zhurbal Tekhnichenkoi Fiziki, 1940.

ANGELLO, DR. STEPHEN J. *An Age of Innovation*. McGraw Hill, 1981.

———. 'Semiconductors – New Vigor in an Old Field', *Westinghouse Engineer*, July, 1950.

ARNAUD, JEAN-FRANCOIS. *Dictionnaire de l'électronique*. Larousse, 1966.

ASHBURN, EDWARD V. *Laser Literature: A Permuted Bibliography, 1958-1966*. Western Periodicals Co., North Hollywood.

ASHKIN, ARTHUR. 'Pression de radiation et lumière laser', *Pour la Science*, issue out of print.

BAKER, W.J. *A History of the Marconi Company*. Methuen and Co., London, 1979.

BALIBAR, F. 'Des électrons pour l'étude des surfaces', *La Recherche*, No. 3, July-August 1970.

——. 'Microscope électronique: la visualisation des atomes', *La Recherche*, No. 31, February 1973.

BARBE, D.F. *Introduction to Very Large Scale Integration VLS*. Springer-Verlag, 1980.

BARDEEN, JOHN. *The Early days of the Transistor*. Urbana Illinois, 1979.

——. *Semiconductor Research Leading to the Point Contact Transistor*. Stockholm, 1957.

BARTHÉLÉMY, CLAUDE. *Evolution industrielle des applications des lasers*. Laboratoires de Marcoussis.

BARTON, DAVID K. 'Historical Perspective on Radar', *Microwave Journal*, August 1980.

BEAUCLAIR, W. DE. *Rechnen mit Maschinen*. Fried-Vieweg und Sohn, Braunschweig.

BEKKER, CAJUS. *Augen Durch Nacht und Nebel*. Willhelm Heyne Verlag, Munich, 1978.

BELLO, FRANCIS. 'The Year of the Transistor', *Fortune* 47, 128, 133, 162, 164, 166, 168, March 1953.

BERNSTEIN, JEREMY. *The Analytical Engines*. New York, Random House, 1964.

BLOCH, F. *Les électrons dans les métaux*. Actualité Scientifique et Industrielle, Hermann, 1934.

BLONDEL, ANDRÉ. *Travaux Scientifiques*. Gauthier-Villars, 1911.

BOETTINGER, H.M. *The Telephone Book*. Bell Telephone Laboratory, 1977.

BOHJLE, G. AND HOFMEISTER, E. *Electronic Semiconductor Components*. Siemens.

BOIS, CHARLES G. DU. 'A Half-Century of Western Electric Achievement', *Western Electric News*, 1919.

BONDELIER, RENÉ. *L'ordinateur à l'hôpital*. Masson, 1970.

BOUTRY, G.A. *La connaissance et la puissance*. Albin Michel, 1974.

BRAILLARD, PIERRE. *L'électricité*. Rencontre, Lausanne.

BRAUN, ERNEST AND MACDONALD, STUART. *Revolution in Miniature*. Cambridge University Press, 1978.

BREDOW, HANS. *Im Banne der Äther Wellen*. Mundus Verlag, Stuttgart, 1954.

BREMENSON, CLAUDE AND PENICAUD, ETIENNE. 'Télécommunication par satellite', *Commutation et transmission*, No. 1, September 1979.

BRIGGS, ASA. *The History of Broadcasting in the United Kingdom*. Oxford University Press, 1965.

BRILLOUIN, LÉON. *Conductibilité électrique et thermique des métaux*. Actualité Scientifique et Industrielle, Hermann, 1934.

——. *Les électrons dans les métaux du point du vue ondulatoire*. Actualité Scientifique et Industrielle, Hermann, 1934.

BROWN, ANTHONY. *Bodyguard of Lies*. Bantam, 1976.

BRUCH, WALTER. *Die Fernseh-Story*. Telekosmos Verlag, Stuttgart, 1969.

BRÜCHE, ERNST, AND SCHWERZER, OTTO. *Begründung der geometrischen elektronenoptik*. AEG Telefunken, 1970.

——, AND MAHL, HANS. *Elektronenmikroscop und Elektronenmikroskopie*. AEG Telefunken, 1957.

BRUMA, MARO. 'Applications scientifiques et techniques des lasers', Palais de la Découverte, 1966 (paper).

BURC, G. AND THE *Fortune* editors. 'Le Monde à l'heure des calculateurs', *Fortune*, 1967.

BUSH, VANNEVAR. *Pieces of the Action*. William Morrow and Co., New York, 1970.

CAMPBELL, L. AND GARNETT, N. *The Life of James Clerk Maxwell*. London, 1882.

CAPRA, FRITJOF. *Le tao de la physique*. Tchou, 1979.

CARNAP, RUDOLF. *Les fondements philosophiques de la physique*. A. Colin, 1973.

CARNEAL, GEORGETTE. *A Conqueror of Space*. Liveright, New York, 1931.

CARPENTER, B.E. AND DARAN, R.W. *The Other Turing Machine*. N.P.L., 1975.

CASIMIR, H.B.G. AND GRADSTEIN, S. (ED). *An Anthology of Philips Research*. Centre Publishing Cy., Eindhoven.

——. 'Thoughts on Integrated Circuits and Microtechnology', paper read before the Electronics Conference of the Hannover Fair, 1968. (Reviewed in *Radio Mentor Electronics*, 34, 366, 1981.)

——. *Les cellules solaires*. Radiotechnique, 1979.

CERN. 'La conférence internationale sur les accélérateurs', *Courrier de CERN*, September 1980.

CHAMPEIX, ROBERT. 'Simple histoire de la T.S.F., de la radiodiffusion et de la télévision', *L'Indispensable*.

CHANSON, CAPITAINE. 'L'optique électronique et ses applications au microscope électronique', Communications à la Société des Radioélectriciens, le 26 mai 1945, published in *l'Onde Electrique*.

CLARK, RONALD. *Einstein: The Life and Times*. Stock, 1980.

——. *Tizard*. Methuen and Co., 1965.

CLARKE, JOHN. 'La cryoélectronique', *La Recherche*, No. 38, October 1973.

COLTMANN, JOHN W.' Resonant Cavity Magnetron', *Westinghouse Engineer*, November 1946. 1946.

Conférence internationale sur les télémanipula-

teurs pour handicapés physiques. I.R.I.A., 1978.
Connaissance de l'électronique. Editions du Tambourinaire, 1955.
COOK, J.S. 'Les câbles optiques', *La Recherche*, No. 45, May 1976.
COUFFIGNAL, LOUIS. *La cybemétique.* P.U.F. Que sais-je? 1978.
CROWTHER, J.G.: *James Clerk Maxwell.* Hermann, 1948.
CR: 'Today and Tomorrow', *Radio Mentor Electronics* 44 n° 9, 334, 1978.
CR: 'Intermetal develops faster', *Radio Mentor Electronics* 44 n° 11, 452, 1978.
CRM: 'Microprocessors for New Jobs', *Radio Mentor Electronics* 33, n° 11, 452, 1978.
CROWLEY-MILLING. M. CERN Publications.
CROZON, MICHEL. *New Projects in Particle Physics,* Endeavour Pergamon Press, 1979.
DAUMAS, MAURICE. *Histoire Générale des Techniques, Vol. V.* P.U.F.
DAVID, PIERRE. *Le Radar.* P.U.F. Que Sais-je?, 1969.
DAVIES, D. W. 'A. M. Turing's Original Proposal for the Development of an Electronic Computer', N. P. L., 1972.
DAVID, REGIS. *L'électronique.* P. U. F. Que saisje?, 1964.
DELORAINE, MAURICE. *Des ondes et des hommes.* Flammarion, 1974.
——, AND REEVES, ALEC. 'The Twenty-fifth Anniversary of Pulse Code Modulation', I.E. E. Spectrum, May 1965.
DEVONS, PROF. S. *Rutherford's Laboratory.* Cavendish Laboratory.
DICKINSON, DALE F. 'Les lasers cosmiques', *Pour la Science*, n° 10.
DRAEGER. *Le laboratoire d'électronique et de physique appliquée.* 1963.
DUARME, PIERRE, AND ROUQUEREL, MAX. *Les ordinateurs électroniques.* P.U.F. Que Saisje?, 1961.
DUMMER, G. W. A. *Electronic Inventions and Discoveries.* Pergamon Press, 1978.
DUPAS, CLAIRE. 'Avec l'holographie électronique: des clichés à l'intérieur des atomes', *La Recherche* n° 49, October 1974.
DUPOUY, GASTON. *La mécanique ondulatoire et ses applications.* Académie des Sciences, 1967.
——.'Le microscope électronique 1,5,MV', *Brit. J. Appl. Phys., séries 2, Vol. 2,* 1969.
——.'Performance and Applications of the Toulouse 3 million volt Electron Microscope', *Journal of Microscopy*, January-March 1973.
——.'La mécanique ondulatoire et ses applications', Académie des Sciences, 1967.
——,AND FRANTZ, PERRIER, 'Microscope électronique fonctionnant sous une tension d'un million de volts', *Journal de microscopie, Vol. 1* n°3-4, 1962.
——'Le microscope électronique à 1.5 millions de volts du laboratoire électronique du C.N.R.S. de Toulouse', *Onde Electrique*, June 1967.
——,AND DURRIEU LOUIS, 'Microscope électronique de trois millions de volts', *Journal de microscopie, Vol. 9*, n°5, 1970.
EAMES, CHARLES AND RAY. *A Computer Perspective.* Harvard University Press, Cambridge Mass., 1973.
'L'Echo des Recherches – CNET', n° 100, May 1980.
EDEN, E.R.C., AND WELCH, B.M. *GaAs Digital Integrated Circuits for Ultra High Speed (LSI/VLSI) Very Large Scale Integration.* Springer-Verlag, 1980.
The Edison Era. Elfun G.E. Hall of History Publication, 1979.
EDGERTON, HAROLD E. *Electronic Flash Strobe.* M.I.T. Press, 1979.
——, AND KILLIAN, JAMES R. JR. *Moments of Vision.* M.I.T. Press, 1979.
Encyclopédie internationale des sciences et techniques. Presses de la Cité, 1971.
'Die Entwicklung des Rundfunkempfängers', *Telefunken.*
L'épopée électrique. Régidée, 1973.
ERICKSON, JOHN. *Radio Location and the Air Defence Problem: the Design and Development of Soviet Radar 1934-1940.* 1972.
ERNEST, J. 'Recherches et réalisations francaises dans la domaine du laser', *L'Onde Electrique*, June 1967.
EVANS, WALTER C. 'Electronics – Prodigy of Electrical Science', *Westinghouse Engineer*, January 1950.
EVE, A.S., C.B.E.D., LLD, F.R.S. *The Life and Letters of the Hon. Lord Rutherford.* Cambridge University Press.
FEYNMAN, RICHARD. *La nature de la physique.* Seuil, 1980.
Fiftieth Anniversary Golden Yearbook. R.C.A., 1959.
Fifty Years of Japanese Broadcasting. N.H.K., 1977.
Five Years at the Radiation Laboratory. M.I.T., 1946.
FLEMING, SIR AMBROSE. Article in *Nature*, June, 1945.
FLEMING, JOHN A. *Memories of a Scientific Life.* Marshall, Morgan and Scott, London, 1934.
FOREST, LEE DE. *Father of Radio.* Milcox and Follett, Chicago, 1950.
FORRESTER, JAY M. *Collected Papers.* M.I.T. Press, 1975.
FORTE, S.T. 'The Technology of Integrated Circuits', *Radio Mentor*, 32, 282, 1965.
FRANC, ROBERT. *Eugène Ducretet.* Tambourinaire, 1964.
FRIEDRICH, V. 'Conditions for and Examples of the Use of Microcomputers', paper read at the Colloquium of the Austrian Tribological Society, Siemens Publication, Vienna, June 1979.
FRITZSCHE, H. 'Les semiconducteurs amorphes', *La Recherche*, n° 6, November 1970.
Fünfzig Jahre Telefunken, May 1953.
Fünfundzwanzig Jahre Telefunken, 1928.
GARUM, VIKTOR. 'Tonfrequenz: Verständertechnik', A.E.G. Telefunken.
GARBRECHT, K. *Microprocessors and Microcomputers. Large Scale Integrated Semiconductor Components. Festkorperprobleme XVII.* Vieweg-Verlag, 1977.
GEDDES, KEITH. *Guglielmo Marconi 1874-1937.* Science Museum, London 1974.
Die Geschichte des Magnettons. Telefunken, 1973.
GILLE, BERTRAND. Gallimard, 1978.
GIRARDEAU, EMILE. *Souvenirs de longue vie.* Berger-Levrault, 1968.
'Le grand débat de la mécanique quantique', unpublished letters between Max Born, Albert Einstein and Wolfgang Pauli, *La Recherche* n° 20, February 1972.
GOLDBLAT, JOSEF.'La course aux armements stratégiques', *La Recherche* n° 37, September 1973.
GOOD, I.J. 'Early Work on Computers at Bletchely', N.P.L. 1976.
GOSLING, W., TOWNSEND, W.G. AND WATSON, J. *Field Effect Electronics.* Butterworths, London, 1971.
GRANIER, JEAN. *L'électron.* P.U.F. Que sais-je? 1958.
GRIVET, PIERRE. *Optique électronique.* Tambourinaire, 1958.
GUIHO, GÉRARD, AND JOUANNAUD, J.P. 'Intelligence artificielle et reconnaissance des formes', *La Recherche* n° 43, March 1974.
GUILLIEN, ROBERT. *L'électronique médicale.* P.U.F. Que sais-je? 1974.
——. *La télévision en couleur.* P.U.F. Que saisje? 1978.
HAG-HAZEN, J.R. *Fifty years of Electronic Components.* Philips Elcoma Division, 1971.
HANDEL, SAMUEL. *Histoire de l'électronique.* Marabout, 1970.
HARREL, MARY ANN. *Those Inventive Americans.* National Geographic Society, 1971.
HARCUP, GUY. *The Challenge of War.* Newton Abbott, 1970.
HEISENBERG, WERNER. *La nature dans la physique contemporaine.* Gallimard, Idées, 1962.
HEMARDINGUEN, P. *Le superhétérodyne et la superréaction.* Chiron, 1926.
HENNING, W. 'Microelectronic Sensors of the Siemens AG', paper read at the Colloquium of the Austrian Tribological Society, Siemens Publication, Vienna, June 1979.
HERBERT, JEAN-LOUIS. 'Histoire et perspectives

du câble sous-marin', I.E.E.E., 1980.
HERMANN, PETER KONRAD. 'Electronische Messtechnik', A.E.G. Telefunken, 1960.
HERTZ, HEINRICH. *Erinnerungen.* Akademische Verlaggesellschaft, Leipzig, 1928.
——. *Schriften Vermischen Inhalts.* Leipzig, 1895.
Historique Thomson. In-house, Thomson-CSF.
A History of Engineering and Science in the Bell System. Bell Telephone Laboratory, 1978.
HOGAN, D.C. LESTER. 'Reflections on the Past and Thoughts about the Future of Semiconductor Technology', Fairchild interface, 1977.
Hommages à Barthélémy, by Strelkoff and Le Duc. Compagnie des Compteurs, 1964.
HOWETH, LINWOOD S. *History of Communications: Electronics in the United States Navy.* Washington, D.C.: Government Press Office, 1963.
HOUSSINI, JEAN-PIERRE. 'Les télécommunications spatiales', *La Recherche* n° 42, February 1974.
IARDELLA, ALBERT B. 'Western Electric and the Bell System: A Survey of Service', 1964.
IIOPOULOS, JEAN. 'Les particules charmées', *La Recherche,* May 1979.
'Information médicale et hospitalière', Symposium de Toulouse, 1968.
JAMMER, MAX. 'Le paradoxe d'Einstein – Podopolsky – Rosen', *La Recherche,* May 1980.
JOHANSEN, ROBERT, VALLÉE, JACQUES, AND SPANGLER, KATHLEEN. *Electronic Meetings.* Addison-Wesley, 1979.
JOHNSON, BRIAN. *The Secret War.* Hamish Hamilton, London, 1978.
JONES, R.V. *Most Secret War.* British Broadcasting Corp, 1978.
KAMPNER, STANLEY. *Television Encyclopedia.* New York Fairchild, 1948.
KAO, K.C. AND HOCKHAM, G.A. 'Optical Communications', *Proc. I.E.E., Vol. 113* n° 7, July 1966.
KITT, HAUPTMANN V. *Die Entwicklung des Sovjetischen Funkmesstechnik.* Militärtechnik Deutschen Militärverlag, East Berlin, 1970.
KURYLO, F. *Ferdinand Braun.* Heinz Moos Verlag, Münich, 1965.
LAFFINEUR, M. 'Radio-Astronomie', *L'Onde Electrique,* June 1955.
LAUNOIS, DANIEL. *L'Electronique quantique.* P.U.F. Que sais-je? 1968.
LAURES, PIERRE. 'Les lasers, principes et applications', Documentation Air Espace, n°112, September 1968.
LAW, SPENCER. *The New Brahmins.* Morrow and Co., New York, 1968.
Leaders in Electronics. McGraw Hill, 1979.
LE DUC, JEAN. *Au royaume du son et de l'image.* Hachette, 1965.
LEE, COLIN K., AND BLACKBURN, HENDLEY N. *Stories of Westinghouse Research.* Westinghouse.
LEFEBVRE, ANTOINE. 'L'espionnage électronique', *Science et Vie,* May 1979.
LENARD, PHILIPP. *Uber Lichtemission und deren Erregung.* Heidelberg, 1909.
LEQUEUX, JAMES. 'l'astronomie en ondes millimétriques', *La Recherche,* June 1980.
LEROY, R. 'Les missiles anti-missiles', *La Recherche* n°3, July-August 1970.
LESSING, LAWRENCE P. *Man of High Fidelity: Edwin Howard Armstrong.* Bantam Books, New York, 1970.
LOBANOV, M.M. *Les débuts de la radiolocation en Union Soviétique.* Radio soviétique, 1975.
LOOMIS, ALFRED. *Physics Today,* November 1975.
LORENZ, G. 'Effects of Microelectrics on Consumer Goods and Medical Systems', Eurocon Lecture, Stuttgart Valvo Publications, March 1980.
LORENZI, J.H., AND LE BOUCHER, E. *Mémoires volées.* Ramsay, 1979.
LOVELL, BERNARD. *Electronics and their Application in Industry and Research.* Pilot Press, London, 1947.
——, AND CLEGG, A. *Radio Astronomy.* John Wiley and Sons, New York, 1952.
LUCAS, PIERRE. 'La commutation électronique', *La Recherche* n°17, November 1971.
LUFF, PETER L. 'Le téléphone électronique', *Pour la Science* n°7.
LYONS, NICK. *The Sony Vision.* Crown, 1976.
MABON, PRESCOTT C. *Mission Communication.* Bell Telephone Laboratory, 1974.
The Magic Crystal . . . how the Transistor Revolutionized Electronics. Bell Laboratories, 1972.
MAILLET, HENRY. *Les applications industrielles des lasers.* Laboratoires de Marcoussis, 1970.
MARCONI, DEGNA. *My Father Marconi.*
MATRA, JEAN-JACQUES. *Radiodiffusion et télévision. P.U.F.* Que sais-je? 1978.
MARTON, L. *Early Hisory of the Electron Microscope.* San Francisco Press, 1968.
MASSAIN, R. *Physique et physiciens.* Magnard, 1966.
MAXWELL, JAMES CLERK. *Traité d'électricité et de magnétisme,* Gauthier-Villars, 1885.
MAXWELL, JAMES CLERK. *A Treatise on Electricity and Magnetism.* Oxford University Press, 1873.
Medical Data Processing, 1976.
Microelectronics, special edition McGraw Hill, 1979.
MILLIKAN, ROBERT ANDREW. *Autobiography.* Prentice Hall Inc., New York, 1950.
——. *The Electron, its Isolation and Measurement and the Determination of Some of its Properties.* University of Chicago Press, 1924.
MILLMAN, JACOB. *Introduction to VLSI Systems.* Addison Wesley Publishing, 1980.
M.I.T. Series published by the Radiation Laboratory.
MORALEE, DENNIS. 'The first ten years at the Cavendish Laboratory'.
MOREL, PIERRE. 'La météorologie de demain', *La Recherche* n°22, April 1972.
NEEDHAM, JOSEPH. *La science chinoise et l'Occident.* Seuil 1969.
——. *Grand Titration.* University of Toronto Press.
NEUMANN, JOHN VON. *The Computer and the Brain.* Yale University Press, New Haven, 1958.
NIEHAUS, WERNER. *Die Radar schlacht.* Motorbuch Verlag, Stuttgart, 1977.
NORA, SIMON, AND MINC, ALAIN. *L'informatisation de la société.* Le Seuil, 1978.
L'onde électrique, special edition July-August 1971.
On the Shoulders of Giants. Elfun G.E. Hall of History Publication, 1979.
OSBORNE, A. *Round-up – in the Whirlpool of Microelectronics.* 1980.
PAGE, ROBERT N. *The Origin of Radar.* Doubleday, New York, 1962.
PAULU, BURTON. *Radio and Television Broadcasting in Eastern Europe.* University of Minnesota Press, 1974.
PELEGRIN, M. *Machines à calculer électroniques.* 1964.
PERCY, J.D. *John Baird.* Television Society, 1950.
PERRIN, JEAN. *Oeuvres scientifiques.* C.N.R.S. 1950.
——. *La science et l'espérance.* P.U.F. 1948.
'Pictorial Resources in the Washington D.C. Area', Compiled by Shirley L. Green with the assistance of Diane Hamilton for the Federal Library Committee, Library of Congress, Washington D.C., 1976.
PONTE, MAURICE, AND BRAILLARD, PIERRE. *L'électronique.* Le Seuil 1964.
——. *L'informatique.* Le Seuil, 1969.
POL, BALTHASAR, VAN DER. *Selected Scientific Papers.* North Holland Public Co., 1960.
POOLE, LYNN AND GRAY. *Electronics in Medicine.* McGraw Hill, 1964.
POPOVSKI, MARK. *L'U.R.S.S. la science manipulée.* Edition Mazarine, 1979.
POWLEY, EDWARD. *B.B.C. Engineering.*
POYEN, JACQUES ET JEANNE. *Le langage électronique.* P.U.F. Que sais-je? 1960.
PRESCOTT, DAISY. *Reminiscences.* Unpublished souvenirs by Marconi's cousin, January 1910.
Presentation. Electronique Marcel Dessault, 1978.
PRICE, ALBERT. *Instruments of Darkness.* MacDonald and Jane's, London 1977.
PRICE J.A. *Electrons: Waves and Messages.* Hanover House, 1956.

PRINCE, J.L. *V.L.S.I. Device Fundamentals Very Large Scale Integration.* Springer-Verlag, 1980.

Proceedings I.E.E special edition, 'Two Centuries in Retrospect', September 1976.

'Proceedings of the Second National Conference and Exposition on Electronics in Medicine', San Francisco, 1970.

RABINOWICZ, ERNEST. 'Les exoélectrons', *Pour la Science* n°1.

'Radar', *Bell Telephone Magazine* 1945-1946.

Radar Series from Radiation Laboratories.

'Radar Stories and The Navy's History of Radar', *Electronics*, June-July 1943.

The Radio and Electronic Engineer, special edition Golden Jubilee of E.R.E., October 1975.

La Radiotechnique 1919-1969.

RANDELL, BRIAN. *On Alan Turing and the Origins of Digital Computers.* University of Newcastle upon Tyne, 1972.

READ, OLIVER, AND WELCH, WALTER Z. *From Tin Foil to Stereo.* Howard W. Sams and Co., 1940.

REICHARD, W. 'Wits versus Chips', Newspaper Seminar of the German Institute for Correspondence Studies at Tübingen University, 1980.

REICHARDT, JASIA. *Robots.* Penguin Books, New York, 1978.

RENARD, BRUNO. *Le calcul électronique.* P.U.F. Que sais-je? 1969.

REUBER, C. 'After 1970 – Integrated Circuits in Entertainment Electronics', *Radio Mentor Electronic* 32, 496, 1966.

——. 'Microprocessor and Consumer Electronics', *Yearbook of Entertainment Electronics* 27-66, 1978.

REUTER, FRANK. *Microprocessor and Consumer Electronics.* Westdeutscher Verlag, Opladen 1971.

Revue Internationale De Défense, 'Les systèmes de défense aérienne', and 'La guerre électronique', two special editions 1979.

REYNER, J.H. *The Encyclopedia of Radio and Television.* Oldhams, London, 1950.

RICARD, PATRICK. *Le livre des Inventions.* Hachette, 1979.

RICHARDSON, O.W. *The Electron Theory of Matter.* Cambridge Press, 1914.

ROBIEUX, JEAN. *Les perspectives ouvertes par l'évolution des recherches dans le domaine du laser.* Académie des Sciences, 1975.

ROBINSON, DONALD. *The 100 Most Important People in the World Today.* Putnam and Sons, 1970.

ROGERS. *L'empire I.B.M.* Laffont, 1970.

ROSE, ALBERT. 'La photoconductivité', Société française de physique et Société des radioélectriciens, October 12, 1953. Published in *l'Onde Electrique.*

ROSENBERG, JERRY. *The Computer Prophets.* MACMILLAN, NEW YORK, 1967.

RÖWENTROP, KLAUS. *Entwicklung der Modernen Reglungstechnik.* R. Oldenburg, Vienna, Munich, 1971.

RUMEBE, GÉRARD. 'Le laser', *Revue de la Découverte*, June 1978.

RUNGE, W.I. *Elektronik ist keine Hexerei.* Econ-Verlag, Düsseldorf and Vienna, 1966.

RYLES, SIR MARTIN. 'Radio Astronomy: The Cambridge Contribution', Cavendish Laboratory, 1970.

SAMPSON, ANTHONY. *The Sovereign State: the Secret History of I.T.T.* Hodder and Stoughton, London, 1973.

SANDERS, FREDERIC. 'Radar Development in Canada', *Proceeding I.R.E. 195-200*, February 1947.

SARNOFF, DAVID. *Looking Ahead.* McGraw Hill, 1968.

SATELLBERG, KURT. *Von Elektron zur Elektronik.* Elitera-Verlag, Berlin, 1971.

SAUVY, ALFRED. *La machine et le chômage.* Dunod, 1978.

SCARLOTT, CHARLES A. 'Television Today', *Westinghouse Engineer*, July 1949.

SCHLESINGER, A. *Principles of Electronic Warfare.* Prentice Hall.

SCHÜLLER, EDUARD. *Das Magnetton.* A.E.G. Telefunken, 1973.

SCHWANKER, ROBERT. 'Der laser 1917-1978', in *Kultur und Technik*, review of the Deutsches Museum in München, March 1979.

SCHWITTERS, ROY F. 'Les particules élémentaires avec du charme', *Pour la Science* n° 2.

SEGALEN, JEAN. 'La guerre électronique', *La Recherche* n° 46, June 1974.

SELME, PIERRE. *Le Microscope électronique.* P.U.F. Que sais-je? 1970.

SHIERS, GEORGE. *Bibliography of the History of Electronics.* The Scarecrow Press Inc., Metuchen, New Jersey, 1972.

SHOCKLEY, WILLIAM. 'Transistor Physics', *American Scientist* 42, 1954.

Six pays face à l'informatisation. Documentation française, 1979.

Sixty Years of Hitachi. 1910-1970.

A Solid State of Progress, Fairchild.

SONNEMAN, ROLF. *Geschichte der Technik.* Aulis Verlag, Derebner, D.D.R., 1968.

The Steinmetz Era. Elfun G.E. Hall of History Publications, 1979.

SOUTHWORTH, BRIAN, AND BOIXADER, GEORGES. *La chasse aux particules.* Tribune editions C.E.R.N., 1978.

STREETLY, MARTIN. *Confound and Destroy.* MacDonald and Jane's, London, 1978.

STUMPERS, F.L.M.H. 'L'oeuvre scientifique de Balthazar Van der Pol', *Revue technique Philips*, 1960/61, Vol. 22, n°2. 'Some notes on the correspondence between Sir E. Appleton and Balthazar Van der Pol.'

SÜSSKIND, CHARLES. *The Encyclopedia of Electronics.* Reinhold, New York, 1974.

——. 'On the Origin of the Term electronics. Coming of Age and Some More', *Proc. of the I.E.E.E.*, September 1976.

——. 'On the First Use of the Term Radio', *Proc. I.R.E.*, March 1962.

TELLEGEN, BERNARD. 'Some Milestones in Electronics', *Wireless World*, February 1979.

TERMAN, F.E.'From Wireless to Radio to Electronics', Stanford University, 1964.

TERRIEN, JEAN. 'La cellule photoélectrique', P.U.F. Que sais-je? 1965.

THOMSON, GOERGE PAGET. *J.J. Thomson and the Cavendish Laboratory in His Day.* Nelson, 1964.

THOMSON, J.J. *Recollections and Reflections.* Bell and Sons, 1936.

——. *Cathode Rays.* Royal Institution of Great Britain, April 30, 1897.

Toshiba Review, N° 100, November-December 1978.

TRENKLE, FRITZ. *Die Deutschen Funkmessverfahren bis 1945.* Motor Buch Verlag, Stuttgart, 1979.

——. *Die Deutschen Funk-Navigations und Funk-Führungsverfahren bis 1945.* Motor Buch Verlag, Stuttgart, 1978.

TRICOT, JEAN. 'Histoire de l'informatique', *Science et Vie*, April-May-June 1979.

TRINQUIER, JACQUES. 'Les microscopes électroniques géants', *La Recherche* N° 13, June 1971.

TULKAY, EDGAR. *Astronomy Transformed.* John Wiley and Sons, 1976.

TURING, SARAH. *Alan Turing.* Cambridge, Heffer, 1959.

TYNE, GERALD F. *Saga of the Vacuum Tube.* Howard W. Sams and Co. Inc., Indianapolis 1977.

VARIAN. *Twenty-five Years.* 1974.

VASSEUR, ALBERT. *De la T.S.F. à l'électronique.* Editions Techniques et Scientifiques françaises 1975.

WARNECKE, R, AND GUENARD, P. *Les tubes électroniques à commande par modulation de vitesse.* Gauthier-Villars, 1951.

WATSON-WATT, SIR ROBERT. *Le Radar.* Lecture given at the Palais de la Découverte, 1946.

——. *Three Steps to Victory.* Odham Press, London, 1957.

WATTS, R.K. *Advanced Lithography, Very Large Scale Integration V.L.S.I.* Springer-Verlag, 1980.

WEIHER, SIEGFRIED, AND GOETZELER, HERBERT. *Weg und Wirken der Siemens Werke im Mortsbritt der Elektrochnick 1847-1972.* Siemens, 1972.

Werk und Wirken. Tekade, 1978.

'Westinghouse in World War II: Radio and X-

Ray Divisions', Westinghouse, October 1950.
WESTMIUZEV, V.K. 'Studies on Magnetic Recording', *Philips Research Reports*, April-June-August-October 1953.
WHITE, SARAH. *Guide to Science and Technology in the USSR*, Francis Hodgson, 1964.
WHITE, WILLIAM C. *The Early History of Electronics in the General Electric Company*. Unpublished 1953-1955.
WHITE, W.C. 'Some Events in the Early History of the Oscillating Magnetron', *Journal of the Franklin Institute*, Vol. 254, n° 3, September 1952.
WIEGAND, OTTO. *Passive Bauelemente Siemens, 1881-1974*.
WIENER, NORBERT. *God and Golem Inc.* M.I.T. Press, 1964.
WILDES, KARL L. *The Digital Computer Whirlwind*. Unpublished M.I.T.
WILDING-WHITE, T.M. *Jane's Pocket Book II – Space Exploration*. MacDonald and Jane's, London, 1976.
WILKES, PROF. *A History of Computing*. Academic Press, New York, 1979.
——. 'Early Computer Developments at Cambridge: The Edsac'.
WILLIAMS, TREVOR. *A History of Technology, Vol. 11*. Clarendon Press, Oxford, 1978.
WINDELL, DR. PETER. 'John Baird', *New Scientist*, November 11, 1976.
WOODBURY, DAVID O. *Battlefronts of Industry: Westinghouse World War II*. John Wiley and Sons, New York, 1951.
YAVIV, AMNON. 'L'optique des guides de lumière', *Pour la Science* n° 17.
ZIEL, A. VAN DER. *Solid State Physics*. Prentice Hall, 1966.
ZUSE, KONRAD. *Der Computer Mein Lebens-Werk*. Munich, 1970.
ZWORYKIN, V.K. 'Some Prospects in the Field of Electronics', Government Printing Office, Washington D.C., 1952.
'Zworykin at 89', *Communicate,* the R.C.A. Magazine, 1978.
——, AND WILSON. *Les cellules photoélectriques et leurs applications*. 1934.
——, AND MORTON, G.A. *Television*. John Wiley and Sons, New York, 1940.

ENGINEERING

ARMYTAGE, W. H. G. *A Social History of Engineering*. Faber and Faber, London, 1961, 2nd edn., 1967.
ARUP, OVE. 'Reinforced Concrete', *Architects Year Book*, 1945.
BANHAM, R. *Theory and Design in the First Machine Age*. Architectural Press, London, 1960.
BIRD, ANTHONY. *The Motor Car, 1765-1914*. Batsford, London, 1963.
BARKER, T. C. AND ROBBINS, MICHAEL. *A History of London Transport, Vol. I, The Nineteenth Century*. Allen and Unwin, London, 1962.
BARMAN, CHRISTIAN. *An Introduction to Railway Architecture*. Art and Technics, London, 1950.
BARTON, D. B. *The Cornish Beam Engine*. Barton, Truro, 1965.
BELLISS, J. EDWARD. 'A History of G. E. Belliss and Company and Belliss and Marconi, Ltd.', *Newcomen Society Transactions, Vol. XXXVII,* 1964-5.
BILL, MAX. *Robert Maillart*. Pall Mall.
BOWLEY, M. *The British Building Industry*. Cambridge University Press, London, 1921.
BURTT, FRANK. *Cross Channel and Coastal Paddle Steamers*. Richard Tilling, London, 1934.
BILLINGTON, DAVID P. *Robert Maillart's Bridges: The Art of Engineering*.
CHADWICK, GEORGE F. *The Works of Sir Joseph Paxton 1803-1865*. The Architectural Press, 1961.
CHALONER, W. H. 'Aron Manby, Builder of the First Iron Steamship', *Newcomen Society Transactions Vol. XXIX,* 1953-5.
CLARK, RONALD H. *The Development of the English Traction Engine*. Goose, Norwich, 1960.
——. *Savages Limited, a Short History*. Modern Press, Norwich, n.d.
COLEMAN, TERRY. *The Railway Navvies*. Hutchinson, 1965, Penguin Books, 1968.
COLLINS, A. R. *Structural Engineering, Two Centuries of British Achievement*. Tarot Print Ltd., 1983.
CROMPTON, R. E. B. *Reminiscences,* Constable, London, 1928.
DAVEY, N. *A History of Building Materials*. Phoenix, London, 1961.
DERRY, T. K. AND WILLIAMS, T. L. *A Short History of Technonology*. The Clarendon Press, Oxford, 1960.
DICKINSON, H. W. AND TITLEY, A. *Richard Trevithick*. Cambridge University Press, Cambridge, 1934.
——. *A Short History of the Steam Engine*. Cambridge University Press, Cambridge, 1938; 2nd edn. Cass, London, 1963.
DUCKWORTH AND LANGMUIR. *Clyde and Other Coastal Steamers*. Brown Ferguson, Glasgow, 1939.
DURHAM, JOHN. *Telegraphs in Victorian London*. The Golden Head Press, Cambridge, 1959.
ELLIS, HAMILTON. *British Railway History*, 2 Vols. Allen and Unwin, London, 1954, 1959.
ESHER, L. *A Broken Wave. The Rebuilding of England 1940-1980*. Allen Lane, 1981.
FLETCHER, SIR BANNISTER. *History of Architecture*. Athlone Press, 18th edn.
Forth Road Bridge Superstructure, Forth Road Bridge Joint Board, The Board, 1964.
FRANCIS, A.J. *The Cement Industry 1796-1914: A History*.
GALE, W.K.V. *The British Iron and Steel Industry*. David and Charles, Newton Abbot, 1967.
——. *The Black Country Iron Industry*. The Iron and Steel Institute, 1966.
GIBBS-SMITH, CHARLES A. *The Invention of the Aeroplane, 1799-1909*. Faber and Faber, London, 1966.
GIEDION, S. *Space, Time and Architecture*. Harvard University Press, 5th edn., Cambridge, Mass., 1967.
GORDON, J. E. *The New Science of Strong Materials*. Penguin, Harmonsworth, 1976.
Great Exhibition of 1851, The. A Commemorative album compiled by the Victoria and Albert Museum.
GUERITTE, T. J. 'Recent Developments in Prestressed Concrete Construction with Resulting Economy in the Use of Steel', *Institution of Structural Engineers Journal*, July 1940.
HAJNAL-KONYI, K. 'Prestressed Concrete', *Architects Year Book No. 3,* 1949.
HALSTEAD, P. E. 'The Early History of Portland Cement', *Newcomen Society Transactions, Vol. XXXIV,* 1961-2.
HAMILTON, S. B. 'Sixty Glorious Years, the Impact of Engineering on Society in the Reign of Queen Victoria', *Newcomen Society Transactions, Vol. XXXI*, 1957-9.
——. 'A Note on the History of Reinforced Concrete in Buildings', *National Building Studies, Special Report No 24,* HMSO, London, 1956.
——. 'A Short History of the Structural Fire Protection of Buildings', *National Building Studies, Special Report No. 27*, HMSO, London, 1958.
HAMILTON, DR. S.E. O.B.E. *Why Engineers Should Study History*. Newcomen Society, 1956.
Henry Maudslay and Maudslay Sons & Field. The Maudslay Society, London, 1949.
HOBHOUSE, CHRISTOPHER. *1851 and the Crystal Palace*. John Murray, London, 1950.
HOBSBAWM, E. J. *Industry and Empire, An Economic History of Britain Since 1750*. Weidenfeld and Nicolson, 1968.
HODGE, MAJOR W. J. 'The Mulberry Invasion Harbours', *Institution of Structural Engineers Journal*, March 1946.
HOPKINS, H. J. *A Span of Bridges*. David and Charles, Newton Abbot, 1970.
HULL, H. V. 'Developments in the Design of Welded Steel Structures', *Institution of Structural Engineers Journal,* August, 1945.
JACKSON, A., AND CROOME, DESMOND. *Rails Through the Clay, a History of London's Tube Railways*. Allen and Unwin, London, 1962.

JEAFFRESON, J. C. *The Life of Robert Stephenson F.R.S.* Longmans, 1864.
JOHNSON, H. R. AND SKEMPTON, A. W. 'William Strutt's Cotton Mills, 1793-1812', *Newcomen Society Transactions, Vol. XXX*, 1955-7.
KING, G. A. 'How the Oil Engine Became Efficient', *Engineering Heritage*. Heinemann, London, 1963.
KINGSFORD, P. W., AND LANCHESTER, F. W. *The Life of an Engineer*. Arnold, London, 1960.
KLAPPER, CHARLES. *The Golden Age of Tramways*. Routeledge and Kegan Paul, London, 1961.
KLINGENDER, FRANCIS D. *Art and the Industrial Revolution*. Noel Carrington, London, 1947; new edn., ed. Sir Arthur Elton, Evelyn, Adams and Mackay, London, 1968.
LANCHESTER, G. H. 'F. W. Lanchester, L.L.R., F.R.S., His Life and Work', *Newcomen Society Transactions, Vol. XXX*, 1957.
MACKAY, J. *Life of Sir John Fowler*. John Murray, London, 1900.
MACDERMOT, E. T. *History of the Great Western Railway*, 2 Vols. Great Western Railway, London, 1927, 1931.
MCNEIL, IAN. 'Hydraulic Power Transmissions', *Engineering Heritage*. Heinemann, London, 1963.
'Manchester Ship Canal, The', *Engineering*, January 26, 1894.
MAINSTONE, ROLAND J. *Developments in Structural Form.*
MARE, E. DE. *Bridges of Britain*. Batsford, London, 1975.
MICHAELS, LEONARD. *Contemporary Structure in Architecture*. Reinhold, New York, 1950.
MOCK, ELIZABETH B. *Architecture of Bridges*. Museum of Modern Art, New York, 1949.
NASMYTH, J. *Autobiography*. (Ed. Smiles). John Murray, London, 1883.
NOCK, O. S. *Steam Locomotive*. Allen and Unwin, 1975; new edn., 1968.
NORRIE, C. M. *Bridging the Years, a Short History of British Civil Engineering*. Arnold, London, 1956.
A Note on the History of Reinforced Concrete in Buildings to the year 1900. Institution Library.
OSBORN, FRED M. *The Story of the Mushets*. Nelson, London, 1952.
PAAR, H. W. *The Severn and Wye Railway*. David and Charles, Newton Abbot, 1963.
PENFOLD, A. (ED.) *Thomas Telford: Engineer*. Thomas Telford, London, 1980.
PHILLIPS, P. *The Forth Bridge*. Grant, Edinburgh, 1899.
PIKE, E. ROYSTON. *Human Documents of the Industrial Revolution*. Allen and Unwin, London, 1966.
PUGSLEY, SIR A. (ED.)*The Works of Isambard Kingdom Brunel*. Institution of Civil Engineers and University of Bristol, London, 1976.
——. 'Aircraft Structures', *Institution of Structural Engineers Journal*, January 1941.
RAISTRICK, ARTAUR. *Quakers in Science and Industry*. David and Charles, Newton Abbot, 1950; 1968.
RIDDING, ARTHUR. *S. Z. de Ferranti, Pioneer of Electric Power*. A Science Museum Booklet, HMSO, London, 1964.
ROLT, L. T. C. *Isambard Kingdom Brunel*. Longmans, London, 1957; Penguin, 1970.
——. *Thomas Telford*. Longmans, London, 1958; Penguin, 1979.
——. *George and Robert Stephenson*. Longmans, London, 1960; Penguin, 1978.
——. *Great Engineers*. Bell, London, 1962.
——. *The Mechanicals, Portrait of a Profession*. Heinemann, London, 1967.
——. *Thomas Newcomen, the Prehistory of the Steam Engine*. David and Charles, Dawlish, 1963.
——. *James Watt*. Batsford, London, 1962.
——.*Tools for the Job, A Short History of Machine Tools*. Batsford, London, 1965.
SAMUELY, F. J. 'Space Frames and Stressed Skin Construction', *Royal Institute of British Architects Journal,* March 1952.
——. *Some Recent Experience in Composite Precast and Insitu Concrete Construction with Particular Reference to Prestressing*. Institution of Civil Engineers, Structural Paper No. 30, February 1952.
SIMMONS, JACK. *The Railways of Britain*. Routeledge and Kegan Paul, London, 1961.
SINGER, HOLMYARD, HALL AND WILLIAMS, editors, *A History of Technology*. Relevant chapters in *Volume IV (1750-1870)* and *Volume V (1850-1900),* Oxford University Press, 1958.
SKEMPTON, A.W. 'Portland Cements, 1843-87', *Newcomen Society Transactions, Vol. XXXV*, 1962-3.
——.'The Boat Store, Sheerness and its Place in Structural History,' *Newcomen Society Transactions, Vol. XXXII*, 1959-60.
——. (ED.) *John Smeaton F. R. S.* Thomas Telford, London, 1981.
——. *Evolution of the Steel Building Frame*. Imperial College Library.
SMILES, S. *Industrial Biography*. John Murray, London, 1883; new edn., David and Charles, Newton Abbot, 1967.
——. *Lives of the Engineers.*
SMITH, E.C. ENG. CAPT. 'Some Pioneers of Refrigeration', *Newcomen Society Transactions, Vol. XXIII*, 1943.
SPRATT, PHILIP H. *The Birth of the Steam Boat*. Charles Griffin, London, 1958.
STRAUB, H. *A History of Civil Engineering*. Translated by E. Rockwell, Leonard Hill, London, 1960.
The Structural Engineer: The Jubilee Issue, Institution of Structural Engineers, London, July 1958.
SUTHERLAND, R. J. M. 'The Introduction of Structural Wrought Iron', *Newcomen Society Transactions, Vol. XXXVI*, 1963-4.
TIMOSHENKO, S. P. *History of Strength of Materials*. McGraw-Hill, New York, 1953.
VERNON-HARCOURT, L. F. *Rivers and Canals, Vol. II*. The Clarendon Press, Oxford, 1896.
WALKER, CHARLES. *Thomas Brassey: Railway Builder*. Frederick Muller, 1969.
WALKER, THOMAS A. *The Severn Tunnel, Its Construction and Difficulties*. Richard Bentley, London, 1891.
WATKINS, G.M. 'The Vertical winding Engines of Durham', *Newcomen Society Transactions, Vol. XXIX*, 1953-5.
WHITE, R. B. 'Prefabrication. A History of its Development in Great Britain', *National Building Sudies, Special Report No. 36*, HMSO, London, 1965.
WHITTICK, ARNOLD. *European Architecture in the 20th Century*. Grosby Lockwood.
WILLANS, K. W. 'Peter William Willans', *Newcomen Society Transactions, Vol. XVIII*, 1951-3.
WILLIAMS, L. PEARCE. *Michael Faraday*. Chapman and Hall, 1965.

INVENTIONS

ABBOTT, C. G. *Great Inventions*. Smithsonian Institution, Washington, 1932.
AITKEN, W. *Who Invented the Telephone*? Blackie, London, 1939.
APPLEYARD, R. *Pioneers of Electrical Communications*. MacMillan, London, 1930.
BARNOUW, E. *A History of Broadcasting in the United States*. Oxford University Press, 1968.
BRIGGS, ASA. *The History of Broadcasting in the United Kingdom, Vols. 1 and 2*. Oxford University Press, 1965.
BRYN, E. W. *The Progress of Invention in the Nineteenth Century*. Munn, New York, 1900.
CARR, L. H. A., AND WOOD, J. C. *Patents for Engineers*. Chapman & Hall, London, 1959.
CARTER, E. F. *Dictionary of Inventions & Discoverers*. Fdk. Muller, London, 1966.
CHASE, C. T. *A History of Experimental Physics*. Von Nostrand, 1932.
CROWTHER, J. G. *Discoveries & Inventions of the 20th Century*. Routeledge & Kegan Paul Ltd., London, 1966.
——. *British Scientists of the 19th Century*. MacMillan, London.
CRESSY, E. *Discoveries & Inventions of the Twentieth Century*. George Routledge, New York. E. P. Dutton, 1914.
DARROW, F. L. *Masters of Science and Invention*. Harcourt Brace, New York, 1923.
DIBNER, BERN. *Heralds of Science as represented*

by 200 Epochal Books and Pamphlets selected from the Bundy Library. Bundy Library, Norwalk, Conn., 1955.
FAHIE, J. J. *History of Electric Telegraphy to the Year 1837*. F. N. Skoon, London, 1884.
——. *History of Wireless Telegraphy*. Wm. Blackwood, Edinburgh and London, 1899.
FLEMING, J. A. *Fifty Years of Electricity*. Wireless Press, 1921.
GOLDSTINE, H. H. *The Computer from Pascal to von Neuman*. Princeton University Press, 1972.
HAWKS, E. *Pioneers of Wireless*. Methuen & Co., London, 1927.
JAFFE, BERNARD. *Men of Science in America*. Simon & Schuster, New York, 1944.
JEWKES J., SAWERS, D. AND STILLERMAN, R. *The Sources of Invention*. MacMillan, London, 1958.
JOHNSON, P. S. *The Economics of Invention and Innovation*. Martin Robinson, 1975.
LARSEN, EGON. *A History of Invention*. J. M. Dent & Sons, London, and Roy Publishers, New York, 1971.
MACLAURIN, W. R. *Invention & Innovation in the Radio Industry*. MacMillan, New York, 1949.
MOORE, C. K. AND SPENCER, K. J. *Electronics – a Bibliographical Guide*. Macdonald, London, 1965.
MOTTELEY. *Bibliographical History of Electricity & Magnetism*. Griffin, London, 1922.
PIERCE, J.R. *The Beginnings of Satellite Communications*. San Francisco Press, 1968.
PLEDGE, H.T. *Science Since 1500*. HMSO, 1946.
Proceedings of the Royal Society, London.
RHODES, F.L. *The Beginnings of Telephony*. Harper, New York, 1929.
RIDER, K.J. *The History of Science and Technology*. Library Assoc. of London, 1967.
ROUTLEDGE, R. *Discoveries & Inventions of the Nineteenth Century*. G. Routledge, London, 1891.
SHIERS, G. *Bibliography of the History of Electronics*. The Scarecrow Press, Metuchen, N.J., 1972.
SINGER, C.J. *A Short History of Science to the Nineteenth Century*. Clarendon Press, 1941.
Timetable of Technology. Michael Joseph, London, 1982.
TRICKER, R.A.R. *Early Electrodynamics*. Pergamon Press, Oxford, 1965.
WHETHEM, W. *A History of Science*. Cambridge University Press, 1929.

HISTORY OF THE STEEL INDUSTRY

ALLEN, G. C. *Industrial Development of Birmingham and the Black Country*. Allen and Unwin, 1929.
BELL, SIR I. L., BT. *The Iron Trade of the United Kingdom*. BITA, 1886.
——. *Principles of the Manufacture of Iron and Steel*. Routeledge, 1884.
——. *Notes on a visit to Coal and Iron Mines and Ironworks in the United States*. ISIJ, 1873.
——. *The Chemical Phenomena of Iron Smelting*. ISIJ, 1871-2.
BESSEMER, H. Autobiography, *Engineering*, 1905.
BOUCHER, J. N. *William Kelly: A True History of the So-called Bessemer Process*. 1924.
BRANDT, D. J. O. *The Manufacture of Iron and Steel*. English Universities Press, 1953.
BROOKE, E. H. *Chronology of the Tinplate Works of Great Britain*. Lewis, Cardiff, 1944; and Appendix, 1949.
BURN, D. L. *The Economic History of Steelmaking, 1867-1939*. Cambridge University Press, 1940.
BURNHAM, T. H. AND HOSKINS, G. O. *Iron and Steel in Britain, 1870-1930*. Allen and Unwin, 1943.
CLAPHAM, SIR JOHN H. *An Economic History of Modern Britain, Vols. I-III*. Cambridge University Press, 1934.
——. 'Work and Wages', in *Early Victorian England*. Ed. G. M. Young, Oxford University Press, 1934.
DENNY, WM. *On the Economical Advantages of Steel Shipbuilding*. ISIJ, 1881.
DUNSHEATH, P. (ED.). *A Century of Technology*. Hutchinson, 1951.
GARRETT, WM. *A Comparison between American and British Rolling Mill Practice*. ISIJ, 1901.
HARBORD, VERNON. *The Basic Bessemer Process. Some Considerations of its Possibilities in England*. ISIJ, 1931.
HEWITT, ABRAM S. *The Production of Iron and Steel in its Economic and Social Relations*. Selected writings, Ed. Allan Nevins, Columbia University Press, 1943.
HEXNER, E. *The International Steel Cartel*. North Carolina University Press, 1943.
JONES, J. H. *The Tinplate Industry*. P.S. King, 1914.
LANGE, E. F. 'Bessemer, Gorransson and Mushet', *Manchester Philosophical and Literary Society Memoirs, Vol. LVII., No.7*, 1913.
LINTON, D. L. (ED.). *Sheffield and Its Region*. The British Association, 1956.
LORD, W. M. 'Development of the Bessemer Process in Lancashire, 1856-1900', *Newcomen Society Transactions, Vol. 25*.
MACKENZIE, T. B. 'The Life of James Beaumont Neilson, F. R. S.: Inventor of the The Hot Blast', West of Scotland ISI *Transactions*, 1928.
MAYNARD, H. N. *On the Use of Steel in the Construction of Bridges*. ISIJ, 1874.
MUSHET, R. F. *The Bessemer-Mushet Process, or Manufacture of Cheap Steel*. 1883.
PALMER, CHARLES MARK. *On Iron as a Material for Shipbuilding and its Influence on the Commerce and Armament of Nations*. ISIJ, 1870.
PEASE, HOWARD. *Sir David Dale: A Memoir*. Murray, 1911.
POLLARD, S. *Three Centuries of Sheffield Steel*. Marsh Brothers, 1954.
PUGH, SIR A. *Men of Steel*. Iron and Steel Trades Confederation, 1951.
SCRIVENOR, H. *A Comprehensive History of the Iron Trade*. Smith, Elder, 2nd Edn., 1854.
SIEMENS, SIR WM. *On the Manufacture of Iron and Steel by Direct Process*. ISIJ, 1873.
——. *Some further Remarks regarding Production of Steel by Direct Process*. ISIJ, 1877.
SMILES, S. *Industrial Biography*. Murray, 1863.
STEWARTS AND LLOYDS. *Review*, Jubilee Number, 1903-1953.
WALTON, MARY. *Sheffield, Its Story and Its Achievements*. Sheffield Telegraph and Star Ltd., 2nd edn., 1949.

Acknowledgements

It is with considerable pleasure that I thank the individuals, companies, institutions, galleries and practices who have given so much time and effort to ensure the quality of content and visual material for the exhibition (RCA November 1987-January 1988) and this commemorative book produced in celebration of the 150th Anniversary of the Royal College of Art. Neither book nor exhibition would have been possible without the contribution of our principal sponsors, Plessey, Balfour Beatty and British Steel, who have been as generous with time and preparation of materials as they have financially. We would also like to thank the National Westminster Bank and Shell UK who also provided sponsorship for the exhibition presentation. Two of the most significant contributions have come from Julia Elton and Jan Walker, who have researched the material for both exhibition and book with great dedication, skill and perseverance. Jan has also made significant contributions to the design, layout and production of the book. The steering group for the publication and exhibition, Professor Edmund Happold (representing Engineering Institutions), Julia Elton and Jan Walker (Research), Professor William Gosling and Michael Aplin (Plessey), Derek Allen (British Steel), John Gayner (Balfour Beatty) and Norman Manners and Linda Anderson (Norman Manners Limited) have been diligent and constructive in shaping the language and presentation of a complex subject with an overwhelming choice of visual material. The Rector of the Royal College of Art, Jocelyn Stevens, has been enormously supportive to my team throughout the long period of preparation. He was central to the establishment of the concept and very active in seeking sponsorship for the exhibition. In my Department at the Royal College of Art, I would like to thank my friend and colleague James Gowan who has contributed to the book and been extremely helpful with material and Gloria Billingham, my secretary, who has administered the promotional aspects with Ingrid Bleichroeder and nobly assisted the research endeavours, particularly biographies and bibliography. In my practice I would like to thank David Reddick, my partner, Roger White who assisted with drawing material, and Elizabeth Hutchinson who sub-edited the text. Finally I would like to thank in particular Andreas Papadakis of Academy Editions for his encouragement throughout the preparation of the book, Frank Russell, Editor-in-Chief and the following members of Academy Editions who have contributed to the preparation of the book, namely Lisa Adams, Mario Bettella, Fi Boyle, Ann Cheatle, Sally Green, Nick Jones and Martin Simpson.

The following institutions have helped us greatly with exhibition and photographic material: Architectural Association; Ben Weinreb Architectural Books Limited; Elton Engineering Books Limited; Institution of Civil Engineers (David Grimes, Librarian); Institution of Structural Engineers; Ironbridge Gorge Museum Trust (David DeHaan, Curator); Science Museum. Other individuals, companies and practices who have helped with the provision of material are: Anthony Hunt Associates (Anthony Hunt); Balfour Beatty (Don Holland and John Gayner); British Steel (Derek Allen and Jim Jamieson); Büro Happold (Edmund Happold and Michael Dickson); Dr. Dennis Smith; English Heritage (Robert Thorne); Dr. Edmund Hambly; Foster Asociates (Norman Foster and Gordon Graham); Felix J. Samuely and Partners (Frank Newby); Francis Pugh; Freeman Fox and Partners (Bernard Wex); Harris and Sutherland (James Sutherland); Ove Arup and Partners (Jack Zunz and Roger Kemp); Pilkington Glass Limited (Sir Alastair Pilkington and David Button); Plessey (Dr. John Bass, Professor William Gosling and Michael Aplin); Richard Rogers Partnership (Richard Rogers and John Young); Royal College of Art (Christopher Frayling and James Gowan); Shell UK Limited (Chris Bullock, Richard Nye and Barry Dugdale).

Photographic credits

The publishers would like to thank the following collections, galleries, organisations and individuals who have been so generous in providing material for 'The Great Engineers': Academy Editions 152, 153; Architectural Association 158, 159, 160, 161, 162, 163, 164, 165; Ben Weinreb, Architectural Books Limited 53, 54, 57, 58, 61, 62, 105; BICC 228; Boulton and Watt Collection, City Library, Birmingham 81; British Architectural Library, RIBA, London 71, 72; British Petroleum 169, 170, 171, 173, 176; British Steel Corporation 112, 168, 174; Büro Happold 24, 25, 26, 27, 28, 29, 30, 31, 32, 35, 38, 39, 82, 216, 237, 239, 240, 241, 242, 244, 245, 263; Channel Tunnel Group 192, 198, 199; Conoco 173, 175; Crombie Taylor 52; Derek Walker Associates 1, 10, 13, 14, 32, 34, 37, 52, 56, 64, 87, 200, 212, 214, 215, 238, 243; Denis Smith 118, 119, 120, 121, 122, 123, 124, 125, 126, 127; Elton Engineering Books 1, 12, 15, 19, 36, 40, 42, 59, 73, 80, 83, 84, 85, 86, 88, 92, 93, 94, 102, 103, 107, 110, 111, 138, 139, 140, 146, 201, 270; Felix J. Samuely and Partners 19, 157, 228, 229, 230, 231; Foster Associates 5, 18, 20, 138, 140, 178, 179, 180, 181, 182, 183, 184, 185, 186, 187, 188, 189, 190, 191, 200, 206, 212, 217, 218, 219, 267, 271; Francis Pugh 40, 72, 74, 75, 79; Freeman Fox 153, 220, 221, 222, 223, 224, 225; Greater London Record Office 63, 66, 67, 118; Hulton Picture Library 92; Institution of Civil Engineers 73, 82, 83, 115, 118, 136, 137; Institution of Structural Engineers 265, 269, 271; Ironbridge Gorge Museum Trust 8, 27, 106, 113, 115, 116, 117, 118, 139, 194, 196; Ironbridge Gorge Museum Trust (Elton Collection) 8, 13, 14, 15, 16, 17, 62, 89, 91, 92, 94, 95, 96, 98, 99, 100, 101, 102, 103, 104, 105, 106, 107, 156, 197; Maritime Museum, Greenwich 34; Marylebone Public Library 69; Mobil Oil

169; Ove Arup and Partners 11, 18, 21, 38, 141, 143, 144, 145, 146, 147, 148, 149, 151, 155, 204, 205, 233, 260, 261, 265, 268; Philips Petroleum 172; Pilkington Glass Limited 207, 208, 209, 210, 211, 213; Plessey 41, 42, 44, 45, 46, 49, 50, 51, 246, 247, 248, 250, 251, 252, 253, 254, 256, 257; Public Record Office 69; Richard Rogers Partnership 6, 21, 22, 23, 36, 155, 204, 205, 232, 234, 235, 236; Robert Howlett 263; Royal Artillery Institution, Woowich 76, 134, 135; Science Museum, London 70, 81, 90, 110; Shell 142, 166, 167, 168, 171, 172, 174, 177; Sotheby's 68; Thames Water Authority 125; Victoria and Albert Museum 54, 55, 56, 59, 63, 65, 66, 129, 130, 132, 133; Welsh Arts Council 78-79; Westminster City Library 68; Windsor Castle Archives 66, 67; Windsor Royal Library 73. Photographers and artists: Ben Johnson 17, 33, 41, 203; Bernard Vincent 235, 269; Cedric Price 155; Edmund Happold 30; Farbewerk Hoecht 237; Foster Associates 179, 180, 182, 183, 184, 186, 187, 188, 191, 231 (Ian Lambot, Richard Davies, John Nye, photographers); Futagawa, Y. 235; Sam Lambert 268; Harry Sowden (Ove Arup and Partners) 144, 147, 148, 149, 150, 151, 156, 260, 261, 266; Institute for Lightweight Structures 216; James Gowan 154, 226, 227; James Sutherland 108, 109, 114; Jan Walker 12, 15, 40, 42, 52, 57, 58, 61, 62, 73, 76, 78, 79, 83, 84, 85, 86, 88, 100, 103, 107, 116, 138, 139, 140, 201, 226, 230, 234; John Donat 265, 267; Ken Kirkwood 10, 212, 217, 219, 238, 242; Max Dupain 150; Ove Arup 233; Richard Einzeig 228, 236; Richard Rogers Partnership 155, 205, 232, 234, 235, 236 (Martin Charles, Richard Bryant, photographers).

Index

Figures in bold refer to biographical entries; figures in italics refer to pages containing illustrations

Abbey Mills Pumping Station 124, *124, 126*
ABK Architects *38*
Adams, George 77
Adams, W. Bridges 63
Agricola, Georgius 76, *80*
Ahlström, Göran 84
Ahm, Povl **261**
Aird and Son 122
Airship R. 100. *115*
Air Supported Structure, 58° North *244*, 263
'Albert Edward' 123
'Alexandra' 123
Algonquin Radio Telescope 220
American Pavilion, Montreal *38*
Analogue Telephone 48
Anthony Hunt Associates *41, 212, 217*
Appleby, John 273
Archigram 154, 155
Architectural Association 154, 155, 227
Architectural Review 56
Architectural Quarterly Review 65
Arctic City 35, 239
Argentine Central Railway 220
Arkwright, Sir Richard 12, **261**, 263, 272
Armengauld *74*
Armstrong 88
Ashby, Sir Eric 23, 84
Artisans' and Labourers' Dwellings Act 126
Art Journal 129
Arup, Ove 23, 144, 216, **261**,
Aspdin, Joseph **261**
Aspen Design Conference 12
Association of Consulting Engineers 142
Aston University Conference Centre 263
Atkins, William **261**
Atomic Bomb 273
Atomic Nucleus 273
Atomic Particle Accelerator 273
Auckland Harbour Bridge 220, 269
Audion 49
Automatic Test System *257*
Ayrton, Maxwell 160, 161, *162*

Babbage, Charles 135
Bacon, Francis 273
Bage 113
Baghdad Metro 220
Baird, John Logie 49, **261**
Baker, Sir Benjamin *19, 88, 116, 197*, 216, **262**, 265
Baker, Mathew *78-79*
Baldwin, Stanley 15
Bangkok Expressway 220
Bank of England 264
Banting, William 132
Bardeen 50
Barker 218
Barlow, William Henry *105*, **262**
Barry, Sir Charles 65, 68
Baths of Caracalla *72*
Battersea Power Station 157
Bauhaus 155
Bazalgette, Sir J. W. *118*, 119, *119*, 120, *120*, 121, 122, 123, 124, 125, 126, 127, *127*, 216, **262**
Beattie 138
Beaumont, Colonel F.E.B. 193
Beckman 218
Beckton Reservoir 124
Bedford to London Railway 262
Bell and Tainter *48*
Bell, Alexander Graham 48, *48*, **262**
Bell, Patrick 272
Bell Telephone Laboratories 247
Bennet, Mr. 86
Berkley, G. 264
Bessemer's Converter 111
Bessemer's Sir Henry 114, **262**, 273
Bessemer Process *112*, 114, 262, 273
Betts, E.L. 262
Bidder, G.P. 83, 87, 121
Biosensor Chip *255*
Birmingham National Exhibition Centre 266
Bishop Fox *71*
Blatchford Viaduct *97*
Blaug, Mark 88
Bletchley Park 49
Blissworth Cutting *99*
Blumlein, Alan 49
BOAC Airport Hangars, London 164, 271
BOAC Maintenance Headquarters *164*, 165, *165*
Board of Ordnance Drawing Room 78
Board of Trade 132
Boabrowski, Jan 32, **262**
Bolton, Arthur T. 154
Booth's Gin Distillery 133
Boots 'Drys' and 'Wets' Buildings 155, *158*, 162, 163, 164, 165,
Bosphorus Bridge 197, 220, *221*
Boulsover, Thomas 272
Boulton and Watt Archive 80
Boulton and Watt Manufactory 80
Box Tunnel *96*
BP's Buckhan Floating Platform *173*
Brady, Francis 193
Bramah, Fox and Company 63
Bramah, John Joseph 63, **262**, 272
Bramah Lock 262
Bramante 72
Brassey, Thomas 124, 138, **262**
Brattain 50
Braun, Karl Ferdinand 266
Brearley, Harry 273
Brent A Platform 173
Brent Oilfield *142*
Bridgewater Canal 27
Briggs, Asa 91
Brindley, James 27, *28*
Bristol and Exeter Railway 264
Bristol Cathedral 227
Britannia Bridge *14*, *82*, 91, *100*, *110*, *111*, 113, *115*
British Broadcasting Company 49, 262
British Channel Tunnel Company 193
British Consultants Bureau 266
British Empire Exhibition, Wembley 160
British Rail *209*
British South African Chartered Company 264
British Standards Committee 263
British Standard Specification 160
British Steel Corporation 168, 178, 197
Brown Daltas Associates *229*
Brown, Joyce M. 114
Brown William 197
Brunel, Isambard Kingdom 12, *14*, 27, *34*, *40*, 45, 68, 69, 79, *81*, 83, 86, *86*, 87, *87*, *89*, *90*, 91, *91*, *92*, *94*, *96*, *97*, *103*, *107*, *110*, 144, 153, *156*, 157, **262**, *263*, 264, 265
Brunelleschi 71
Brunel, Sir Marc Isambard 79, 87, 114, 157, 262, 268, 272
Brunless, Sir James 264
Buchanan, R.A. 84, 87
Buckthorpe 239
The Builder 56
Büro Happold *25*, *38*, *216*, 239, 263, *263*, 265, *266*
Burrell Collection *229*
Burton, Decimus *17*, *33*
Butler, Edward 273

Calgary 32
Cambridge History Library *203*
Cambridge Library 227
Campbell, Captain James 272
Campilo, Francisco Salva 42
Canadian Grand Trunk Railway 262
Canadian National Research Council 263
Candela, Felix, 32, *35*
Cannon Street Railway Station 264
Caracalla Baths 72
Carlisle 42
Cartensen 69
Carlton House Terrace 113, 117
Cartwright, Edmund 12, **263**, 272
Cast Steel Process 273
Casteels, Maurice 157
Castle Howard 11, 12
Cayley, Sir George 263, 273
Central Argentine Railway 264
Central Cormorant Oil Field 177
Chadwick, Edwin 119, 120
Chadwick, George F. 60
Chambers, William 75
Channel Tunnel 265
Channel Tunnel Developments Limited 196
Chapman, Hendrik 77
Chappe, Claude 40

Charing Cross Station *105*
Chelsea Embankment 262
Chermayeff 227
Chesapeake Bay Project 197
Christmas Card 272
Churchill, Sir Winston 264
Cinematograph Exhibitors' Association 265
'City of Rome' 114
Civil Engineer Corps 138
Clark, Arthur C. 50
Clark, E.F. 83
Clarke, J.F. 114
CLASP 60
Clement, Joseph 15
Clifton Suspension Bridge 155, *156*, 262
Clutch 272
Coalbrookdale Ironbridge 28, 29, *30*, 112, 272
Coalbrookdale Ironworks 27, 269
Coates 227
Cochrane, John 67
Cockerell, Sir Christopher **263**, 273
Cockroft, Sir John **263**, 273
Coherer 48
Cole, Henry 128, 129, 130, 132, 133, 134, 135, 272
Colossues 49
Compact Disc 51
Compression Ignition Engine 273
Concorde *209*
Concrete Society 262
Conder, F.R. 83
Connell 227
Conoco 177
Conway Bridge 264
Coode and Partners 142
Cooke, Sir William Fothergill *42*, 44, 45, 46, 47, 48, 271, 272
Coriolis Force *154*, 157
Corlett, Ewan 114
Cormorant A Platform 177
Corrugated Iron 272
Cort, Henry 111
Coventry Cathedral 261, 265
Cowper, Charles 55
Cowper, Downes *64*
Cowper, Edwin A. 63
Cragside 88
Craigellachie Bridge *73*
Crank-driven Bicycle 272
Crimean War 138
Crompton, Samuel 272
Cromwell Road Cencrete Bridge, London 273
Crooke, William 87
Crossness Pumping Station *121*, *124*
Crossness Reservoir 126
Croydon Waterworks *16*
Crozer, Claude 76
Crumlin Viaduct *83*, *102*
Crystal Palace, London *12*, *13*, 16, 28, 32, *32*, *36*, 52, *52*, *54*, 55, 56, *56*, 60, *60*, *62*, 63, *63*, *64*, 65, *65*, *66*, 67, *67*, 68, 69, *82*, 113, 135, 157, 216, 264
Crystal Palace, New York 69
Cubitt, William 264
Czernowitz *138*

Daily Express Buildings, London and Manchester *158*, 161, 162, 164 271
d'Alembert 76
Darby, Abraham 111, 272
Darwin, Charles 88,
Davenport, Professor Allan 184
da Vinci, Leonardo 72, *73*, 75
Davy, Sir Humphry **263**, 272
Dee Bridge 113, 114
de Forest, Lee 49
de Gamond, Thome 193
Delamotte *66*
De la Rue *58*
de Mottray, Tessier 193
Denton, Trevor *214*
Deptford Creek 122
Deptford Power Station *43*, 264
Derek Walker Associates *10*, *238*, *243*
de Rothschild, Baron James 268
Dewar, Sir James 273
Derwent Timber Frame 60
Descartes, R. 76
Deutsche, Rudolf Manuel 76
Dickson, Michael 239, **263**, *263*
Diderot 76
Differential Gear 273
Disc Brake 273
Dollis Hill Synagogue *161*, 164
Dome of Discovery *153*, 157
Dorchester Hotel 161, 162
Dorman Long 265
Dougan, David 88
Downes, Charles 55
Dowson, Philip 265
Duke of Wellington 128, *129*, *131*, 132, *132*, 133, *133*, 135
Duke of York 134
Dulwich Gallery 153
Dumas, Alexandre 40
Dunican, Peter **263**
Dunlop, John **263**, 273
Dunlop Rubber Company 263
Dunwood 273
Duplicating Machine 272, 273
Duralumin 115
Durenner Metalworke 115
Durham Footbridge 232
Dynamo 272

East India Company 141
Ecole Polytechnique 76
Eddystone Lighthouse 76, *77*, 269, *269*
Edinburgh News 65
Edinburgh School of Engineering 88
Edison, Thomas Alva 270
Effra Pumping Station 124
Eiffel Tower 157
Electric Arc Lamp 272
Electric Telegraph *42*, 272
Emberton 227
EMI 49, 262
Emmerson, George S. 87
Emperor Pedro II of Brazil 262
Empire Pool, Wembley *161*, 163, 165
Erskine Bridge 220
Euclid 79
Euston Station *106*

Faber, Oscar **264**
Fairbairn, Sir William 113, **264**
Faraday, Michael 45, 87, **264**, 272
Fedida, Sam 273
Feedback and Systems Stability Theory 49
Felix Samuely and Partners *155*, *157*, *203*, *212*, *215*, 265, 268
Fenestration 2000 Project 215
Ferranti, Sebastian *43*, **264**
Festival of Britain *19*, *157*, 216, 271
Findhorn Bridge 161, *162*
Finniston Committee 85
Five Needle Telegraph 45
Fixed Railway Signals 272
Fleming 49
Float Glass *207*
Float Plant *208*
Float Process 210
Flowers, T. H. 49
Flying Shuttle 272
Folkstone UK Channel Tunnel Terminal Site, *192*
Ford 163
Forth Road Bridge *19*, *88*, *116*, 220, *221*, 262, 265, 266, 269, 271
Foster Associates *18*, *20*, *138*, 178, *178*, *212*, *217*, 266
Foster, Norman 216
Fowler, Sir John *19*, 87, 88, *88*, 116, 123, *197*, *201*, 262, **264**
Fox, Francis Beresford 87, 220, **264**
Fox Henderson and Company 55, 60, 63, 65, 67, 68, 69, 216, 264
Fox, Sir Charles Beresford *12*, *32*, 60, 63, 216, 220, **264**, 265
Fox, Sir Douglas 220, **264**, 269
Frayn, Michael 202
Freeman Fox and Partners *153*, 157, 197, 216, 220, *265*, 266, 269, 271
Freeman, Sir Ralph 220, *220*, 264, **265**, 269
Freemason's Council Chamber 154
French Channel Tunnel Company 193
French, Yvonne 52
Frethun France Channel Tunnel Terminal *193*
Friese-Greene, William **265**
Frischmann, Willie 197
Fry, Roger 28
Fuel Cell 273
Fuller, Buckminster 32
Fullerton, Arni *244*
Fun Palace *155*

Galbraith 24
Gallium Arsenide Device *251*
Gargarin, Major Yuri 273
Garrard's 130
Gas Street Lighting 272
Gasson, Barry *229*
Gauss 44
General Electric Company 211
German and Swiss Standard Specification, 1887, 115
Gestetner, David 283
Giedion, Siegfried 56
Gilbey, W.A. 269
Gildermeister 69
Glider, (first manned flight) 273
Gimbrede, Thomas 76
Gloag and Bridgwater 113
Gooch, Sir Daniel 87, **265**
Goodhart – Rendel 227
Gosling, William 19
Gosse, Philip 87
Gover Viaduct *89*
Graf Zeppelin Airship *24*
Grand Junction Railway 262
Graves Michael *153*
Great Engineers Exhibition 12
Great Exhibition 12, *13*, 52, 60, 128, 134, 264
Great Western Railway 45, *86*, 91, *107*, 262, 265
Gropius, Walter 155
Grove Cell 44
Gutbrod, Rolf *38*
Hamilton, S.B. 114
Hammam, Conrad 227
Hampstead Road Cutting *106*
Hancock Associates 265
Handyside *136*
Hanwell Viaduct *103*
Happold, Edmund 12, 16, 19, *38*, *216*, *237*, 239, *265*, **265**
Harbord, David *214*
Hargreaves, James 12, 261, 272
Harrison, John 272
Hartley 49
Harwell Atomic Energy Research 263
Hawkshaw, John 63, *105*, 193
Hawksley, T. 121
Hayward Gallery 216
Hazelhurst, Colonel *154*, 157
Helsby, Cyril 227
Henderson, John 63, 67
Henkel 218
Hennebique 114
Henry, Joseph 46, 271
Hersent 193
Hertz, Heinrich 48
Hewlett Packard 253

High Marnham Power Station 220
Highpoint I & II, Highgate 261
Hill, Octavius 132, 133, 134
Hills System 60
Hitler, Adolf 28
HMS 'Atlanta' *78-79*
HMS 'Iris' 114
Hobbs, Ronald 218, 263, **265**
Hobhouse, Christopher 52
Hockham, George 50
Hock Park Scheme *38*
Hodgkinson, Eaton 113
Hologram 273
Hologram Production Facility *210*
Honeywell 205
Hong Kong and Shanghai Bank 16, *20, 178, 179, 180, 181, 182, 183, 184,* 184,*185, 186, 187, 188, 189,* 189, *190, 191, 200, 271*, 271
Hong Kong and Shanghai Banking Corporation 178, 184, 189, 232
Hong Kong Aviary 35
Hong Kong Metro 220
Hong Kong Tunnel 220
Hooke, Robert 32
Hoover, Herbert 32
Hopewell Centre, Hong Kong 268
Hopkins, Michael 266
Horeau, Hector 193, *196*
Horner, William George 272
Horseguards Audience Chamber *132*
Horseguards Parade 133
Horsley, John Calcott 272
Horticultural Society 268
Hovercraft 263, 273
HRH Prince Albert *95*, 128, 132, 135
HRH Prince of Wales 39, 123
HRH Queen Elizabeth I 141
HRH Queen Victoria 15, 127, 128
Hughes, Professor D.E. 273
Humber Bridge 197, 220, *222, 224, 225*, 265
Hungerford Bridge 157, 262
Hunt, Anthony 216, **265,** *265*
Huntsman 111
Hutton Field 177
Hydraulic Jack 272
Hydraulic Press 262, 272

IBM Personal Computer 189
IBM's Travelling Museum 232
Illustrated London News 12, *54, 129, 131*, 133, 141, *196*
Imperial Measures 272
Indented Bar and Concrete Engineering Company 159
Indian Army 44
Indus Basin Scheme 141
Inmos Microelectronics Factory 266
Inmos Microprocessor Factory *41*
Institution of Civil Engineers 84, 112, 119, 127, 142, 232, 262, 264, 270
Institution of Engineers of Ireland 84
Institution of Structural Engineers 227, 239, 261, 263, 266, 268
Institution of Mechanical Engineers 84
Intergrid 60
International Association of Bridge and Structural Engineering 266
International Meridian Conference 273
Iron Boat 272
Iron-Framed Building 272
Iron Smelting 272
Isle of Dogs Pumping Station 124
ITT 49
Ivy Bridge Viaduct *89*

Jacquard Looms 129
James, J.G. 112
James Watt and Company 123, 124
Jeaffreson, J.C. 113
Jencks, Charles 155
Jenkin, Fleeming 88
Jenkins, Ronald 218, 265, **266**
Jet Engine 273
Johnson, Ben *203*
Johnson, H.R. 113
Joint Contracts Tribunal 265
Jones, Owen *68, 69, 105*, 134
Journal of Design and Manufactures 128, 130

Kao, Charles 50
Kappafloat 214
Kay, John 272
KDKA 49
Kelly 111
Kelly, Dr. Anthony 117
Kennard, William *83, 102*
Kennedy, Lt. Colonel J. *140*
Keogh, Maria 119
Kerensky, Oleg Alexander **266**
Kerr, Robert 56
Kevlar 117
Kier, J.L. 216, 227
Kilby, Jack 50
King's College London 45
Kingsgate Footbridge, University of Durham *233*, 266
King Fahad Medical City 261
Kowloon Leisure Park *243*, 263
Krier 155

Lake Huron 138
Lambda System 261
Lambot 114
Lanchester, Frederick 273
Landseer 87
Langley, Batty 76
Lansdowne Bridge 141
Lasdun 227
Lawson, Brigadier General Robert *76*
Le Blanc *74*
Le Corbusier 154
Leicester University Engineering Building *203, 226*, 227, *227*
Leoni, Gillcomo 75
Lesseps 87
Lewis, Michael 218, **266**
Liddell, Ian *38, 216, 237*, 239, *266,* **266**
Lieberman, Ralph 52
Lincoln's Inn Fields, London *152*
Linear Induction Motor 272
Littlewood, Joan *155*
Liverpool and Manchester Railway 216, 269, 270
Lloyd's Building *21*, 23, 154, *155*, 202, *204, 205, 232, 233*
Lloyd's Register 114
Locke, Joseph *83*, 87, *92*, 262, **266**
'Locomotion' 136
London and Birmingham Railway 45, 91, *98, 104*, 216, 264
London to Birmingham Iron Bridge *101*
London County Council 127
London Docklands 261
London Zoo Aviary 227, 230-31
Lord Snowdon *230-31*
Lukin, Lionel 272
Lutine Bell 23

Macdonald 239
MacGregor, Ian 197
Mackenzie, William 138
Macmillan, Kirkpatrick 272
Magnus Platform *169*, 173
Manasseh, Leonard 157
Mannheim Garden Centre *38, 237, 244, 265*, 266
Mansion House 11
Marconi Company 49
Marconi, Guglielmo *47*, 48, 262, **266**, 273
Marine Chronometer 272
Marlborough House 132, 134
MARS Group 227
MARS Plan of London 227
Marsh, Richard 272
Martin, John L. 155, **266**
Martin, Leslie 155
Mass Production Manufacturing System 272
Mathematical Information Theory 50
Mathieu, Albert 193
Maudslay Sons and Field *81*
Maudslay and Field 262
Maudslay, Henry *268*
Maureen Steel Gravity Platform *172*
Maxwell, James Clerk 48
Maybach, William 273
Mayhew, Henry 65
McAlpine, Malcolm 162
McDermott 12
McNeill, Sir John 119
Mechanics' Institute 129
Mediaeval Court 135
Meikle, Andrew 272
Menai Suspension Bridge 27, *34*, 262, 264, 270
Mendlesohn, Erich 227
Menil Gallery, Houston 232, 269
Merecedes Benz 39
Mersey Tunnel 220, 264
Metcalfe, Sir Charles 264
Metropolis Management Act 120
Metropolitan Board of Works 120, 121, 122, 123, 124, 125, 126, 127, 262
Metropolitan Commission of Sewers 120, 121
Metropolitan District Railway 123, *201*, 262
Metropolitan Tramway Company 159
Michael, Duncan **268,** *268*
Microphone 273
Midland Railway 262
Military Survey of Scotland 79
Miller, Horace 272
Milton Keynes Shopping Building *215*, 227
Ministry of Defence 193
Mississippi River 28
MKDC (Walker, Mosscrop, Woodward) *212*
Mobil's Beryl B Platform *169*
Modern Movement 261
Modular Construction 273
Monge, Gaspard *72*, 76, 77
Monier 114
Montrose Bridge *163*
Morse, Samuel 46, 47, 271
Mosscrop, Stuart *215*
Mouchel 115
Moya, Hidalgo 157
Mulberry Harbour 264
'Mule' Spinning Machine 272
Munich Aviary *25*, 35, *245*, 263, *263*
Murdoch, William 272
Murray 40

Napoleon 40, 60, 76, 134
Nasmyth, James *15*, 272
National Art Training School 135
National Building Agency 263
National Exhibition Centre, Birmingham 266
Nervi, Pietro Luigi 32, *35*
New Street, Birmingham 63
Newby Frank 12, 216, 227, *230, 268,* **268**
Newcastle High Level Bridge *73*, 91, *97, 111*, 113
Newcomen, Thomas 27, *27*, 272
Newton-Einstein House Project *154*
Nicholson 42
No. Eight Machine Shop 63
No. Four Slip 63

Oakhanger Satellite Ground Station *49*
Oersted 43
Off-line Coating Plant *211*
Oldroyd, D.R. 88
Olivetti 227

One-Needle Telegraph 47
Ordish, R.M. *137*
Otto Beit Bridge 220
Otto, Frei *25,* 32, *38, 237,* 239, *244,* 269
Ove Arup & Partners *10, 11, 18, 20, 21, 23, 36, 38, 136, 141, 143, 144, 145,* 155, 156, 178, *178, 200, 204, 206,* 218, *218,* 232, 239, 261, *261,* 263, 265, 266, 268, *268, 269,* 271

Paddington Station 68, *68,* 69
Palladio, Andrea *72,* 75
Palm House, Kew *17, 33*
Palmer 141
Paris Exhibition 48
Parkes, Alexander 273
'Parkesine' 273
Parsons, Sir Charles 12, **268**, 273
Pascal 76
Pasley, General 114
Patscenter 232
Paxton, Sir Joseph 12, *12, 32,* 55, 56, 60, 63, 65, 67, 68, 69, 135, 144, 145, 157, 216, 264, **268**
Payerne, Dr. Prosper 193
Penguin Pool, London Zoo 155, 227, 261
Penney Committee 265
'Penny Black' 272
Perrault, Claude 75
Persig 28
Peter Jones 115
Peter the Great 112
Peto, Sir M. 138, 262
Petts 138
Photophone *48*
Piano and Rogers *23, 155, 234-36,* 269
Pickard, James 272
Pilichowski 227
Pilkington *206,* 208, 210, 214
Pilkington Fibre Optics Factory 206
Pilkington, Sir Alastair **268**
Pink Floyd Umbrellas *216, 240, 266*
Piston-operated Steam-engine 272
Playfair, Lyon 85
Plessey Company 41, *49,* 49, *50, 246, 250,* 253
Plessey Microship *246*
Plessey Research Centre *46*
Pneumatic Tyre 272, 273
Point Contact Transistor 273
Polished Plate Process 211
Pompidou Centre *22, 23,* 32, *36,* 154, *155,* 232, *234-36,* 239, *261,* 265, 269, *269*
Portland cement 261
Powell and Moya *19, 154, 157, 228*
Powell, Philip 157
Power Loom 263, 272
Price, Cedric 155, *155, 230-31,* 268
Priestley, J.B. 157
Primrose Hill Tunnel *102*
'Prince Consort' 123
Private Eye 19
Proton 273
Puddling Furnace 111
Pugin 135
Pugsley, Sir Alfred 87
Pulse Code Modulation (PCM) 49
Pyro-electric Sensors *253*
Pyro-electric Thermograph *253*

Quantrill, Malcolm 11

R. 100 Airship 270
Railway Electrification 273
Raphael 72
Ravenna 114
Railway Locomotive 272
Reaping Machine 272
Redgrave 130, 132, 134, 135
Reeves, Alec 49
Reflectafloat 213
Reinforced Concrete 264
Reliance Controls 216
Renault Building *206*
Renault Centre 206, 271
Rendel, James 141
Rennie, John 86, 272
RIBA 56
Rice, Peter 218, **268**
Richard Rogers Partnership *21, 41, 155, 204, 232*
Richardson 273
Rindl, Sven 216
Ritz Hotel 114
River Conway 270
River Euphrates 142
River Green Crescent 142
River Indus 141
River Indus Suspension Bridge 220
River Jumma 141
River Nile 142
River Ouse 165
River Severn 204, 270
River Spey *73,* 161, 165
River Tigris 142, 261
Riyadh Council of Ministers 239
Riyadh Diplomatic Club *38*
Roberts, Sir Gilbert **269**
Rogers, Richard 23, 154, 266
Rolf Gutbrod/Frei Otto *241-2*
Rolls Royce *209*
Rolt, L.T.C 79, 83, 87, 113, 153
Ronalds, Francis 42, 43, 49
Rope-making Machine 272
Rothwell and Company 124
'Rotamahana' 114
Royal Academy 128, 135
Royal Albert Bridge 91, 262
Royal College of Art 12, 15, 134, 136, 154, 263
Royal Institute of British Architects 161, 261, 264
Royal Military Academy, Woolwich 79
Royal Society of Arts 117
Rowe A.P., 270
Ruskin 28, 65, 154
Russell, John Scott 70, 85, 87
Russell, William 138
Rutherford 273

Saarinen 268
Safety Glass 213
Sainsbury Centre, Universtiy of East Anglia 216, *217, 218, 219, 265,* 266
Saint, Andrew 60
Saint-Cloud Glass House *68*
St. Andrew's Halls of Residence 227
St. Anne's, Liverpool 112
St. Catherine's College, Oxford 261
St. Paul's Cathedral 108, 133, 134, 220
St. Pancras Station 262
St. Peter's Cathedral 72, 108
St. Sophia 108
Salomons, Sir David 273
Saltash Bridge *94*
Sama Bank, Jeddah *229*
Samuely, Felix James *19, 154,* 155, 157, 216, 227, **269**, *269*
San Diego Aviary 35
Sandby, Paul 78
Satellite Broadcast 273
Savery, Captain *27*
Schematics and Biosensor Operation *254*
Schneider 193
Schönberg, Isaac 49
School for New Woodland Industries 35
School of Design 128, 129, *131,* 134, 135
School of Machine-tool makers 262
Schum Peter, J.A. 88
Schweigger, Johann 43, 44
Science Museum 157
SCORE 50
Scotia Steelworks *112*
Scots Magazine 40
Scott, Norman 88
Seed Drill 272
Segal, Walter 155
Semi-conductor Laser *252*
Semper, Gottfried 130, 132, 134
Separate-Condenser Steam-engine 272
Severn Bridge 220, 265, 266, 269, 271
Severn Tunnel 265
Severud 268
Severud, Elstad and Krueger Associates 265
Sextant 272
Shakespeare *61*
Shannon, Claud 49
Sheffield and Manchester Railway 270
Shell/Esso's Brent C Platform *167*
Shell/Esso's Cormorant A Platform *172*
Shockley 50
Shuttleworth, John *119*
Siemens-Martin Open-hearth Process 111
Siemens Schukert 227
Silicon Chips *248*
Silicon Wafer *249*
Silver Plate 272
Simmons, Jack 83
Simon Bolivar Bridge 220
Simpsons, Picadilly 269
Skempton, A.W. 75, 84, 113, 115
Skylon *19,* 157, *157, 228,* 269
Slaughter, Gruning and Company 122
Slide-rest Lathe 268
Smeaton, John 27, *26, 27,* 72-75, 76, 7, *77,* 84, 86, 239, *269,* **269**
Smiles, Samuel 128
Smythe 218
Snow, C.P. 157
Snowdon and Price 227
Snowdon Aviary *230-31,* 268
Snowdon Mountain Railway 220, 264
Soane, Sir John *152,* 153, 154
Society of Civil Engineers 84
SOM 227
Somerset House 128, 129
South Bank Festival of Britain 265
South Eastern Railway Company 193
South India Railway Company 264
South Devon Railway *97, 107*
South Wales Railway 264
Southern, John 80, *81*
Southwark Bridge *111*
Spacecraft 273
Spectrafloat 213
Speight, Sadie 155
Spey Bridge *162*
Spinning Jenny 272
Sputnik I 50
Sputterer *256*
Stray Carburettor 273
Sports Stadium, Jeddah *38*
SS 'Great Britain' *34, 81, 87,* 91
SS 'Great Eastern' *40, 70,* 75, *90, 92, 139,* 262, *263,* 265
Stainless Steel 273
Standard Telecommunications Laboratories 254
Stansted Airport *18*
STC Laboratories 50
Steam Hammer 272
Steam Turbine 273
Stephenson, George 87, 136, **269**, 270, 272
Stephenson, Robert *14, 72, 82,* 83, *84,* 86, 87, 88, 91, *92, 97, 98,* 99, *100, 102, 103, 110,* 113, *115, 136,* 144, 239, 264, **269**
Stepensons, George and Robert 27
Stevens, R.L. 270
Stirling and Gowan *203, 227*
Stirling, James *203,* 227
Stoney 273
Storr, F. 114
Strowger 48

'Structures 3' 239
Strutt 113
Stuart, Herbert Akroyd 273
Stuttgart Staatsgalerie 232
Suez Canal 87
Super Lattice *251*
Sutherland, James 63, 114
Swan, Sir Joseph **270**
Swenarton, Mark 155
Swindon Engine House *110*
Sydney Harbour Bridge 216, *220, 222,* 265, 269
Sydney Opera House *11, 143*, 144, *144*, 145, 232, *261*, 265, 266, *268, 271*

Taff Vale Railway Bridge 96
Taptee Viaduct *140*
Tatlin Tower 216
Tay Bridge 262
'Team 4' 216
Tecton *156*
Teddington Lock 120
Telegraph Construction Company 265
Telford, Thomas 27, *34*, 84, 85, 86 112, **270**
Telstar 50
Temple Bar 132, 133, 134
Temple of Olympian Zeus, Agrigento 108
Terry 155
Tewkesbury Bridge 112
Thames Bridge 220
Thames Embankment 262
Thermionic Valve 273
Thompson, J.J. 273
Thomson, Robert William 264, 272
Thorne, James 123
Three-wheeled Electric car 273
Threshing Machine 272
Time Zones 273
The Times 138
'Titania' 87
Tithebarn Street, Liverpool 63
Tower of London 78
Transformer 272
Trans-Siberian Railway 27
Transatlantic Telephone Cable 273
Treatise on Electricity and Magnetism 273
Trevithick, Richard 27, *39*, 269, **270**, 272
Triode 49, 273
Trissino, Giangiorgio 72
Tritton, William 141, 273
Trussed Concrete Steel Company 160
Tubbs, Ralph *153*, 157
Tull, Jethro 262, 272
Tunnelling Machine 272
Turbine-driven Vessel 273
Turing, Alan Mathison 49, **270**, 273
Turner, Richard *17, 33*
Twain, Mark 28
Two-needle telegraph 47
Tyne and Wear Metro 261

Underwater Manifold Centre (UMC) 177
United States Court 57
University of East Anglia 216
Universtiy of Western Ontario 184
Ure, Andrew 135
U.S. Embassy 227
Utzon, Jorn *11*, 145

Vacuum Flask 273
van der Rohe, Mies 154
Vertue, William *71*
'Victoria' 123
Victoria and Albert Museum 128, 157
Victoria Embankment *120, 122*, 123, *123*
Victoria Falls Bridge 220, *223*, 265
Videotex 273
Vignoles, Charles B. 83, 266, **270**
Vignoles, K.H. 83
Vitruvius 72, 75
Volta 42
Volta River Bridge 269
Voltaic Pile 42, 44
von Canstatt, Schilling 44, 45, 46
von Ohain, Hans 271

Walker, Derek *214, 215*
Walker, John 272
Wallis, Sir Barnes Neville **270**
Walton, Ernest 263, 273
Wachsmann, Konrad 157, 216, 268
Wansford Bridge 161
Ward and Lucas 227
Waterfield, Giles 153
Waterloo *129*
Water-powered Spinning Machine 272
Watford Tunnel *99*
Watkin, Sir Edward 193
Watson, J.W. *214*
Watson-Watt, Sir Robert **270**
Watt, James 80, *81, 84*, 272
Weber 44
Webster, William 122, 123, 124
Welch, M. 76
Wells 227
Wembley Exhibition 163
Wembley Stadium 155, 160
West Point Military Academy 76, 77
Western Pumping Station 124
Wex, Bernard Patrick **271**
Wheatstone, Sir Charles *42*, 45, 46, 47, **271**
Wheeler's 130
Whistler 27
Whistler's Father *31*
Whistler's Mother *31*
Whittaker, Mr. 132, 133
Whittle, Sir Frank **271**, 273
Wiener, Martin J. 12, 19, 88
Wiener Thesis 15, 88
Wilkinson, John *81*
Wilkinson, W.B. 114
Willes, Mr. 132, 133
Wilm, Alfred 115
Williams, Captain M. 63
Williams, Sir Owen 115, 159, 160, 162, 163, 164, 165, 216, 271, *271*
Willis Faber Dumas *212*, 216, 266
Wills, W.D. and H.O. 227
Winchester Cathedral *71*
Windsor Castle 125, 127
WonderWorld Theme Park *10, 238*, 263
Wireless Transmission 273
Wood, Geoffrey **271**
Wood, John Crewe 273
Woodward, Christopher *215*
World War I 75, 115, 157, 160
World War II 32, 49, 115, 142, 157, 263, 264
Wyatt, Matthew Digby 68
Wye Bridge 220
Wylson, James 193

York Station 11

Zambesi River 220
Zunz, Jack 12, 218, 263, 266, **271**